ACS SYMPOSIUM SERIES **778**

Nuclear Site Remediation

First Accomplishments of the Environmental Management Science Program

P. Gary Eller, EDITOR
Los Alamos National Laboratory

William R. Heineman, EDITOR
University of Cincinnati

American Chemical Society, Washington, DC

Library of Congress Cataloging-in-Publication Data

Nuclear site remediation : first accomplishments of the Environmental Management Science Program / P. Gary Eller], editor, William R. Heinemann, editor.

    p. cm.—(ACS symposium series ; 778)

    Papers presented at a symposium held at the 218th National Meeting of the American Chemical Society in New Orleans from Aug. 22–26, 1999"

    Includes bibliographical references and index.

    ISBN 0–8412–3718–2

    1. United States, Dept. of Energy, Environmental Management Science Program—Congresses. 2. Cleanup of radioactive waste sites—United States—Congresses.

    I. Eller, P. Gary. II. Heinemann, William R. III. Series.

TD898.118 .N827   2000
628.5′2—dc21   00–58617

The paper used in this publication meets the minimum requirements of American National Standard for Information Sciences—Permanence of Paper for Printed Library Materials, ANSI Z39.48–1984.

Foreword

THE ACS SYMPOSIUM SERIES was first published in 1974 to provide a mechanism for publishing symposia quickly in book form. The purpose of the series is to publish timely, comprehensive books developed from ACS sponsored symposia based on current scientific research. Occasionally, books are developed from symposia sponsored by other organizations when the topic is of keen interest to the chemistry audience.

Before agreeing to publish a book, the proposed table of contents is reviewed for appropriate and comprehensive coverage and for interest to the audience. Some papers may be excluded in order to better focus the book; others may be added to provide comprehensiveness. When appropriate, overview or introductory chapters are added. Drafts of chapters are peer-reviewed prior to final acceptance or rejection, and manuscripts are prepared in camera-ready format.

As a rule, only original research papers and original review papers are included in the volumes. Verbatim reproductions of previously published papers are not accepted.

ACS BOOKS DEPARTMENT

Contents

Overview

Radionuclide and f-Element Chemistry

v

Separations and Treatment Chemistry

Geochemistry

Physical and Radiation Chemistry

Analytical Chemistry

Biochemistry

Preface

From 1943 until the end of the cold war in 1989, the United States and the former Soviet Union developed massive stockpiles of nuclear weapons. Because the development of these formidable arsenals was considered to be in the supreme national interest of both countries and because of technological limitations, short-term production concerns sometimes overrode long-term environmental considerations. Both countries incurred large-scale contamination of soils, water, air, and facilities and amassed huge quantities of wastes in currently unacceptable forms as a result of the nuclear weapons missions. Monumental cleanup efforts in both countries now are addressing these cold war legacies.

The U.S. Department of Energy (DOE) has responsibility for addressing the nuclear weapons legacy in the United States. As outlined in Chapter 1, this program is the most expensive environmental remediation program ever attempted. The challenges are technically, administratively, financially, and politically daunting. For many of the remedial challenges, cleanup technology is either unacceptably expensive or risky at present, or simply unknown. In recognition of this situation, in 1996 the U.S. Congress funded and DOE instituted an Environmental Management Science Program (EMSP) to develop the science base for breakthrough technology that can address these difficult problems. To date almost 300 scientific investigations have been funded under the EMSP program.

Because the initial three-year funding cycle for the first group of EMSP projects was about to end and chemistry is at the heart of many of these projects, Dr. Lester Morss of Argonne National Laboratory (the 1999 chair of the American Chemical Society (ACS) Division of Nuclear Chemistry and Technology (DNCT)) proposed that it was timely and appropriate for the ACS to sponsor a symposium highlighting the scientific achievements of the program to date. After the concept was endorsed by the Executive Committee of the DNCT, Dr. Morss honored us by asking us to organize such a symposium. As a result, a symposium was held at the 218th National ACS Meeting in New Orleans from August 22–26, 1999. Sixty-three plenary, tutorial, and topical scientific lectures and fifty-four posters were presented over a period of five days. Technical sessions were organized around the subject areas of actinide, separations, analytical, physical, materials, geo/biochemistry, and inorganic chemistry. Five ACS Divisions as well as DOE provided sponsorship in the form of administrative and financial support. The outstanding response to this symposium encouraged the consideration of pub-lishing a symposium proceedings volume. Enthusiasm among symposium partici-pants to this idea likewise was overwhelmingly positive. To meet the ACS require-ments for this book, we chose a limited number of presentations from the symposium to be developed as chapters that would give broad coverage of this subject. The volume you are now reading is the result.

Many individuals and organizations contributed to the success of this effort. First and foremost, DOE provided the funding for each project discussed at the ACS symposium. DOE also provided generous financial support that allowed a large number of undergraduate, graduate, and postdoctoral students to present their results. The support by Mr. Mark Gilbertson (DOE Office of Environmental Management) and Dr. Roland Hirsch (DOE Office of Science) is particularly recognized in this regard. The DNCT provided the primary ACS sponsorship for this symposium, with cosponsorship by the ACS Divisions of Analytical Chemistry, Industrial and Engineering Chemistry, Environmental Chemistry, Inc., and Physical Chemistry. These Divisions contributed both administratively and financially. Neither the symposium nor proceedings would have occurred without the enthusiastic support, encouragement, and guidance of Dr. Lester Morss, chair of the DNCT. The Glenn T. Seaborg Institute branch at Los Alamos National Laboratory graciously provided skillful administrative support by Ms. Susan Ramsay. Similarly, Ms. Kim Carey provided essential administrative support courtesy of the University of Cincinnati, Department of Chemistry. The successful conduct of the symposium and preparation of this proceedings volume would not have been possible without the commitment of all these people.

In the final analysis, however, the success of this symposium and proceedings must be attributed to the outstanding scientific skill and knowledge of the principal investigators and their colleagues. We hope this symposium proceedings volume succeeds in providing a compact forum for the world to learn of their efforts.

P. GARY ELLER
Nuclear Materials Technology Division
Los Alamos National Laboratory
Los Alamos, NM 87544

WILLIAM R. HEINEMAN
Department of Chemistry
University of Cincinnati
Cincinnati, OH 45221–0172

Overview

Chapter 1

The Environmental Management Science Program: One Facet of a Balanced Solution-Based Investment Strategy

Gerald Boyd[1], Mark Gilbertson[2], Roland Hirsch[3], and Arnold Gritzke[2]

[1]Office of Science and Technology, Office of Environmental Management, Department of Energy, 1000 Independence Avenue, Washington, DC 20585
[2]Office of Basic and Applied Research, Office of Science and Technology, Office of Environmental Management, Department of Energy, 1000 Independence Avenue, Washington, DC 20585
[3]Medical Sciences Division, Office of Biological and Environmental Research, Office of Science, Department of Energy, Germantown, MD 20585

DOE's Office of Environmental Management (EM) is engaged in the cleanup of the facilities used in nuclear materials production. The science and technology to support many aspects of this mission do not exist. To address this, EM formed the Office of Science and Technology. In 1996, EM began reevaluating the notion that the cleanup would take over 30 years. *Accelerating Cleanup: Paths to Closure which resulted,* established EM's strategy for cleanup most of the sites by 2006, expediting cleanup of the remaining sites, and implementing long-term stewardship. At the same time, Congress mandated development of a basic science program.

Introduction

The DOE's Office of Environmental Management (EM) is now engaged in the cleanup of the facilities used in over fifty years of nuclear materials production. EM was established with the mission of cleaning up the legacy of environmental pollution at DOE weapons complex facilities, preventing further environmental contamination, and instituting responsible environmental management. However, a coordinated effort in the conduct of scientific research and technology development to support many aspects of this mission did not exist. To address this need for effective and affordable solutions to the daunting environmental problems it faced, EM formed the Office of Science and Technology (OST) to conduct a national basic scientific and technology research, development and deployment program. OST's mission is to ensure the availability of cost-effective, technically better, and safer technologies to achieve the complex cleanup. The Environmental Management Science Program provides a key facet of the development of new cleanup methods and the improvement of existing methods by providing the scientific understanding of the environment and the chemical, biological and physical aspects of cleanup processes.

In 1996, EM began to reevaluate the notion that the cleanup of the former weapons complex would take over 30 years to accomplish. The result of this reevaluation was *Accelerating Cleanup: Paths to Closure, (1)* which articulated EM's vision to cleanup and transition the majority of the operating sites by the year 2006, expedite the cleanup of the remaining sites, and implement a long-term stewardship program to address the complicated and intractable environmental management problems. This new planning document established a path forward, which allowed major cleanup activities to be completed in the near term, while leaving recalcitrant technical problems for the longer term. This staging of the cleanup effort allows the research efforts within OST to be focused on areas where there is a high technical need and allows adequate time for the required science and technology to be developed and tested.

At the same time, Congress mandated the establishment of an environmental science program within EM, *(2)* to be managed in collaboration with the DOE Office of Science and the Idaho Operations Office. The objective of this program is to focus the nation's scientific research infrastructure on EM's science needs in order to provide technical scientific underpinnings for the long-term cleanup effort. The first of the Environmental Management Science Program Projects (EMSP) have just completed their initial 3-year grants.

EM's Environmental Management Problem

EM is engaged in the cleanup of the radioactive, hazardous, and mixed wastes left from over 50 years of U.S. nuclear weapons production. These materials are in a variety of forms (e.g., stored as waste in tanks and drums, disposed in landfills and trenches, released to the surface, contaminants in buildings and other structures, and as contaminants in soils and groundwater.) To assess the magnitude of the problem and to coordinate the cleanup effort DOE has organized the cleanup by problem areas.

Total life-cycle costs for the cleanup are estimated to be approximately $147 billion from 1997 to 2070. This currently represents the largest cleanup effort in the world.

High Level Waste

One of the largest problems DOE faces both in terms of cost and scientific challenge is the cleanup of the waste contained in storage tanks. Within the DOE complex, over 335 underground storage tanks have been used to process and store radioactive and chemical mixed waste generated from weapon materials production and manufacturing. Collectively these tanks hold over 340 million liters of high-level (HLW) and low-level radioactive liquid waste in sludge, saltcake, and as supernate and vapor (3). Very little of this material has been treated and/or disposed of in final form. Most of the waste is alkaline and contains a diverse portfolio of constituents including nitrate and nitrite salts, hydrated metal oxides, phosphate precipitates, and ferrocyanides along with a variety of radionuclides. The tanks are located at five DOE sites and many of these tanks have known or suspected leaks. Major challenges exist in the areas of characterization and separations or pretreatment.

Mixed Waste

An estimate of the inventory of mixed waste inventory shows approximately 167,000 cubic meters of over 1,400 different types of mixed waste located at 38 sites in 19 states (3). This inventory, however, is increasing with newly generated waste resulting from ongoing processes and environmental restoration, facility decontamination, and facility transition activities. After appropriate treatment, much of this waste is expected to be placed in the Waste Isolation Pilot Project facility (WIPP). The first shipments of TRU have already been sent to WIPP. However, challenges in characterization, separations and stabilization still exist before all of the mixed waste can be disposed.

Deactivation and Decontamination

DOE is addressing the problem of deactivation transformation of over 7,000 buildings and decommissioning over 900 buildings and their contents. This represents more than 6 million square meters of buildings containing contaminated concrete, equipment, machinery, and pipes *(3)*. Therefore, substantial quantities of metal and concrete must be decontaminated and nearly 200,000 tons of scrap metal must be disposed. Deactivation and Decommissioning activities across the complex will likely require decades to complete. Research is needed to develop or improve methods for the characterization, deactivation, decontamination, monitoring, and certification of facilities, as well as removal, control, treatment, and stabilization of Deactivation and Decommissioning derived waste. Scientific breakthroughs in these areas could reduce the schedule and associated long-term costs.

Subsurface Contaminants

Soil and ground water contaminated with radionuclides, heavy metals, and dense, non-aqueous phase liquid (DNAPL) contaminants are found at most DOE sites. This consists of more than 50 million cubic meters of contaminated soil and 5,700 ground water plumes, which have contaminated over 2 trillion liters of water. In addition, there are numerous landfills that contain nearly 4 million cubic meters of radioactive and hazardous buried waste, some of which has contaminated surrounding soils and ground water *(3)*. Currently available cleanup technologies are often unacceptable due to excessive costs, increased risks, long remediation schedules, or production of secondary wastes. Research is needed to improve characterizing and delineating contamination, removal and remediation of contaminants, separation of radionuclides from hazardous compounds, and prediction of contaminant migration.

Nuclear Materials

Certain nuclear materials, such as plutonium, can be dangerous even in extremely small quantities, particularly if ingested or inhaled. In addition, finely divided plutonium dust may spontaneously ignite when exposed to air above certain temperatures. Extreme precautions are required in storing, handling, and transporting such materials. DOE's plutonium and other nuclear materials exist in a variety of forms -- from acid solutions, to rough pieces of metal, to nearly finished weapons parts *(3)*. In addition, unknown amounts of plutonium have collected on the surfaces of ventilation ducts, air filters, and gloveboxes at some DOE facilities. Metal and chemical wastes containing nuclear materials typically are stored in drums and monitored, pending ultimate disposition.

Methods are needed to package and stabilize the hundreds of metric tons of metals and oxides, residues, and other processing intermediates that resulted from weapons manufacturing, laboratory samples, and research reactors both domestic and international.

Accelerating Cleanup: Paths to Closure Objectives

The vision of the *Accelerating Cleanup: Paths to Closure* strategy document is to set forth a path for the completion of the cleanup of the DOE complex. With this vision, a sequencing of projects has been established that calls for the cleanup of most of the 53 remaining sites by 2006 (Figure 1, below). At the 10 remaining sites, including the five largest sites, treatment will continue beyond 2006 for the "legacy" waste streams. This vision will drive budget decisions, subsequent sequencing of projects, and the actions needed to meet program objectives. Meeting this vision will require a sustained national commitment and sufficient funding. It will also require the use of world class technologies, improved efficiency, national integration, and university and industry partnerships.

Office of Science and Technology Mission

OST was formed in 1989 to provide a full range of science and technology resources and capabilities to meet EM's cleanup and long-term stewardship problems. This includes not only the development of technologies but also includes overseeing the targeted basic research needed to successfully develop new technologies and the support activities the sites may need to assure that these technologies are deployed.

When OST was formed, there was a realization that the cleanup of many of the sites in the DOE complex could not be accomplished without major advances in science and technology. At that time, either no technical solutions existed or currently available solutions were exorbitantly expensive or may have posed high risk to workers, the public, or the environment. OST's role is to provide scientific or technical alternatives, which will overcome these obstacles, on a schedule that will meet DOE schedules and goals. In recent years, the OST program has focused on deploying technologies to reduce the near-term cost and promote faster and more efficient cleanup. At the same time, OST, through the Environmental Management Science Program, is investing in long-term basic targeted research to improve and reduce the cost and risk of post 2006 cleanup and environmental management activities.

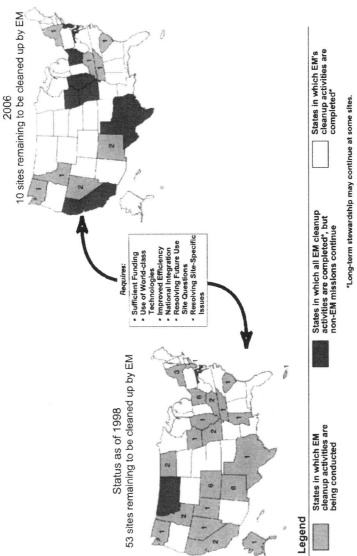

Figure 1. 2006 cleanup goal of the Accelerating Cleanup: Paths to Closure

8

OST Objectives

OST has recently completed a Strategic Plan *(4)*, an EM R&D Program *(5)* Plan, and a Management Plan *(6)*. These documents establish the following four major goals for OST activities and define indicators of success in meeting these goals: meeting high priority needs, reducing the cost of EM's major cost centers, reducing EM's technological risk, and accelerating the deployment of innovative technologies. These OST goals support the 7 objectives delineated for EM in the EM Research and Development Portfolio.

Meet High Priority Needs

EM's science and technology investments are end-user driven; they address the highest priority needs identified by the sites (Figure 2, below). This includes those needs on the critical path to site closure and those that represent major science and technology gaps that threaten project completion. Needs are rank ordered to ensure that EM focuses first on the problems with the biggest impact. Over 500 science and technology needs have been identified. Forty-six percent address highest risk problems, such as, stabilizing/securing plutonium, HLW tanks, etc. EM intends to bring more than 100 technologies to bear on these needs in the next 3 to 5 years.

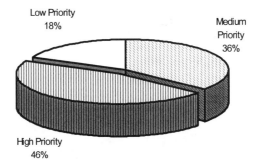

Figure 2. Distribution of Needs as Related to Site Critical Paths

Reduce the Cost of EM's Major Cost Centers

Current life-cycle cost estimates for the cleanup of the former weapons complex total nearly $150 billion in constant 1998 dollars for all projects through 2070 (Figure 3, below, shows the distribution of costs by Problem Area). EM's five Focus Areas target 80% of these cleanup costs, with almost half of it devoted to high-level waste. Over 300 opportunities to help meet EM's cost reduction goals have been identified. The potential cost saving associated with these opportunities is in excess of $12 billion with the largest opportunities in the area of HLW tanks.

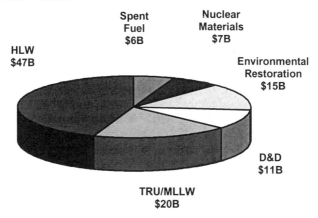

Figure 3. Life-Cycle Cleanup Cost by Problem Area

Reduce EM's Technological Risk

Many of the cleanup plans at the sites rely on enabling or breakthrough science and technology assumptions that have not been fully validated. Science and technology investments are targeted at these areas to help reduce the risk that the critical science and technology will not be available when needed or will not perform as expected. EM has identified 50 Project Baseline Summaries that present medium to high technological risk and that are on the critical path to closure (Figure 4, below). Most of these projects are managed by the Savannah River, Oak Ridge, Albuquerque, Idaho, and Richland Field Offices. OST specifically targets the technologies associated with these projects to ensure that site closure goals are met.

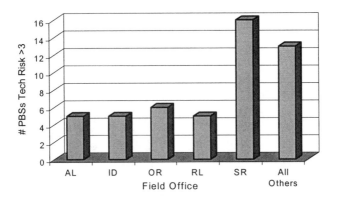

Figure 4. Number of Project Baseline Summaries with Technical Risk Greater than 3 by Field Office

Accelerated Technology Deployment

Throughout the investment in science and technology projects, the Focus Areas work with the sites to facilitate and support accelerated deployment and acceptance of the new technologies. In close coordination with the end-user community, Focus Areas provide technical advice and assist in development of site-specific deployment plans. Additionally, the Focus Areas assist in gaining regulatory acceptance of the new technologies, ensure that the national technology development program addresses site-specific requirements, and evaluate the progress and readiness of technologies under development (Figure 5, below). In fiscal year 1998, the target for deployment of new technologies was 49. The actual level achieved was 113 new deployments.

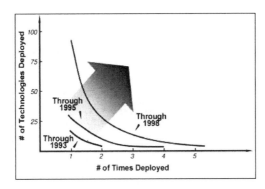

Figure 5. Number of Technologies Deployed 1993 through 1998

OST Investments

As noted above, OST addresses problems through a "Focus Area Centered Approach." The OST Focus Areas are aligned with DOE's problem areas (Figure 6, below, shows the OST investment by Problem Area) and are led by Field Offices, i.e., Richland is the lead for HLW tanks; Idaho is the lead for MLLW and shares the leadership of Nuclear Materials with Albuquerque; Savannah River is the lead for Subsurface Contaminants; and Oak Ridge is the lead for Deactivation and Decommissioning. Not all problem areas are represented by a Focus Area. However, many needs and solutions applicable to one problem area can also be used to address other problem areas e.g., HLW and Mixed Waste solutions are applicable to Spent Fuel and Plutonium problems. Aspects of the transportation and health/ecology/risk problem areas are encompassed by all of the focus areas.

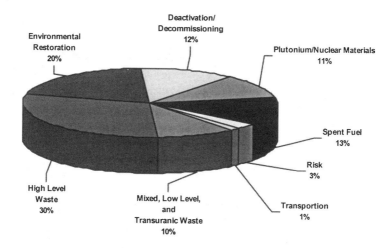

Figure 6. To-Date OST investment by Problem Area

EMSP Role

The EMSP was established in response to a mandate from Congress in the fiscal year 1996 Energy and Water Development Appropriations Act. Congress directed the Department to "provide sufficient attention and resources to longer-term basic science research which needs to be done to ultimately reduce cleanup costs, ...develop a program that takes advantage of laboratory and university expertise, and ...seek new and innovative cleanup methods to replace current conventional approaches which are often costly and ineffective". This mandate followed similar recommendations from the Galvin Commission to the Secretary of Energy Advisory Board. The EMSP also has received valuable input from the National Academy of Sciences, regulators, citizen advisory groups, and other stakeholders.

The goal of the EMSP is to develop and fund a targeted, long-term basic research program that will result in transformational or breakthrough approaches for solving the Department's environmental problems. The purpose is to provide the basic scientific knowledge that will lead to reduced remediation costs, schedule, or risk, or that will help alleviate otherwise intractable problems. EMSP research is focused on Department's cleanup problems and is linked to problem holders, including technical staff, managers, and stakeholder advisory groups at the sites. The program supports research that could:

- Lead to significantly lower cleanup costs and reduced risks to workers, the public, and the environment over the long-term.
- Bridge the gap between broad fundamental research that has wide-ranging applicability, such as that performed in the Office of Science, and needs-driven applied technology development conducted by the EM Office of Science and Technology Focus Areas.
- Serve as a stimulus for focusing the nation's science infrastructure on critical national environmental management problems.

Research projects are solicited and awarded according to program needs of the DOE sites and the degree to which those needs can be influenced by scientific findings. Awardees interface with OST Focus Area representatives during the projects. Research results are integrated into technology development activities of the Focus Areas and site end-users through a number of facilitated interactions, such as topical and site-specific workshops, national workshops, other EM program meetings and other national meetings such as the American Chemical Society Symposium that is the focus of this proceedings volume.

EMSP Organization

EMSP's relationship to other programs is shown in Figure 7, below. The EM/OST Office of Basic and Applied Research (OBAR) is the lead organization for the EMSP. It provides policy and programmatic guidance, solicits research needs, ensures research is applicable to DOE cleanup problems, and communicates research results. DOE-Office of Science (SC) provides input into program policy development, manages the solicitation of research applications, oversees the scientific review process, and manages the scientific aspects of the Program. DOE-ID is the lead Field Office for program execution, which includes assisting OBAR in analyzing science needs, providing procurement services, integrating research results, and managing grant funding. DOE-ID also serves as the interface between the EM Focus Areas, Crosscutting Programs and DOE Field Offices.

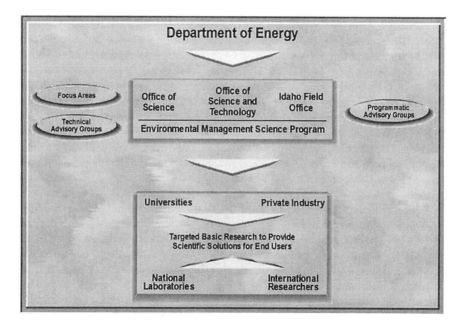

Figure 7. EMSP Organization and Interactions

The EMSP has a number of technical advisory groups providing valuable input on both technical and programmatic matters. In the technical area, EM's Site Technology Coordinating Groups (STCGs) identify science and technology needs and opportunities associated with all field sites and cleanup projects. The Focus Areas work with EMSP and the STCGs to evaluate site needs and set basic research directions. The EM Integration team identifies ways to improve efficiencies and cost savings throughout EM and identifies additional science priorities associated with critical waste streams.

Several programmatic advisory groups assist the EMSP with strategic and policy recommendations. The EM Federal Review Board and the EM Advisory

Board (EMAB) advise and evaluate overall Program execution. The EMAB Science Committee reviews the processes used to select projects, provides recommendations on Program direction, and advises on EM science policy. Similarly, the Biological and Environmental Research Advisory Committee of the Office of Science has provided input on the EMSP. DOE's Strategic Laboratory Council also advises the EMSP on planning and execution processes in order to ensure programmatic relevance and successful utilization of research results. The National Research Council/National Academy of Sciences provides periodic external policy recommendations.

EMSP Participation

Since its inception in Fiscal Year 1996, EMSP has invested over $220 million in support of 274 research projects. This investment has provided funding for researchers at 90 universities, 13 national laboratories, and 22 other governmental and private laboratories. Research is being conducted in 38 states and the District of Columbia, two Canadian provinces, Australia, Russia, the United Kingdom, and the Czech Republic. Forty percent of funding to date has gone to universities and other non-Federal labs.

There are 202 projects started prior to 1998, which included over $160 million distributed across all 13 science categories. From 1998 onward, the grants have become focused on particularly intractable problems. In Fiscal Year 1998, 33 projects were funded to respond to needs in the areas of high level radioactive waste and deactivation and decommissioning. Drawing on expertise available from abroad, particularly in high level waste, the 1998 awards also included grants to researchers in the Czech Republic and Russia.

Projects by Problem Areas

The 1999 awards for 39 projects focused on subsurface contamination and low dose radiation effects. This latter area is likely to have an impact on the cleanup activities in most of the problem areas. The number projects anticipated, depending on available funding, is 31 projects in Subsurface Contaminants/Vadose Zone (Subsurface Contamination Problem Area) and 8 projects in Low Dose Radiation Effects (Health/Ecology/Risk Problem Area)(Table 1, Below, shows the annual distribution of projects by problem area). SC has funded and additional seven projects in the Low Dose Radiation Effects area. Five additional projects may be funded by EMSP in FY2000, depending on funding levels and SC may fund up to nine more projects in FY2000, again depending on funding levels.

	FY96	FY97	FY98	FY99
High Level Waste	26	22	20	-
Spent Nuclear Fuel	-	5	-	-
Mixed Waste	22	9	-	-
Nuclear Materials	2	6	-	-
Subsurface Contamination	72	11	-	31
Health/Ecology/Risk	10	8	-	8
Decontamination and Deactivation	4	5	13	-
TOTAL	136	66	33	39

Table 1. EMSP Projects by Year by EM Problem Area

EMSP Successes

While most EMSP projects are funded on a three year basis, useful results have emerged from many projects before their completion. These research results may be instrumental in developing not only new tools for environmental cleanup, but also in improving the understanding of scientific principles that underlie conventional cleanup methods. The following are examples of some early EMSP results that could provide significant benefits to EM's cleanup effort. Many others are highlighted in the remaining chapters of this volume.

One research project in the high level waste problem area, led by Pacific Northwest National Laboratory, focuses on a key problem in processing radioactive wastes where insoluble sludges clog transfer lines or interfere with solid-liquid separations. In order to control the problem, better understanding of the characteristics and mechanisms for controlling the physical properties of sludge suspensions, which consist of submicron colloidal particles, is needed. The project focuses on the factors controlling colloidal agglomeration, determining how agglomeration influences slurry rheology, and developing strategies to control agglomeration. Early results are pointing the way to optimization of waste processing conditions for retrieval, transport, and separation of the waste currently in tank storage.

A research project in the mixed waste problem area, being conducted at Los Alamos National Laboratory, is developing a high-flux neutron source for nondestructive assay of containerized TRU waste at DOE sites. This research could lead to a more stable and powerful neutron source, which would improve assay results. Some potential applications are the characterization of TRU wastes for mixed waste residues prior to stabilization and disposal, cemented or vitrified wastes, spent nuclear fuel, and high level wastes.

Basic research at the University of Georgia, addressing the remedial action problem area, deals with mercury contamination in soils. This work uses genetic engineering to transfer properties from soil bacteria to plants which would allow them to degrade mercury into a less toxic form. These plants can then be grown in contaminated areas where they could withdraw mercury from the soil and reduce its toxicity. This work is expected to lead to the development of a passive method for remediation of mercury and possibly other contaminants.

Another project in the area of subsurface contamination deals with the problem of water flow processes in mixed soil or fractured rock environments in the vadose zone (that is, above the ground water table). Researchers at Lawrence Berkeley National Laboratory have developed equations that describe the pattern of fractures in basalt and the trajectory of flow paths in the basalt. The results of this research may change the approach used to predict flow and transport of groundwater and contaminants in fractured media. With this information in hand, the environmental restoration program stands the best chance of stopping contaminants before they seep into the water table.

In the problem area of mixed waste, the mixture of toxic chemicals, heavy metals, halogenated solvents and radionuclides in many DOE waste materials presents the challenging problem of separating the different species and disposing of individual contaminants. A microbiological treatment system is an attractive possibility for separation and treatment. A project is being conducted at the University of Washington to develop organisms capable of detoxifying metals, such as mercury, and halogenated organics in mixed waste. This research involves cloning these beneficial properties from other bacteria into a bacteria strain that is highly resistant to radioactivity.

Incorporating Science Results into Technology Development

The results from many of the projects of the EMSP have been or will be incorporated directly into ongoing cleanup projects, baseline technologies, or innovative technologies under development by OST. Others will require further development before they can be utilized in cleanup.

OST uses a "stage-gate" model for managing its investment portfolio. Research and development activities are categorized into 7 distinct "stages" with a review process at the "gates" for advancing a project to the next stage. This includes technical merit review and go/no-go decisions (e.g., American Society of Mechanical Engineering reviews at gate 4) and the programmatic relevance review each year by Focus Areas. Scientific merit reviews and relevance reviews are used before the initiation of basic research, i.e. gate 0.

With the emphasis that has been placed on demonstrations and deployments to achieve short-term goals of the *Accelerating Cleanup: Paths to Closure* and the initiation of new basic research projects to meet the *Accelerating Cleanup: Paths to Closure* long-term goals, there has been a lack of emphasis on applied research and exploratory development. As the first of the 3-year grants in the EMSP reach the end of their primary funding period steps are being taken to move this basic research into the applied research stage. This is shown conceptually in Figure 8, below, where the valley associated with applied research funding is less pronounced in 2001. Due to the flat or decreasing overall funding for EM-OST the increase in applied research funding will be accomplished at the expense of basic research and projects at more mature stages. In planning for the transitioning of EMSP sponsored basic research into later stages of the stage-gate approach used by OST, many of the same evaluation criteria are used that are used for evaluation and ranking of more mature technologies. These parameters include meeting needs, high probability of use, potential to treat currently untreatable waste, and potential for cost savings. Pressures on the EM budget inevitably have not allowed EM to build the applied research program as quickly as desired, but progress is being made toward this goal.

Conclusions

EM must invest in a balanced portfolio ranging from basic research to deployment assistance for the sites. Demonstrations and deployment assistance are needed to achieve the goal of closure of most sites by 2006. Basic and applied research is needed to meet long-term goals. EM's cleanup mission remains heavily dependent on achieving basic scientific breakthroughs that will allow the cleanup of intractable problems. EMSP therefore is integral to the success of the overall OST mission. The EMSP is how EM reaches out to the

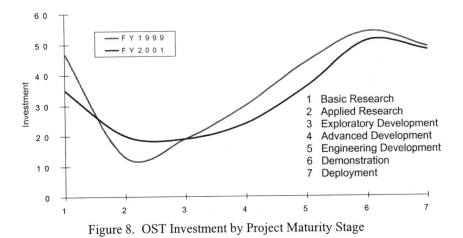

Figure 8. OST Investment by Project Maturity Stage

basic research community and is an integral part of the balanced portfolio strategy. EM is now addressing the difficult task of transitioning the basic science projects that reached the end of their primary funding period in FY 1999 into the applied research stage.

Fortunately, the Department has many of the best scientists in their fields participating in the EMSP. These scientists have a strong interest in helping to solve the problems facing the Nation's cleanup effort. They also possess a firm commitment to close collaboration with each other and the technologists and managers at the sites seeking solutions to the many currently intractable problems. The staff of the Office of Environmental Management and Office of Science is grateful for the dedication of the scientists who participated in the symposium from which this volume is derived. The organizers and participants deserve congratulations for this impressive display of accomplishments and vision for the future.

References

1. Accelerating Cleanup: Paths To Closure; United States Department of Energy, 1998
2. H.R. Conference Report No.293, 105th Congress 1st Session, 1996.
3. DOE Research and Development Portfolio: Environmental Quality; Volume 3: United States Department of Energy.
4. Environmental Management Strategic Plan for Science and Technology, United States Department of Energy, 1998.
5. Environmental Management Research and Development Program Plan, United States Department of Energy, 1998.
6. Office of Science and Technology Management Plan, United States Department of Energy, 1999.

Chapter 2

Challenges and Scientific Issues for the Department of Energy Weapons Complex Cleanup: High-Level Waste and Spent Nuclear Fuel

William L. Kuhn

Pacific Northwest National Laboratory, 3350 Q Street, K7–15, Richland, WA 99352

This paper provides a perspective of challenges and scientific issues associated with the U.S. Department of Energy's mission to remediate tanks containing high-level waste and facilities containing stored spent fuel. It is directed toward the broad community of engineers and scientists capable of addressing these challenges. The topic is too broad to discuss comprehensively in this paper. Instead, we provide a few examples and discuss the diversity of scientific issues and the cultural difficulty of infusing innovative technologies or ideas into "baseline" projects through which waste or spent fuel are treated and disposed are discussed.

Introduction

A broad technical community of engineers and scientists working on behalf of sites managed by the U.S. Department of Energy (DOE) have resolved many issues important to safe treatment and disposal of stored nuclear waste. This waste includes notably high-level waste in tanks and spent nuclear fuel. For these wastes, many scientific and institutional challenges still remain that need to be addressed by the technical community. To do this it needs to understand the breadth and depth of the problem, the cultural difficulty of introducing new technical approaches into a large project, and the nature of the scientific challenges. Here we attempt to address each of these aspects in turn, not comprehensively, which is beyond the scope of this paper, but conceptually and by presenting useful examples. First, we summarize the current status of stored tank waste and spent fuel and the means of treating the tank waste. Next, we describe cultural issues associated with introducing new technical approaches into "baseline" plans. Finally, we discuss the diversity of the scientific challenges and provide some specific examples of such challenges and of recent promising scientific work.

High-Level Waste in Tanks

Remediation of tanks containing highly radioactive waste is a major technical and programmatic challenge for the U.S. Department of Energy (DOE) (1). The DOE system currently stores about 340 million liters of waste containing more than 700 million Curies (MCi) in 282 tanks at five major sites:

- The Savannah River Site (SRS) near Aiken, South Carolina, has 51 tanks storing 125 million liters of waste containing about 400 MCi of radioactivity.
- In Washington State, the Hanford Site has 177 tanks that store 208 million liters of waste containing about 200 MCi of radioactivity.
- The Idaho National Engineering and Environmental Laboratory (INEEL) near Idaho Falls, Idaho, has 11 tanks with 5.3 million liters of liquid waste containing 520,000 Ci of radioactivity and 3.8 million liters of calcined (a granular powder) waste with 24 MCi of radioactivity stored in seven bin sets.
- The Oak Ridge Reservation (ORR) in Oak Ridge, Tennessee, has about 1.6 million liters of waste containing 47,000 Ci of radioactivity in 40 tanks. ORR also annually adds approximately 56,000 liters of active waste containing 13,000 Ci of radioactivity to 13 of their tanks.
- The West Valley Demonstration Project (WVDP) near West Valley, New York, has retrieved and vitrified approximately 95% of the 2.3 million liters of reprocessing waste that was stored in 3 tanks.

Most of these wastes originated from processing spent nuclear fuel to recover plutonium or uranium. Spent fuel is reprocessed by dissolving the fuel in acid and extracting (or precipitating, in earlier processes) the plutonium. The raffinate, designated high-level waste, which includes most of the fission products, is therefore highly acidic. To save costs and because stainless steel was not readily available during World War II, reprocessing waste was neutralized by adding an excess of sodium hydroxide to create a highly
alkaline solution that could be stored in carbon steel tanks. Sodium nitrite was added as a corrosion
inhibitor. Thus, wastes at Hanford, Savannah River, and Oak Ridge contain mainly sodium nitrite, sodium nitrate, and sodium hydroxide and include a host of non-radioactive metal oxides and radioactive fission products.

Spent fuel is also reprocessed at INEEL to recover uranium. INEEL chose to calcine the resulting acidic waste to accommodate the high fluoride concentration, to reduce the waste volume, and to create a dry waste form. However, some of the acidic waste has a high sodium concentration that complicates calcination. Consequently, this high sodium-bearing waste has been stored in stainless steel tanks as an acidic liquid. INEEL plans ultimately to calcine this waste also.

Alkaline High-Level Wastes at Hanford and Savannah River are both separated and immobilized into a relatively small volume of high-level waste glass and a relatively large volume of a low-level waste form to be disposed of on site. At Savannah River, strontium is removed in-tank using ion exchange materials that subsequently settle to become part of the sludge, which is then vitrified. Tetraphenly borate was going to be used to precipitate cesium, but this process is being reevaluated to consider also silicotitanate ion exchangers and solvent extraction. The remaining low-level waste is being grouted (as "saltstone") and disposed of on site. At Hanford, cesium and strontium will be removed in ion exchange columns from tank supernatant and vitrified along with sludge from the tanks. The remaining low-level waste will be vitrified and disposed of on site.

Stored Spent Fuel

The DOE also stores and manages spent fuel. The fuel originates at a variety of sources, including production reactors at Hanford and Savannah River, research reactors at the INEEL, U.S. Navy submarine and surface-ship propulsion reactors, the decommissioned Fort St. Vrain power plant in Colorado, the damaged core from the Three Mile Island power plant in Pennsylvania, and university and foreign research reactors. The U.S. also accepts spent nuclear fuel from foreign research reactors to reduce the threat of nuclear terrorism in the world; it contains uranium that was enriched in the U.S. and initially exported in the "Atoms for Peace" program. The resulting diversity of spent fuel forms has been described (2) and is shown in Figure 1. The boxes in the figure show the form of the uranium and the type of reactor from which it has been retrieved. For most of the forms, uranium is alloyed or combined microstructurally with other metals to provide desired properties, or it is dispersed in a matrix providing such properties. Additionally, the uranium may be non-metallic, such as in the familiar oxide form or as a hydride, nitride, or carbide. A salt matrix was used in the MSRE was the Molten Salt Reactor Experiment, which inherently required a molten matrix to circulate uranium dissolved in it as the tetrafluoride. It is apparent that DOE spent fuels include chemically reactive metallic uranium and in this respect differ prominently from reprocessing wastes.

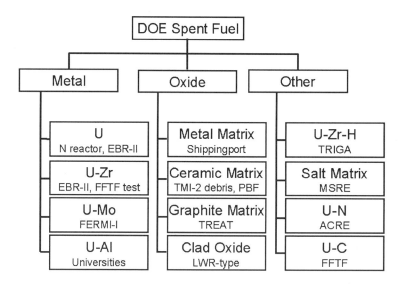

Figure 1. Types of DOE Spent Fuel (Karell et al. 1997)

The Challenge Of Infusing New Ideas Into Baseline Approaches

As the technical community conceives new technical solutions through investments in directed science, the solutions can only be inserted within appropriate projects through which high-level waste or spent fuel are treated and disposed or stored. To appreciate the challenge this represents, consider the perspective of a project manager responsible for meeting the project objectives on time and within budget. Before altering the project's "baseline" technical approach, the manager must be persuaded that proposed new solutions overall decrease the risk of not meeting those objectives.

Ideally, all elements of the project remain unchanged over the life of the project and are completed according to the original schedule. At any given moment, the project identifies "open technical issues" that, until resolved, are significant challenges to the success of the baseline technical approach. The project usually schedules and funds work—which may be scientific in nature—to "close" these open issues. However, if science-based solutions become available in time to meet the needs of the project, they would likely be incorporated even if developed outside the project.

The issue, then, is whether new solutions actually meet the needs of a project. Many of the elements in a project, such as those directly addressing operations, regulatory drivers, financial incentives, integration with the existing infrastructure, etc., are not subject to change by new technologies or technical information. Thus, project managers tend to be skeptical about anticipated benefits of new technical approaches until it is apparent there is a net improvement considering all affected elements in the project. Where scientist envision a breakthroughs, project managers may envision delays caused by sorting out many practical and mundane aspects of incorporating something new, including safety analyses and approvals, readiness reviews, operator training, and fitting equipment into an existing site infrastructure.

On the other hand, project managers worry about unanticipated problems with the baseline technical approach for which no technical solution or responsive information exists. This could cause major delays until new technical approaches or information resolving the problem can be provided. Such problems can be discovered though a revelation, which tends to clarify (albeit belatedly) scientific work needed to support the project, or through an "asymptotic" realization of such needs. If the latter case the solution seems to be only a year or so away for many years, and hence funding work on more innovative solutions seems to be unjustified.

An "asymptotic" realization of a problem confronts technical innovators with a "window of opportunity" barrier. At any given time it appears that an innovative technical approach must be available within a brief "window of opportunity" because resolving problems with the baseline approach seems to be imminent. This perception can continue for years. In retrospect we see the situation depicted fancifully in Figure 2. Here the ordinate is the predicted time of completion of a project and the abscissa is the time at which that prediction is made. Eventually we reach the "tomorrow has come" barrier where the project is supposed to be completed per the promised completion date. The project manager needs to cross this barrier into the promised land. Thus, the ideal is the horizontal dashed line, where the predicted time remains the same up to the actual completion. However, unanticipated problems arise (the exclamation mark). When this is realized suddenly, one sees the need for innovative solutions and recognize the resulting shift in the schedule (the upper line). But when the realization is gradual, the schedule for providing alternatives seems to be compressed to the narrow window of opportunity between the "asymptotic" schedule and the "tomorrow has come" barrier. Actually, there may be a much wider window of opportunity--between the ideal (dashed line) and the revealed shift in the baseline schedule (uppermost line), but of course we do not act on this unless we see the true shift in the baseline schedule.

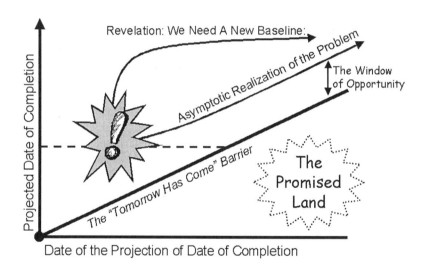

Figure 2. Retrospective View Of The "Window Of Opportunity" Barrier

If, as an act of faith, we make prudent investments in scientific investigations that provide information or technologies seemingly too late to fit into a project's window of opportunity, then potential solutions to unanticipated problems will be available if needed. They serve essentially as insurance against the need to shift baseline schedules when unanticipated problems arise. Of course, budgets are limited and determining what is "prudent" is difficult: are we insuring against a plausible or a highly unlikely problem, or against a major or a minor threat?

The opportunity to infuse new technical approaches into a project depends on the stage of a project, as depicted conceptually in Figure 3. The ordinate depicts the progress of the project (upward) through major stages. The abscissa depicts conceptually the departure of a technical approach from the baseline (the baseline being the ordinate). In the very early stages of the project many technical approaches might be considered as alternatives to the nominal baseline. However, as formal design phases ensue, the project must define its baseline to progress at all. Given the effort spent on design reviews, the need to accommodate truly new technical approaches should decrease rapidly as the project progresses. The project will invest in closing open technical issues, and make prudent investments (if funding allows) in managing technical risks (insurance), and these are the principal opportunities to infuse new technical approaches into the project. There may also be new technical approaches funded outside the scope of the project, i.e., by programs such as DOE/EM focus areas or the EM Science Program.

If these are seen by the project as reducing overall technical risks, they will probably be incorporated. The most likely impact of programs outside the project is to provide information needed by the project to resolve or address regulatory or safety issues or to define or demonstrate the acceptability of technical risks.

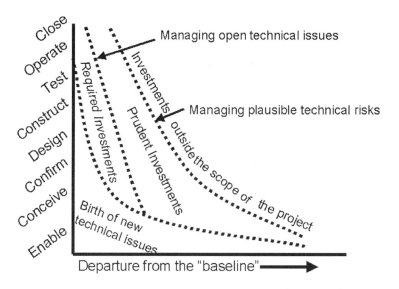

Figure 3. Conceptual depiction of the decreasing opportunity to infuse new technical approaches as a project progresses.

Scientific Issues

Many scientific issues have been articulated as science needs. DOE's Office of Science and Technology (OST) maintains a Needs Management System to receive and document work needed to support projects (required investments) at sites as well as opportunities (prudent investments) to support the sites that are identified by DOE/EM sites. The needs are cataloged according to technology vs. science needs, by site, and by the focus areas that identify and respond to them. Tables I and II summarize the number of needs recently published over the internet (http://emNeeds.em.doe.gov/Current/Home/frameset.asp) by OST. Only needs corresponding to spent fuel and high-level waste in tanks are shown, and only for sites that have identified a significant number of needs in those areas. The totals for the focus areas and the sites are shown for comparison. For example, Idaho has identified 22 needs related to SNF, 31 related to HLW, and 112 needs overall.

Table I. Technology Needs and Opportunities

SITE	Spent Nuclear Fuel National Program	Tanks Focus Area	All Focus Areas
Albuquerque	0	0	79
Idaho	22	31	112
Hanford	6	24	143
Savannah River	2	18	85
All Sites	30	86	637

Table II. Science Needs and Opportunities

SITE	Spent Nuclear Fuel National Program	Tanks Focus Area	All Focus Areas
Albuquerque	2	1	3
Idaho	1	0	15
Hanford	0	27	60
Savannah River	1	2	3
All Sites	4	30	114

The total number of technology needs is about six times the number of science needs; this corresponds roughly to the situation depicted in Figure 3. Technology needs are more closely associated with project baselines than are science needs because the former are relatively short-term and the latter relatively long-term. The large number of needs pertinent to tank waste compared to spent fuel reflects the relative maturity of DOE-funded R&D on tank waste, and conversely reflects the relative novelty of R&D on treatment of defense spent fuel. The lower number of needs identified by DOE's Albuquerque office and Savannah River site reflect the lesser inventories of these wastes in the first case and the relative maturity of the stage of the projects in the second case.

Diverse Physicochemical And Spatial Dimensions

The diversity of science issues, and hence the diversity of experience, capabilities, and interests in the scientific community that need to be focused on spent fuel and high-level tank waste problems, can be depicted roughly in terms of the diversity of spatial dimensions and physical vs. chemical domains implied by the issues. Selected prominent issues pertinent to treating spent fuel from high-level tank waste are shown in Figure 4. Of course, the arrangement shown is subjective, and it is not comprehensive, but it does serve to illustrate the diversity that characterizes scientific issues that need to be understood to manage technical risks. The spatial domain could be extended to smaller, "electronic" dimensions to illustrate the growing importance of computational chemistry to understand and predict both physical and chemical phenomena. Such modeling cuts across virtually all technical issues that occupy the "molecular" length scale; other experts need to then interpret and thereby extend this understanding across the spectrum of spatial domains.

Examples of Spent Fuel Science Issues

Metallic uranium spent fuel is stored underwater at Hanford. Defects in the fuel cladding, or non-uniform corrosion of the cladding, admit water, which reacts with the uranium. The resulting hydrated oxides expand relative to the metal and can rupture the cladding and occasionally virtually dismantle fuel rods. The situation is depicted in Figure 5. The stored spent fuel includes some fraction that has been breached and hydrated. Water moves along the interface between the fuel and the cladding. Uranium is oxidized by water to release hydrogen, and water can also corrode the inside of the cladding.

Safe storage requires drying the hydrated spent fuel; the planned method for this is vacuum drying. This will remove some portion of the water, including presumably any free water. One of the important remaining scientific issues is the long-term stability of the hydrates: will their chemical or physical form change significantly? Chemically-bound water may remain in cracks and bound to surfaces even following proposed drying steps, leading to possible long-term corrosion of the containers and/or fuel rods themselves, generation of H_2 and O_2 gas via radiolysis, and reactions of pyrophoric uranium hydrides (3).

Diverse Physicochemical and Spatial Dimensions

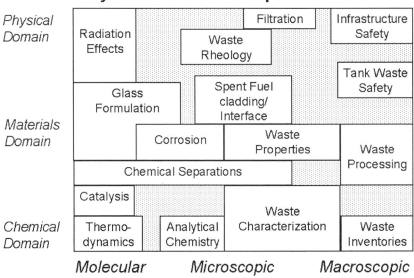

Figure 4. Conceptual Distribution of Science Needs Over Physicochemical and Spatial Domains.

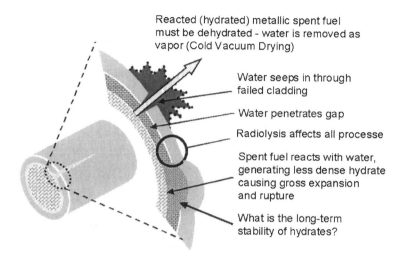

Reacted (hydrated) metallic spent fuel
must be dehydrated - water is removed as
vapor (Cold Vacuum Drying)

Water seeps in through
failed cladding

Water penetrates gap

Radiolysis affects all processe

Spent fuel reacts with water,
generating less dense hydrate
causing gross expansion
and rupture

What is the long-term
stability of hydrates?

Figure 5. Hydration of Metallic Fuel Underwater in K-Basin at Hanford

A different issue has been identified at INEEL, as depicted in Figure 6. There, spent fuel is being moved from wet to nominally dry storage, but still there is concern that bacterial "biofilms" could retain or absorb water and then, because of heterogeneous growth, enable non-uniform corrosion such as pitting (4). The potential for this and the behavior of bacteria in such biofilms is a significant scientific issue.

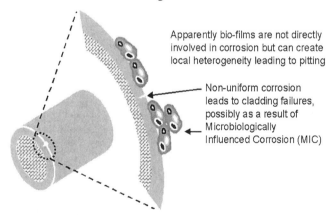

Apparently bio-films are not directly
involved in corrosion but can create
local heterogeneity leading to pitting

Non-uniform corrosion
leads to cladding failures,
possibly as a result of
Microbiologically
Influenced Corrosion (MIC)

Figure 6. Non-uniform Corrosion of "Damp" Spent Fuel Is a Concern at INEEL

Other issues pertaining to spent fuel that are not discussed here include:
- Stabilization: pyrophoricity, combustion parameters
- Characterization/nondestructive examination
- Corrosion, radionuclide release mechanisms
- Accelerated degradation of containers and the effects of microbes
- Characterization of water clarity of basins
- Alternative spent nuclear fuel processes & waste forms
- Development of methods to remove moisture without damage to the fuel elements

Examples of High Level Tank Waste Science Issues

A comprehensive discussion of the many science needs pertaining to high-level tank waste is beyond the scope of this paper. We present only a few selected examples that illustrate the nature and diversity of the issues, and hence many important issues, such as vitrification or characterization, happen not to be discussed here. The interested reader is referred to OST's Technology Management System, described above, and to summary publications of OST's Tanks Focus Area, which are available over the internet at http://www.pnl.gov/tfa/program/index.stm.

Solution Thermodynamics

The liquid phase in high-level waste tanks is often a solution saturated with respect to sodium nitrite, nitrate, and often aluminate, as well as including several molar sodium hydroxide. The result is ionic strengths of twenty molal or more. In solutions such as this, water accounts for little more than half of the moles, as illustrated in Figure 7. Much of the water hydrates sodium cations: as little as two water molecules may be available per cation. Consequently, the solution is more like a loose crystal with all ions associated with each other and simultaneously hydrated than a set of ions each independently hydrated by water and separated by "free" water. Sophisticated thermodynamic models, such as Pitzer's, are required to predict phase equilibria or heat effects in such liquids. Physical chemists have modeled only some of the associations among ions that are important in tank wastes, which are significantly more complex than depicted here. For example, although sulfate appears in much lower concentration than nitrate or nitrite (i.e., among the "other" species), it can have a profound effect on vitrification of the waste because ultimately it is sparingly soluble in the glass. Therefore, it is important to model accurately the solubility of sulfates in tank wastes, and therefore important to model the ion associations involving the less concentrated anions.

Effects of Generation and Storage of Gas in Waste Sludges

A portion of tank waste exists as solids either because it is insoluble or because even high solubility limits are exceeded when the liquid is concentrated. The solids form a particulate sludge layer that often includes organic species and in which ordinary thermal and also radiolytic chemical reactions generate flammable gases such as hydrogen and nitrogen oxides. The storage and release of resulting bubbles in and from the sludge is poorly understood, but the consequences can be spectacular. A classic example is Tank SY-101 at the Hanford site. In this tank, the sludge exhibits a yield stress which, together with capillary forces, holds small gas bubbles that cause gas to accumulate to the point that the sludge becomes less dense than the saturated liquid above it. Until recently, this situation caused cyclic "roll-overs" of the sludge, thereby releasing the stored flammable

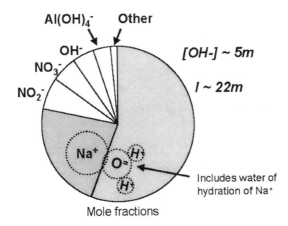

Figure 7. Principal Components In Concentrated Alkaline Tank Waste Solutions.

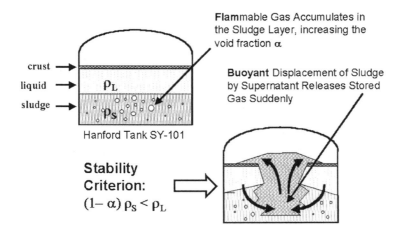

Figure 8. Buoyancy Instability Causing Cyclic Releases of Flammable Gas

gas suddenly as depicted in Figure 8 (5). The geometry, growth, and release of the bubbles need to be understood on a microscopic scale to be able to predict when such problems can occur, for example as different wastes are blended during waste treatment operations.

Heterogeneous Radiolysis

Because high-level tank waste usually includes fine solids in a liquid, there is a large interfacial area at which heterogeneous radiolytic processes can occur in addition to homogeneous processes in the liquid phase. The situation near a surface being bombarded by a secondary electron from absorption of radiation is depicted in Figure 9. Radiolysis generates oxidizing species (NO_x, O^-, O_2) that degrade organic complexants and initiate gas generation reactions. These processes are poorly understood compared to homogeneous radiolysis. Understanding them helps us understand the aging of high-level wastes and to resolve questions about storing them safely.

Recent Accomplishments of the EM Science Program

OST's Environmental Management Science Program (EMSP) has led to notable advances addressing important issues pertaining to radiochemical separations in high-level tank wastes. Three examples are:

- Technetium in HLW had been thought to exist as pertechnetate, a very soluble, anionic, heptavalent form. However, studies of the performance of anion exchangers have revealed that some technetium does not exist as pertechnetate. Technetium was introduced into the tank wastes as the pertechnetate anion, thus separations schemes have been based on this. However, years of thermal and radiolytic digestion in the presence of organic material may have produced stable, reduced technetium chelate complexes (6, 7). Existing Tc separations are selective only for pertechnetate anion; the complexed form is not separated and unfortunately is not easily oxidized. This revelation has spawned research that may lead to alternative separation methods.

- Solvent extraction has been explored at INEEL for separation of fission products and transuranic elements from dissolved calcine (if one were to process the calcine) and from the high-sodium acidic liquid wastes. The stability of extractants in these highly acidic solutions can be a problem. Candidate extractants include cobalt dicarbollide, for which considerable experience exists in Russia and the Czech Republic. A cesium extractant was developed at Argonne National Laboratory as part of an integrated fission product separations process (8). More recently, an alternative extractant, more stable in acidic solutions, has been developed at Oak Ridge National Laboratory (9) using a crown ether.

- The use of crown ethers is usually limited by the need for polar environments, weak efficiency, and dependence on matrix anions. However, modern computational chemistry techniques have been used to predict the behavior of proposed new extractants combined with spectroscopic and separation techniques to test predictions. This led to an extractant stable in strong acids as an option for separations at INEEL.

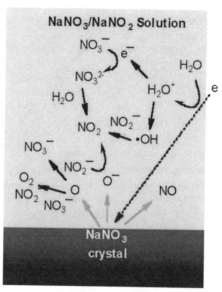

Figure 9. Heterogeneous Radiolysis in Tank Waste Near a Sodium Nitrate Crystal

- Crystalline silico-titanate (CST) has emerged as a commercially available material for exchanging cesium from alkaline tank waste solutions, but the ultimate disposal of the exchanger can be difficult if the nominal waste form is a glass because the solubility of the titanium in glasses is limited. Recent work has shown that heat treated crystalline silico-titanate (IE-911) with no additives is a viable waste form itself and is more Cs leach resistant than baseline borosilicate glass formulations (*10*). Heat treated IE-911 is multi-phase; the exchanged cesium appears to reside in a crystalline silicate phase, $Cs_2XSi_3O_9$. Water and hydroxyls are removed during heat treatment, eliminating the possibility of hydrogen generation during short-term or long term storage. Coarsening during heat treatment eliminates fines, and the thermal conversion process reduces the CST volume by 40%.

Conclusions

As the technical community conceives new technical solutions through investments in directed science, the solutions must be inserted within appropriate projects through which high-level waste or spent fuel are treated and disposed or stored. If problems with baseline technical approach are recognized only gradually, then it may seem that innovative solutions have only a brief "window of opportunity" to be inserted because resolution of problems with the baseline approach seems to be imminent. Actually, there may be a much wider window of opportunity, but we cannot act on this without the revelation that the problems requires a change in the baseline technical approach. If, as an act of faith, we make prudent investments in scientific investigations that provide information or technologies seemingly too late to fit into a project's window of opportunity, then potential solutions to unanticipated problems will be available if needed, serving essentially as insurance against the need to shift baseline schedules when unanticipated problems arise.

The diversity of science issues, and hence the diversity of experience, capabilities, and interests in the scientific community that need to be focused on spent fuel and high-level tank waste problems, are reflected in the diversity of spatial dimensions and of physical vs. chemical domains implied by the issues. A comprehensive discussion of the many science needs pertaining particularly to high-level tank waste is beyond the scope of this paper, but selected examples illustrate the nature and diversity of the issues, such as:

- The liquid phase in high-level waste tanks is so concentrated it is more like a loose crystal with all ions associated with each other than a set of ions each independently hydrated by water and separated by "free" water. Sophisticated thermodynamic models, such as Pitzer's, are required to predict phase equilibria or heat effects in such liquids.

- Because high-level tank waste usually includes fine solids in a liquid, there is a large interfacial area at which relatively poorly understood heterogeneous radiolytic processes occur that effect the aging of high-level wastes and associated issues such as safe storage.

- The geometry, growth, and release of the bubbles in tank waste sludge need to be understood on a microscopic scale to be able to predict when sudden gas releases can occur.
- There is a large interfacial area in tank waste sludges at which heterogeneous radiolytic processes can occur in the liquid phase and that, although important, are poorly understood compared to homogeneous radiolysis.

References

1. Stewart, T. L., J. A. Frey, D. W. Geiser, and K. L. Manke. 1997. *Overview of U.S. Radioactive Tank Problem.* Presented at the 1997 Annual American Chemical Society, Las Vegas, Nevada.

2. Karell, E.J., K. V. Gourishankar, and C. C. McPheeters, 1997. *Application of the Electrometallurgical Treatment Technique to Long-Term Disposition of DOE Spent Fuel,* presented at the American Nuclear Society 1997 Winter Meeting, November 16-20, Albuquerque, NM

3. Marschman, S. C. et al. 1998. *Radiolytic and Thermal Process Relevant to Dry Storage of Spent Nuclear Fuels.* 1998-1999 Progress Report, Project 60392, Environmental Management Science Program, Office of Science and Technology, Department of Energy, Washington, D.C. Accessible via the internet at http://www.etde.org/html/em52/60392.html

4. Defense Nuclear Facilities Safety Board. 1995. *Stabilization of Deteriorating Mark 16 and Mark 22 Aluminum-Alloy Spent Nuclear Fuel at the Savannah River Site.* Report DNFSB/TECH-7. Accessible via the internet at http://www.dnfsb.gov/techrpts/tech-7.html

5. Meyer, PA, ME Brewster, SA Bryan, G Chen, LR Pederson, CW Stewart, and G Terrones. 1997. *Gas Retention and Release Behavior in Hanford Double Shell Waste Tank,* PNNL-11536, Rev.1, Pacific Northwest National Laboratory, Richland, Washington.

6. Schroeder, N.C. et al. 1995. *Technetium Partitioning for the Hanford Tank Waste Remediation System: Anion Exchange Studies for Partitioning Technetium from Synthetic DSSF and DSS Simulants and Actual Hanford Wastes (101-SY and 103-SY) Using Reillextm-HPQ Resin.* LA-UR-95-4440, Los Alamos National Laboratory, Los Alamos, NM

7. Schroeder, N. C., K. R. Ashley, G. D. Whitener, and A. P. Truong. 1996. LANL Pretreatment: Technetium Removal Studies. LA-UR-96-4470, Los Alamos National Laboratory, Los Alamos, New Mexico.

8. E. P. Horowitz, M. L. Dietz, and M. P. Jensen. 1996. *A Combined Cesium-Strontium Extraction/Recovery Process.* In Value Adding through Solvent Extraction: Proceedings of ISEC '96, International Solvent Extraction Conference, Vol. 2, Eds. D. C. Shallcross, R. Paimin, and L. M. Prvcic, p. 1285.

9. Moyer, B.A et al. 1998. *Fission Product Solvent Extraction.* Proceedings of the Efficient Separations and Processing Crosscutting Program 1998 Technical Exchange Meeting, Augusta, GA, Mar. 17-19, 1998.

10. Balmer, M. L. and B. C. Bunker, BC. 1995. *Waste Forms Based On Cs-Loaded Silicotitanates.* Presented at the 97th annual meeting of the American Ceramic Society, Cincinnati, OH. PNL-SA-26071, Pacific Northwest National Laboratory, Richland, WA

Chapter 3

The Subsurface Contaminants Focus Area: Innovative Technology Development for Environmental Remediation

Van Price

TRW/WPI Support Team, 223 Gateway Drive, Aiken, SC 29803

Introduction

The purpose of this paper is to give a brief overview of soil, ground water, and buried waste issues across the United States Department of Energy (DOE) Complex, as these are the subject area for technology development by the Subsurface Contaminants Focus Area (SCFA). The intended audience is the science and technology community within universities and industry who may have something to contribute to DOE's efforts to reduce the environmental impact of past defense production activities. In these few pages it is not possible to present extensive detail, but references, including world-wide-web addresses, are provided for the reader who wants more information.

The United States Department of Energy is the successor organization to the Atomic Energy Commission (AEC) and the Energy Research and Development Administration (ERDA). Activities of these organizations spanned nuclear raw materials, reactor operations, chemical processing, weapons production, and research and development. Each of these areas produced a unique set of by-products, waste, and environmental effects.

While defense programs are still an important part of DOE's overall mission, this discussion is limited to environmental cleanup only. The objective of environmental cleanup is to reduce risk of injury and illness from the waste and contamination inherited as a legacy of past activities. One thing to keep clearly in mind is that none of our forebears set out to poison their children or harm our environment. Almost all of their activities were consistent with standard and prudent industrial practices of their time. Risks that we perceive today from industrial chemicals and radioactive materials were simply not well known in the past.

DOE's current cleanup activities span everything in the defense legacy resulting from uranium mining through weapons testing. Many details of the process are given in "Linking Legacies"(*1*). Table I is adapted from information in that document.

The Subsurface Contaminants Focus Area (SCFA)

DOE's Environmental Management (EM) organization includes a research and development arm, the Office of Science and Technology (OST or EM-50). Field activities of OST are organized into 'focus areas,' where "areas" means sets of similar problems. To assure that the focus areas are really in touch with cleanup problems in the field, DOE moved in 1995 to decentralize operation of the focus areas from headquarters to its field offices. SCFA is staffed as part of the EM organization in the DOE Savannah River Field Office (DOE-SR), but serves EM clients across the DOE Complex. Information on each of the focus areas is available through the DOE web site at: http://ost.em.doe.gov/ifd/ost/programs.htm.

At each DOE site, environmental restoration leading to cleanup and closure is a major activity. The total annual budget for restoration is on the order of six billion dollars. Within that, science and technology (OST) has been allocated $352M, $274M, and $220M in fiscal years 1997, 1998, and 1999 respectively.1 In FY99, science and technology received about four percent of the restoration budget, with approximately ten percent of this, or $24M going to SCFA. Of the $24M, $3.6M is available for development of new technology, and $11.2M is devoted to deployment assistance for previously-developed technology. $3.6M is used for technical assistance to bring expertise to bear, on a case-by-case-basis, on difficult environmental restoration problems. The balance of the $24M

---

1 These data were obtained from the DOE world-wide-web site at: http://www.em.doe.gov/budget/sect7.html on November 15, 1999.

Table I. Summary of Defense Uranium Cycle.

Activity	Description	Typical Legacy Waste
Mining of Uranium	Uranium ore is mined and treated to produce uranium oxide (yellow cake) and ground up rock waste	Mill tailings
Enrichment	Isotopic separation by diffusion of gaseous UF_6 through membranes or in centrifuges	Isotopically heavy UF_6 and yellow cake. Industrial solvents such as carbon tetrachloride. Hg
Reactor Operations	Uranium consumed, plutonium produced, tritium produced	Activation products - (Co-60, others), U, Pu, Am, H-3
Separation	Reactor fuel and target materials chemically separated	Tritium, Fission products (Sr-90, Cs-134,137, others), Pb, Hg, Cd, Tc-99, acids, caustics, solvents
Fuel and target fabrication		U, Cr, Li, Solvents – chlorinated organics
High Explosives (HE) manufacture	HE produced and machined	HE – TNT, RDX, Pb, Ba, nitrate
R&D for process and technology development	DOE labs are located across the country	Small amounts of everything
Fabrication	U, Pu machined	U, Depleted U (DU), Pu
Testing of nuclear warheads and detonation systems	DU mock-ups crashed, exploded	DU, HE, Ba, Pb
Testing of weapons	Weapons tested	U, Pu, Fission products, Tritium
Waste Disposal	Landfills, Seepage basins, Cribs, High level waste tanks, spills, leaks	Fission and activation products, caustics, solvents, tritium

goes to special projects and program administration. A portion of the EM Science Program budget, about $7.5M in FY99, is also allocated to remediation research.

SCFA's mission, simply put, is to provide DOE site problem holders with innovative solutions to their most pressing and difficult problems related to subsurface contamination. Subsurface includes any buried waste and fugitive contaminants in natural media such as soil, rock, and underground water. 2

DOE's Remediation Problem and SCFA's Scope

A number of estimates have been made of the extent and cost of the remediation problem for which DOE is responsible. The numbers here represent an order-of-magnitude estimate. Generally these are from Linking Legacies (*1*) and BEMR (*2*). About 9,000 'release sites' have been tabulated where some contaminant has potentially been released to the environment. About 475 billion gallons of ground water are in 5,700 migrating plumes. About 3 million cubic meters of buried waste are in DOE landfills, and DOE sites include over 70 million cubic meters of contaminated soil. Total remediation costs have been estimated to be between 200 and 300 billion dollars (*2*).

Most remediation is being handled with conventional technology. Projects are designed and executed with input from federal and state regulators and the local community. Occasionally problems arise which are not amenable to solution through conventional technology. These are elevated through the Site environmental restoration (ER) organization and brought to a Site Technology Coordinating Group (STCG) for study. STCGs in some cases may be able to alert the ER project manager to viable solutions. In other cases, STCGs develop problem descriptions called Technology Need and Opportunity Statements, which are the official mode of requesting Focus Area attention to the problems. These statements also include names and addresses of persons who can give further information about the problems.

About 200 such need statements fall within SCFA's purview. They change from time-to-time as new problems are found, better defined, or elevated in importance, and as old problems are solved. Generally the problem descriptions

---

2The ground-water surface-water interface and fluvial transport are emerging scope

are written at a high level of abstraction, but overall they are an excellent source of information about the problem solving needs of DOE. They may be accessed through links to STCG home pages at: www.em.doe.gov.

Each year SCFA performs an analysis of the site need statements. Data on contaminants, geologic settings, performance requirements for solutions, schedule, and other factors useful in determining types of solutions are extracted. These data are combined to define and defend the types of projects that SCFA will fund in the upcoming budget years. In addition to the data analysis, SCFA conducts site visits in which a team of technical experts is sent to each site to discuss problems with field project managers - the owners of these problems who initiate the individual STCG need descriptions.

During discussions of need statements, it often becomes apparent that there are logical, but unstated extensions to the need. An additional objective of site visits is to elicit emerging issues from site personnel. For example, validation, verification, and long-term monitoring of containment and treatment was identified as a key issue through site discussions. When faced with questions about in-situ subsurface remediation, most problem holders acknowledge that proof of effectiveness is an important component of remediation, and furthermore, the effectiveness must be for a long period of time. The issue, however, had not been fully identified in need statements from the sites.

Based on results of needs analysis and site visits, SCFA has defined five strategic work areas and eleven work packages. The strategies are:
- *Locate* and quantify subsurface contamination accurately
- *Contain* or stabilize leaks and buried waste "hot spots" in situ
- *Treat* or destroy mobile contaminants in situ
- *Remove* "hot spots" not subject to in situ treatment
- *Validate* and verify system performance.

Table II summarizes the current SCFA work packages. A work package is a segment of the overall scope of SCFA's effort, which may or may not be populated with actual funded projects in a given fiscal year. These were originally derived by rolling up key word summaries of site needs into groups with common features. Thus all site needs track to one or more work packages. (Some need statements include, for example, characterization, treatment, access, and validation components.)

Table II. SCFA Work Package Summary

Work Package #	Work Package Name	Objectives	Typical Technology Solutions	Typical Contaminants
SS-01	Characterization, Monitoring, Modeling, and Analysis	Location, Delineation, Characterization, Prediction	Geophysics, Chemical Sensors, "Smart Sampling," Models	Buried Waste, Spills Plumes
SS-02	Containment Barrier Systems	Prevention of Migration	Grout Walls, Permeation Grouts	Buried Waste, Spills Plumes
SS-03	Stabilization Technologies	Prevention of Migration	REDOX, ISV, Stabilization grouting	Spills, Plumes, Buried Waste
SS-04	Long-Lived Caps	Prevention of Migration	Caps, Covers, ALCD, Surface Barriers	Buried Waste
SS-05	In-Situ Passive-Flow Treatment	Contaminants on the Move	Reactive Barriers	Cr, U, DNAPL Chemicals
SS-06	Biological Treatment Systems	Dispersed Contaminants	Engineered Plants, Microbial Systems	DNAPL Chemicals, Hg
SS-07	Vadose Zone Treatment Systems	Contaminated Soil	Soil Mixing (KMnO4), Soil Flushing, Vapor Extraction, Gas REDOX	Metals, Rads, DNAPLs
SS-08	Saturated Zone Treatment Systems	Contaminated Aquifer	REDOX, Vapor Stripping, Co-solvent / Surfactant Release	Metals, Rads, DNAPLs
SS-09	Access and Delivery Systems	Delivery of Characterization Tools and Treatment Materials	Improved Cone Penetrometer, Horizontal Drilling, Sonic Drilling, Sampling methods	Metals, Rads, DNAPLs, especially in fractured rock or coarse gravels
SS-10	Hot Spot Removal	Contamination Source	Grout-to-Remove, Dig-face Characterization, Volume reduction methods	Buried Waste, Spills
SS-11	Validation, Verification, and Long-Term Monitoring of Containment and Treatment	Proof of Effectiveness for Long-Term Stewardship	Embedded Sensors, Statistical Sampling, Long-term monitoring	All

SCFA's Allied programs

There are several semi-independent sub-programs, which function in support of the Focus areas. SCFA works closely with CMST (Characterization, Measurement, and Sensors Technology), IP (Industry Programs), UP (University Programs), EMSP (Environmental Management Science Program), and ASTD (Accelerated Site Technology Deployment). In addition, there is a more autonomous group at WETO (Western Energy Technology Office in Butte, Montana) that works cooperatively with SCFA to develop and deploy remediation technologies.

CMST manages many projects that fall within SCFA's work packages one and eleven. Development and demonstration of chemical and radioactivity sensors and adaptation of these to novel access systems, such as the cone penetrometer, are done through CMST. CMST participates in SCFA Site Needs analysis visits. Their headquarters is in the DOE-NV Office in Las Vegas, Nevada.

If SCFA determines that a novel, but unproven, technology may be available in the private sector, IP develops a commercial solicitation package and requests bids for a demonstration of the technology.

UP maintains contracts with several universities. The scope of these contracts allows SCFA and other focus areas to access university expertise on selected projects. Both IP and UP are managed through the NETL (National Energy Technology Laboratory) in Morgantown, West Virginia.

EMSP is responsible for feeding the technology development pipeline. As noted above, in FY99 approximately $7.5M of the $224M OST budget funded basic research in environmental remediation projects.

ASTD encourages site cleanup project managers to deploy innovative technologies through cost sharing. Uncertainties associated with non-conventional technology carry over as threats to project budgets and schedules because of an increased risk of failure. Therefore some incentives are provided to reduce the threat. In addition to the potential cost and schedule savings promised by innovative technology, the ASTD program provides half the cost of deploying the innovative technology.

Each of the above programs maintains a www site[3] to which the reader is referred for further information. Most of the programs fund research and development within the DOE Laboratory system. EMSP and IP issue calls for proposals from non-DOE entities.

SCFA solicits and contracts primarily within the DOE laboratory system for technology development, but many of the DOE labs sub-contract to universities or companies. SCFA also solicits from non-DOE sources through IP and provides results of needs analysis to EMSP to assist with their development of language for calls for proposals.

Other SCFA Resources

In addition to funding R&D projects, SCFA also provides technical assistance to DOE sites through several mechanisms. These include ITRD (Innovative Technology Remediation Demonstration), TechCon (Technology Connection), and the SCFA Lead Laboratory.

ITRD works with Site personnel, regulators, and concerned citizens to resolve issues related to selecting an appropriate innovative technology to solve a site problem. Once feasibility issues are identified, ITRD can fund studies and small-scale technology demonstrations that lead to selection of the optimum path to solve a problem.

TechCon maintains a database of remediation technology vendors who have demonstrated their ability to solve difficult remediation problems. TechCon will help site problem holders identify available commercial technology and then sponsor a vendor forum where potential solutions are discussed with problem holders, at the appropriate DOE Site. They maintain a web site at Argonne National Laboratory - http://web.ead.anl.gov/techcon/.

SCFA has funded the Savannah River Technology Center (SRTC) to coordinate efforts of DOE Laboratory personnel across the DOE Complex in a "Distributed SCFA Support Laboratory" (referred to as the SCFA Lead Laboratory above). This virtual lab concept is new, and seems to be functioning well. Recently (November 1999) a "Lead Lab" team representing expertise drawn from across the DOE Complex met at DOE's Paducah, Kentucky, site

---

3 See links at www.envnet.org/scfa/

and developed recommendations to solve a set of problems that have recently gained prominence.

SCFA Project Selection

Each fiscal year, SCFA develops a scope of work for new projects based on performance gaps between STCG needs and available technology. These gaps are screened for problems that are best solved through technical assistance, for those that can be addressed through commercial sources, and for those that require basic research. Solicitation to commercial sources is carried out through IP, and basic research scope becomes the subject matter of EMSP calls for proposals.

SCFA issues a proposal call within the DOE Laboratory system for technology development and demonstration projects. The individual DOE laboratories may enlist university and commercial partners. Proposals received are ranked by a committee using criteria developed over a period of years. The committee consists of DOE employees representing technical expertise drawn from each DOE Field Office and thus representing remediation needs across the DOE Complex. Existing projects are also re-evaluated at this time to assure that they are still on track to solve a problem.

Funding is allocated to projects based on the ranking. The details of the within-DOE allocations process are outside the scope of this paper. Part of the allocation depends on a headquarters (OST) ranking of all Focus Area work packages, based on problem prevalence, cost, and risk data supplied to a OST database by the individual sites.

SCFA's Project Portfolio

SCFA's currently active projects are listed in Table III. Some of these are "Accelerated Site Technology Deployment" (ASTD) projects and are described in detail at: http://id.inel.gov/astd/index.html. Others are described at: www.envnet.org/SCFA.

Table III. Summary of FY99 SCFA Projects

Project Number*	Project Title	Comments	Assigned to: Work Package
AL19SS03	Non Traditional In Situ Vitrification Demonstration	Support for in-situ vitrification	SS-03
AL19SS40	Non Traditional In Situ Vitrification Demonstration	In-situ vit of soil with traces of plutonium. Melt begins at bottom.	SS-03
AL27C221	Environmental Measurement while Drilling ("C" designates CMST)	Gamma detector fitted to drill	SS-01
AL28SS40	Environmental Measurement while Drilling	Drilling support	SS-01
AL29SS41	Permeable Barrier Reactive Material Performance Database	Compilation of treatment barrier data	SS-05
FE06IP01	Industry Programs Technology Development Projects	Several projects	
FE07IP02	University Programs	Several projects	
ID78SS31	Field Demonstration of SEAR-NB	Isopropanol and surfactant used to mobilize DNAPL droplets at 'neutral bouancy'	SS-08
ID78SS32	RDTF Bioremediation	RTDF is a consortium working on bioremediation projects. Funded by DOE, EPA, DoD, others. www.rtdf.org	SS-06
OH18SS40	Mobilization, Extraction, and Removal of Metals & Radionuclides	Pump and treat for uranium to protect Great Miami Aquifer	SS-08
OR09SS30	Reactive Barrier Performance Monitoring & Verification	Follow-up to Oak Ridge treatment walls	SS-05
OR18SS31	In Situ Remediation of DNAPLs in Low Permeability Media	Hydrofracture emplacement of permanganate, iron filings, other treatment material	SS-07
OR18SS32	Field Demonstration of In Situ Chemical Oxidation	Study of effects of treatment zones, including on microbial populations	SS-08
OR18SS35	Intrinsic and Accelerated DNAPL Remediation	Support for RTDF bioremediation studies	SS-06
OR18SS41	In Situ Reactive Barriers at Y-12 Plant	Performance evaluation	SS-05
RL35C223	JCCEM Contaminant Transport Studies	Transport data from Russian nuclear weapons production areas	SS-01
RL38SS42	In Situ chemical Treatment of Soils by Gaseous Reduction	Hydrogen sulfide in soil to reduce and fix chromate	SS-07
SR16C221	SCAPS Logistics	Operation of CPT	SS-09
SR17C221	Innovative DNAPL Characterization Technology	Several characterization devices tied to Cone Penetrometer	SS-01
* First 3 chars =DOE site; 4th=year; if 5th="C", then CMST			

Special Projects

Several special projects are worthy of note here. SCFA is in the process of producing a book designed to capture the state of knowledge, art, and practice in vadose zone cleanup, "Vadose Zone Science and Technology Solutions." (*3*)

SCFA has recently held a workshop on the state of the art and practice in phytoremediation and will post the proceedings of this conference on the SCFA home page.4

International cooperative projects are active in Poland and Argentina.

Summary

SCFA is that part of the DOE Office of Science and Technology charged with developing and advocating innovative solutions to DOE's subsurface environmental problems. These problems are a legacy of the U. S. nuclear defense program, and span processes across mining, metallurgy, research, manufacturing, reactor operations, chemical processing, weapons testing, and waste disposal. They affect soil, ground water, and bedrock in all parts of the United States with climates ranging from cold and dry to wet and warm. SCFA is an umbrella organization with many components and programs, and with close ties to other DOE programs. Programs change to some extent each fiscal year but can be tracked through world-wide-web links given here.

Acknowledgements

This paper was improved by reviews of Lou Middleman, Bruce Erdal, P.G. Eller, and an anonymous reviewer. Opinions and remaining errors are those of the author and do not necessarily represent the opinions of the US Department of Energy.

References

Some World-Wide-Web sources are noted in the text, but not included in the reference list.

---

[4] SCFA home page: www.envnet.org/scfa/

1. BEMR - 1996 Baseline Environmental Management Report. This DOE Headquarters summary presents an assessment of the magnitude of cleanup problems at DOE Sites. It is available online at: www.em.doe.gov/bemr96/

2. Linking Legacies, DOE/EM-0319, 230p., January 1997. This is a good summary of the history and mission of the US department of Energy and its predecessor organizations. It also includes revisions of the contaminated media estimates in BEMR.

3. Looney, B.B., and Falta, R.W., editors, Vadose Zone Science and Technology Solutions (ISBN 1-57577-083-3), In Press, 2000

Radionuclide and f-Element Chemistry

Chapter 4

Actinide Behavior in Neutral Media

G. R. Choppin

Department of Chemistry, Florida State University, Tallahassee, FL 32306–4390

To address the separation and disposal of actinide elements present in storage at USDOE waste sites as well as to proceed with remediation of such sites, it is necessary to understand the chemical speciation of the actinides in the wastes and in the soil and waters to which they may be released. The speciation is a function of a complex interaction of the redox behavior, hydrolytic effects and interaction with residual separation agents in the waste and/or the natural ligands in the environment as well as of the sorptive and solubility characteristics of the various species formed in solution. Uncertainties encountered in using present data to model oxidation distribution of plutonium are discussed. The need to validate such data for use in designing effective separation agents for nuclear wastes and to model properly, the patterns of actinide migration in contaminated soils and waters by use of oxidation analogs is discussed.

Introduction

From the initial experiments on the production and subsequent isolation of transuranium elements, the separation processes have been conducted in acidic media. With the end of the cold war and the consequent shift in emphasis from production of these materials to treatment and disposition of the accumulated by-product radioactive wastes, it has been necessary in the U.S. to develop an adequate knowledge of the behavior of the actinide elements in neutral and basic media. A large amount of the wastes are stored in tanks in solutions of pH $\geq$ 12, so treatment, speciation and separations in basic solutions must be considered a high priority. Equally important is the related need to understand the behavior of the actinides which have been or will be released (as a result of treatment and storage activities) to soils. In such soils, the ground and surface waters are in the neutral range (pH

~ 6-8). In this paper, the behavior of the actinides are discussed for such neutral media.

Much research is being conducted at present on actinide behavior in contaminated soils of former weapon production sites. There is a significant amount of earlier literature on the behavior in natural waters of actinides from above-ground nuclear testing. In evaluating these data, it is useful to divide actinide species in the environment into two classes: source-dependent and source-independent [1]. The source-dependent species are not at thermodynamic equilibrium with their surroundings but remain in the forms in which they were injected into the environment. They may be refractory oxides, kinetically stable complexes and ions occluded in soluble particles. The behavior of these species allows only limited generalizations. Eventually, the actinides may be released from such forms at which time their geochemical behavior becomes more amendable to modeling. Source-independent species have kinetics which are sufficiently rapid that equilibrium (or, at least, steady state) modeling describes the net behavior over a period of time. It is these active species which are useful in laboratory and field studies for developing a geochemical data base for prediction of actinide behavior in aquatic systems. At bomb test sites, the source-dependent species are significant, whereas at reprocessing and waste sites, source-independent species are present and determine the environmental behavior.

Some Chemical Properties of Actinides

A brief review of some relevant chemistry of the actinides is useful. Plutonium is the actinide of greatest environmental interest and also has the most diverse chemical behavior (2). Plutonium can be present in aqueous solutions of the pH and E_H range found in nature in four oxidation states, the III, IV, V, and VI. In its different oxidation states, plutonium behavior resembles closely that of the other actinides in a particular oxidation state. The III and IV states of actinides are simple hydrated cations, while the V and VI states form dioxo cations, AnO_2^+ and AnO_2^{2+}, respectively. For plutonium, the lower oxidation states are stabilized by more acidic conditions and the higher states become more stable with increasing basicity. Such trends may be negated by factors such as complexation, hydrolysis, etc., which can reverse the relative stability of the different oxidation states. Disproportionation reactions are a common feature of the IV and V states of plutonium in acidic solutions, but are unlikely to be of concern at environmental concentrations and pH, It is not uncommon for several oxidation states of plutonium to coexist in the same solution.

Uranium and neptunium can exist in the same 4 oxidation states in solution as plutonium. However, the potentials vary for the different couples much more than do those of plutonium (see Figure 1) with the result that in oxic wet soils, NpO_2^+

50

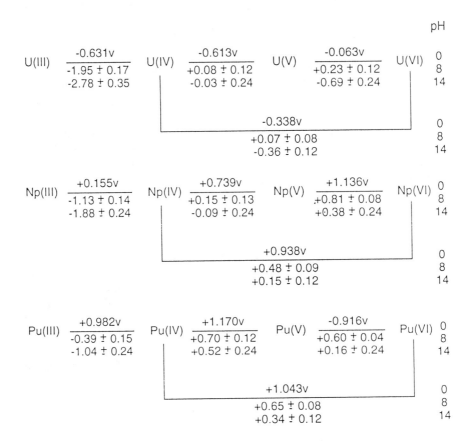

Figure 1. Reduction potentials (v) for U, Np and Pu at pH = 0/8/14.

and UO_2^{2+} are the common forms and Np^{4+} and U^{4+} are found in reducing media.

Actinides in the different oxidation states may have different migration rates since these depend on solubility, sorption/desorption phenomena and interactions with inorganic anions and organic material, all of which are greatly influenced by the actinide oxidation state. For example, in anoxic waters, U(IV) forms insoluble, polymeric, mixed hydroxides and carbonates which sorb in the soil. In oxic waters, U(VI) is capable of forming more soluble species, e.g., $UO_2(CO_3)_3^{4-}$, allowing greater migration rates.

Whatever the oxidation state, actinide cations interact with anionic species by ionic bonding. The effective charge on the plutonium atom in PuO_2^+ and PuO_2^{2+} is ca. +2.2 and +3.2, respectively (3). As a result, the trend in complexation strength is:

$$Pu^{4+} \succ PuO_2^{2+} \succ Pu^{3+} \succ PuO_2^+ \qquad (1)$$

Hydrolysis occurs for actinides of all these oxidation states at the pH values of natural waters and serves to remove them from solution by precipitation or by sorption to suspended particles and to the surface of ionic minerals. For any particular oxidation state, the log of the stability constants for complex formation typically have a linear correlation with the ligand pKa values of the ligands for related ligands of similar charge and dentation. Due to the problems of maintaining a unique oxidation state for plutonium in solution and of competing with hydrolysis, reliable values for stability constants with many anions of environmental interest are lacking. However, with caution, it is often feasible to use data from other actinides (e.g., Am(III), Th(IV), Np(V)) and U(VI) to estimate plutonium values, as discussed subsequently.

The other actinides are somewhat simpler in their chemistry, as their redox behavior is more restricted. Except in very strongly oxidizing systems, americium is present in the trivalent state and thorium in the tetravalent. Uranium is likely to be found either in the IV or VI state, while neptunium more commonly exists in the IV or V state. In natural waters, redox changes may result by interaction with peroxide formed by photolysis of humic substances present in the water. For solutions with higher amounts of radioactive species, radiolysis can result in formation of H_2O_2, ClO^-, etc. which can also produce redox reactions in actinides. Figure 2 is a schematic summary of the reactions to be considered in interpretation of the behavior of plutonium in neutral waters.

Models for Waste Treatment and for Migration in Soils

Predictive models are needed for migration of metal ion species in soils. The major drawback of such models is the large amount of thermodynamic and kinetic data necessary to incorporate the multitude of possible processes. Questions arise

52

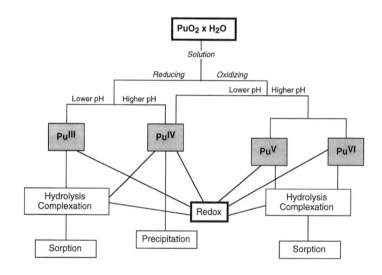

Figure 2. Schematic diagram of possible reaction of plutonium in natural waters.

as to whether all relevant chemical species in an ecosystem have been included in the calculations. Poorly defined humic interactions complicate the system, as do colloidal interactions (which are rarely incorporated). As the models are based commonly on thermodynamic data, most of the speciation calculations neglect kinetic contributions to the speciation in the ecosystem, which may lead to steady state rather than equilibrium conditions. These chemical models may have more value as guides to defining what further experimental analysis is needed rather than for the quality of their predictive attributes [4].

The lability of plutonium with respect to variation of its oxidation state distribution greatly complicates measurement of stability constants due to the uncertainty of the oxidation state present. This difficulty is a major problem in tracer level studies where it is not possible to use spectroscopic signatures to measure the relative concentrations of the different oxidation states. The major factors influencing the oxidation state distribution of plutonium in aqueous solution include E_H and pH, the presence of complexing ligands and radiolytic effects in high activity solutions. A limiting factor of the concentration of plutonium in neutral/basic solutions is the very strong hydrolysis of Pu(IV). For example, in such solutions, the most stable species of dissolved plutonium is PuO_2^+; however, the insolubility of $Pu(OH)_4$ results in extremely low concentration of Pu^{4+} such that the resulting redox equilibrium limits PuO_2^+ to concentrations of ca. 10^{-6}-10^{-8} M.

In conclusion, to perform useful modeling, it is very important to have reliable knowledge of the oxidation state (or states) present under the experimental conditions and to have stability constants for any complexes which may be formed by the metal (e.g., plutonium) ion in these oxidation states. Obtaining such information of plutonium at submicromolar concentrations is a formidable challenge.

Complexation and Hydrolysis Speciation

The difficulty in defining the distribution of plutonium between the four oxidation states which can be present in waters of neutral/basic pH has resulted (due to incorrect assumptions of this distribution) in a wide variation in reported values of plutonium complexation and hydrolysis constants needed for valid modeling of separation and remediation methods and for performance assessment of repositories for nuclear wastes.

Many research groups have used data for oxidation state analogs (i.e., metal ions with similar chemical behavior to the plutonium species of interest) to model the plutonium behavior or to validate the plutonium values to use. Ideally, the chemical behavior should be identical for a particular plutonium state and its oxidation state analog. However, such ideality is rare and, often, not necessary as modeling answers with 20-50% uncertainty are acceptable in many cases.

Validation that the data of particular analogs can be acceptable for use in modeling the chemical behavior of plutonium in its different oxidation states is, therefore, of major value in the use of such data in performance evaluation assessments of nuclear storage sites, separation processes, remediation of contaminated areas, etc. A major advantage associated with the oxidation state analogs is that macro concentrations of the analogs can be studied since the oxidation state distribution is not a function of the concentration as it is, for example, for plutonium. This allows use of such techniques as Raman, luminescence, UV-vis and NMR spectroscopies. For non-radioactive analogs, the absence of redox perturbation due to radiolytic effects is a further advantage, as plutonium behavior at tracer levels is free of such perturbations.

In all four oxidation states of interest, plutonium forms strongly ionic bonds (hard acid cations) and interacts preferentially with hard base donor atoms. Thus, within a specific oxidation state, the relative strength of the resulting bonds between the plutonium cation and any particular ligand is dependent primarily on the cationic radius. Accordingly, the primary requirements for choice of the cations used as oxidation state analogs are that they are in the same oxidation state, are hard acid cations, and have a radius similar to the ion of the plutonium oxidation state of interest.

Actinide Analogs

Analogs for Pu(III)

While Pu(III) is important in strongly acidic and in anoxic media, it is not present in oxic surface waters nor is it a factor in nuclear wastes stored in strongly basic media. For oxidation state III, the correlation in the behavior of Pu(III) with that of other trivalent 4f and 5f elements is very good (5). The 4f analog element cations allow a variety of other techniques to be employed at higher concentrations. For example, for Nd(III) the hypersensitive spectral absorption bands (17373, 19198 cm^{-1}) are sensitive to changes in complexation, symmetry, etc. (6). Eu(III) has very useful luminescence characteristics which can be used to measure the number of binding donor sites of a ligand from the shift in wavelength of the $^5D_0$-$^7F_0$ transition at 17276 cm^{-1} for the hydrated Eu(III) ion (7). Also, the residual number of waters of hydration in the inner sphere can be calculated from the half-life of the luminescence decay. Finally, the number of species present can be obtained from the number of lines in the $^5D_0$-$^7F_0$ excitation spectrum of Eu(III) (8). Both the shifts and the relaxation times of ^1H and ^{13}C NMR spectra induced by the paramagnetism of Eu(III) and Gd(III) can provide information on the structural characteristics of ligands bound to the lanthanide ion (9). Pu(III) cannot be studied by most of these techniques.

Am(III) and Cm(III) are also useful analogs. ^{241}Am is a gamma emitting radionuclide which simplifies sample preparation and counting, while ^{248}Cm is a relatively long-lived nuclide with excellent luminescence properties. Even at 10^{-11}-10^{-12} M concentrations, the luminescence half-life of ^{248}Cm can be used to measure the number of water molecules bound to the Cm(III) in a complex (10). Table 1 compares hydrolysis and carbonate complexation constants for Pu(III) and other 4f and 5f trivalent analogs.

Analogs for Pu(IV)

Few reliable data are available on stability constants for Pu(IV) and, unfortunately, tetravalent plutonium is the cation for which the oxidation state analog approach has the greatest uncertainty. Th(IV) can be used as an oxidation state analog for Pu(IV) but for Th(IV), the complexation is somewhat weaker, as is the hydrolytic tendency. Th(IV) has no useful spectroscopic properties, but its complexes can be studied by potentiometry and NMR (^{1}H and ^{13}C spectra of ligands). The use of Th(IV) as a model requires adjustment of its data to fit Pu(IV) behavior as the ionic radii (5) are relatively different (1.048 Å for Th(IV) vs. 0.962 Å for Pu(IV). U(IV), Np(IV) and Ce(IV) are closer in radii to Pu(IV) but control of their tetravalent oxidation state can be a problem in their use as an analog.

Analogs for PuO$_2^+$ and PuO$_2^{2+}$

Since the concentration of PuO$_2^+$ is limited in acid systems to tracer levels by disproportionation to Pu(IV) and Pu(VI) and in neutral and basic systems by the redox equilibrium with extremely insoluble Pu(OH)$_4$, little data has been reported to provide comparison with analogous NpO$_2^+$ data. The available data for 1:1 complexation does show very good agreement between NpO$_2^+$ values and the few values available for PuO$_2^+$ (usually within 2 σ). Moreover, the lack of analogs is not a major problem as PuO$_2^+$ is a weakly complexing cation whose concentration is predominantly determined by its redox equilibrium with the III, IV and VI oxidation states and in both acidic and basic solutions, it is present only in tracer level concentrations. To the extent that PuO$_2^+$ forms complexes or is hydrolyzed, its behavior can be adequately modeled using NpO$_2^+$ data.

PuO$_2^{2+}$ shows a close similarity to UO$_2^{2+}$ in chemical behavior. The lower specific activity of common uranium isotopes and ^{237}Np allows use of a much broader variety of spectroscopic and other techniques which require 10^{-4}-10^{-3} M concentrations, providing more information on the systems under study. Both NpO$_2^+$ and UO$_2^{2+}$ can be studied by Raman spectroscopy (at 770 cm^{-1} for NpO$_2^+$ and

873 cm^{-1} for UO_2^{2+}). In addition, the fluorescence of UO_2^{2+} provides a useful spectroscopic tool analysis and characterization.

In Table 2 some stability constants of plutonium and its analogs are listed to indicate the degree of similarity in their values for the IV, V and VI states.

Approach to Estimations of log β_n Values

The ionic nature of f-element bonds would suggest that data of analog cations could be refined by correcting for the differences in the radii of Pu(IV) and the tetravalent analogs (e.g., Th) using the equation:

$$\log \beta_n^{corr} = \log \beta^{exp}[d(Th)/d(Pu^{IV})] \qquad (2)$$

Using this equation, log β_1 = 12.9 for $ThOH^{3-}$ formation gives a value of 14.1 for log β_1 of $PuOH^{3+}$. This value compares with that of 13.3 reported in ref. (12) for $PuOH^{3+}$ formation. A similar calculation for $NpOH^{3+}$ formation gives a value of 13.1 for log β_1 for $PuOH^{3+}$.

It has also been proposed that the Born equation can be used to validate reported stability constants (13). In this approach the estimate values are calculated by a modification of the Born equation:

$$\log \beta_n(A) = \log \beta_n(B) \cdot (Z_A/Z_B) \cdot (D_B/D_A)(d_B/d_A) \qquad (3)$$

where: Z_i = "effective" charge on ion i; D_i = "effective" dielectric constant for ion i; d_i = Shannon radius for ion i   (14).

The effective charges are +3 for M(III), +4 for M(IV), +2.2 for MO_2^+ and +3.3 for MO_2^{2+} (3). The effective dielectric constants which have been used successfully in the Born equation to estimate stability constants are 65 for MO_2^+, 57 for M(III), 55 for MO_2^{2+} and 40 for M(IV) systems (15). The radii used were for a coordination number of 8 for M(III) and M(IV) and of 6 for MO_2^+ and MO_2^{2+} (14). Table 3 compares values of log β_{101} calculated by this approach using experimental values of Sm(III) to calculate values of Th(IV) NpO_2^+ and UO_2^{2+} with the experimental values for those ions. The agreement indicates that acceptable values can be calculated for plutonium complexation.

A primary use of these values would be in screening values reported in the literature whose uncertainty is in question due to the researchers' failure to consider adequately experimental problems due to redox, sorption, precipitation, ternary complexation, etc. In lieu of directly measured stability constants, these values may have use, also, in preliminary sensitivity modeling to assess the need for research to obtain more valid constants.

Table 1

Stability Constants for Pu(III) and Analog (12)

Reaction: M + X = MX

MX	OH			CO_3^{2-}		
	$\log \beta_1^0$	$\log \beta_2^0$	$\log \beta_3^0$	$\log \beta_1^0$	$\log \beta_2^0$	$\log \beta_3^0$
Sm(III)	6.8	13.0	17.1	8.6	13.2	-
Am(III)	7.6	13.9	16.3	7.8	12.3	15.2
Pu(III)	7.5	11.5	15.5	7.5	12.4	-

Table 2

Comparison of the log β_{101} Values for Complexation of Pu and Analog Ligands

Metal	F^-	SO_4^{2-}	NO_3^-	HPO_4^{2-}
		a. An(IV)		
Th	7.45	3.25	2.5	10.8
U	7.76	3.50	2.1	-
Np	7.54	3.51	2.2	-
Pu	7.61	3.66	2.6	12.9
		b. An(VI)		
UO_2^{2+}	6.1	3.5	-0.6	7.24
NpO_2^{2+}	5.7	3.4	-0.9	8.18
PuO_2^{2+}	5.7	-	-	8.19

Calculation of "Total" vs "Free" Plutonium

In neutral and basic media, plutonium in the III, IV and VI oxidation states exist in very low concentrations of the uncomplexed, hydrated cations. To reflect the magnitude of the complexed vs "free" plutonium, the effect of hydrolysis and complexation by carbonate anions has been calculated for Pu(III), Pu(IV), Pu(V) and Pu(VI) systems using pH = 8.0, $[CO_3^{2-}] = 10^{-5}$ M and the metal-carbonate stability constants from ref. 12. The ratio of $[PuX]:[Pu^{n+}]$ was calculated for the reaction:

$$Pu^{n+} + mX^{n-} = PuX^{n-mp} \qquad (4)$$

(where $Pu^{n+} = Pu^{3+}$, Pu^{4+}, PuO_2^{1+} or PuO_2^{2+}, $X^{m-} = OH^-$ or CO_3^{2-}) by the equation

$$[PuX_m]/[Pu] = \beta_m[X]^m \qquad (5)$$

The calculated values are listed in Table 4 for species which are in at least 5% relative abundance in the total species concentration. As can be seen, for Pu(III) and Pu(IV), hydrolysis is more significant than is carbonate complexation ($[CO_3^{2-}] = 10^{-5}$ M was chosen as this is the approximate concentration of neutral surface waters), whereas for PuO_2^+ and PuO_2^{2+}, the reverse order is found.

Redox Effects

Although a great amount of study has been given to actinide behavior in neutral/basic media, the experimental difficulties are such that the data still have large degrees of uncertainty about their validity. Much use has been made of Pourbaix diagrams of E_H vs pH ($E_H = E_H° + \{0.059/n\} \cdot \log [Ox]/[Red]$). A typical diagram is shown in Figure 3 which indicates that at pH = 8, $[PuO_2^{2+}]_T \succ [PuO_2^+]_T$, whereas the reverse is observed both in natural, oxic waters and in the laboratory.

It is commonly assumed in redox calculations for plutonium speciation that the E_H half cell value to use is that of the O_2/H_2O couple. In oxic waters, this couple has an $E_H = 0.76v$ at pH 8 and the diagram indicates [Pu(IV)] as the dominant oxidation state species in solution. Using the most reliable half cell values for plutonium at pH = 0 (Figure 3), but ignoring hydrolysis and complexation, gives ratios for $E_H = 0.76v$ equal to:

$$\log[Pu(V)/[Pu(VI)] = 2.71 \qquad (6)$$
$$\log[Pu(V)]/[Pu(IV)] = 25.0$$

Table 3

Estimates from the Born Equation Based on Corrections of log β(Sm)

$$\text{Log } \beta(An(X)) = \log\beta(Sm(III))\frac{Z_x\, d(III)\, D_{III}}{Z_{III}\, d(X)\, D_x}$$

Log β_1 : Estimates vs. Experimental

Ligand	Th(IV)	Np(V)	U(VI)
Acetate	3.8 vs. 3.9	1.1 vs. 0.9	2.5 vs. 2.5
Oxalate	9.5 vs. 8.2	2.7 vs. 6.3	6.3 vs. 6.0
Citrate	15 vs. 12	3.5 vs. 2.5	8.8 vs. 7.4
NTA	22 vs. 13	6.0 vs. 6.8	14 vs. 10

Table 4

Values of $[PuX_m]/[Pu]$ Calculated for pH 8, $[CO_3^{2-}] = 10^{-5}$ M

Pu	Ligand	Species	Ratio	Ligand	Species	Ratio
(III)	OH^-	1:1	$10^{1.6}$	CO_3^{2-}	1:1	$10^{1.5}$
(III)		1:2	$10^{1.9}$		-	
(IV)		1:4	$10^{22.2}$		1:3	10^{15}
(V)		1:1	$10^{-1.4}$		1:1	10^{0}
(VI)		1:1	$10^{2.7}$		1:3	10^{7}
(VI)		1:2	$10^{3.5}$		-	

Ratios of $[Pu]_T/[Pu]_F$		
Pu	pH 8, no $[CO_3^{2-}]$	pH 8, $[CO_3^{2-}] = 10^{-5}$ M
III	$10^{2.1}$	$10^{2.2}$
IV	$10^{22.2}$	$10^{22.2}$
V	$10^{0.02}$	$10^{0.3}$
VI	$10^{3.6}$	$10^{7.0}$

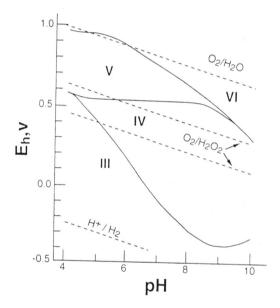

Figure 3. Pourbaix diagram, E_H vs pH, for plutonium at low concentrations in low ionic strength solutions. The solid lines represent equal concentrations of the adjacent oxidation strengths; hydrolysis and carbonate (10^{-5} M) complexation included in concentration calculations.

When hydrolysis is included (calculated from the values in ref. 12), the ratios have the values:

$$\log[Pu(V)]/[Pu(VI)] = -0.7 \qquad (7)$$
$$\log[Pu(V)]/Pu(IV)] = 2.8$$

When both hydrolysis and carbonate (Table 2) complexation are included, the ratios are:

$$\log[Pu(V)]/[Pu(VI)] = -4.0 \qquad (8)$$
$$\log[Pu(V)]/[Pu(IV)] = 3.0$$

These three sets of values of the redox patterns of plutonium do not agree with the prediction from Figure 3 (i.e., $\log[(Pu(VI)] \succ [Pu(V)])$, as experimentally it has been well documented that Pu(V) is the dominant oxidation state of soluble plutonium in waters of pH=8. Figure 3 is based on an E_H value of 0.76v indicating that the O_2/H_2O half cell is not the source of the observed redox patterns in natural oxic waters.

The available data indicates that $[Pu(VI)] \succ 10^{-2}[Pu(V)]_T$ in natural (and sea) waters as well as in laboratory studies in ca. 1 M NaCl solution. If $[Pu(V)_T] = 10^{-2}$ M, an upper limit can be set on the E_H value of 0.41v. We have not attempted to relate this value to any half-cell reaction which might be operative in natural waters.

Estimating an E_H to fit the (V)/(IV) data is more difficult as it is very uncertain what value to use for $[Pu(IV)]_T$. The Pu(IV) concentrations measured are related to the size filter used in the separation procedure - the smaller the filter aperture, the lower the concentration. Also, calculations based K_{SP} values are not reliable. The hydrolysis reaction

$$Pu^{4+} + 4\ OH^- = Pu(OH)_{4\ (am)} \qquad (9)$$

has been measured to have a log $K_{SP} = -56$ (16). For pH = 8, this gives a concentration of $[Pu^{4+}] = 10^{-32}$ M. From the ratio of $[Pu(IV)]_T$ to $[Pu(IV)]_{free}$, a concentration of ca. 10^{-10} M would result for $[Pu(IV)]_T$ when hydrolysis species are included. For concentrations of $Pu(V)_T$ measured in oxic, pH = 8 waters of 10^{-7}-10^{-8} M, from the ratio of $[Pu(V)]/[Pu(IV)]$ at $E_H = 0.76$v, the $Pu(IV)_T$ concentration would be ca. 10^{-10}-10^{-11}, which is in reasonable agreement with the estimate from the K_{SP} and stability constant values.

Conclusions

In summary, even a brief analysis of measured redox speciation of plutonium indicates the need for better measurements of the soluble species and more reliable stability (and hydrolysis) constants for use in modeling analyses in order to separate soluble species behavior from sorption, colloidal transport, etc. effects. In the absence of reliable constants, values of the stability constants of the appropriate oxidation state analogs of Pu(III), PuO_2^+ and PuO_2^{2+} can be used in modeling the behavior of plutonium in natural systems. Corrections based on the difference in radii can provide estimated values which can be used in preliminary modeling. Modeling with reliable values, measured or estimated, can provide valuable understanding of the role of soluble vs colloidal species which, in turn, is strongly influenced by the predominant oxidation state(s) present in the system.

Preparation of this paper was done under an USDOE EMSP contract to Florida State University.

References

1. Nelson D.M., Larsen B.P., Penrose W.R., *Symposium of Environmental Research for Actinide Elements*, Pindar J.E., Alberts J.J., McLeod K.W., Schreckhise R.G., (Eds), DOE86008713, NTIS, Springfield, VA., 1987.
2. Carnall W.T., Choppin G.R., (Eds), *Plutonium Chemistry*, ACS Symposium Ser. 216, Am. Chem. Soc., Washington, D.C., 1983.
3. Choppin, G.R., Rao, L. F. *Radiochemica Acta*, **37** 143, 1984.
4. von Gunten H.R., Benes P., *Radiochimica Acta* **69** 1, 1995.
5. Choppin G.R., Rizkalla E.N.: in *Handbook of the Physics and Chemistry of Rare Earths*, Vol. 18, Gschneidner Jr. K.A., Eyring L., Choppin G.R.., and Lander G.H., (Eds) North-Holland Publ., Chap. 128, 1994.
6. Beitz J.V.: in *Handbook of the Physics and Chemistry of Rare Earths*, Vol. 18, Gschneidner Jr. K.A., Eyring L., Choppin G.R.., and Lander G.H., (Eds) North-Holland Publ., Chap. 120, 1994.
7. Choppin G.R., Wang Z.M.: *Inorg. Chem.* **36** 249, 1997.
8. Horrocks D.W., Albin M.: in *Progress in Inorganic Chemistry*, Vol. 31, Lippard S.J., (Eds) John Wiley & Sons, New York, 1984.
9. Powell D.H., Gonzales G., Tissieres V. Mickei K., Brucher E., Helm L., Merbach A.E.: *J. Alloys Comp.* **207/208** 20, 1994.
10. Beitz J.V., *Radiochim. Acta* **52/53** 35, 1991.
11. Kimura T., Choppin G.R.: *J. Alloys Comp.* **213/214**, 41, 1994.
12. Shibutani S.. et al., JNC Thermodynamic Database for Performance Assessment of High-Level Radioactive Wastes Disposal System, JNC, 1998.
13. Choppin G.R., *Radiochim. Acta* (in press).
14. Shannon R.D. *Acta Cyst.* **A32** 751, 1976.
15. Choppin G.R., Rao L.F.; *Radiochim. Acta* **37** 143, 1984.
16. Rai D., *Radiochim. Acta* **35** 97, 1997.

Chapter 5

The Aqueous Complexation of Eu(III) with Organic Chelates at High-Base Concentration: Molecular and Thermodynamic Modeling Results

Andrew R. Felmy, Zheming Wang, David A. Dixon, Alan G. Joly, James R. Rustad, and Marvin J. Mason

Pacific Northwest National Laboratory, 3350 Q Street, Richland, WA 99352

The solubility of $Eu(OH)_3(c)$ was investigated over a broad range of NaOH concentrations extending to 7.5 molal in the presence and absence of added EDTA, and NTA. Thermodynamic analysis of the experimental data indicates that formation of mixed metal-chelate-hydroxide complexes is occurring in solution. These mixed metal-chelate-hydroxide complexes increase the effectiveness of the chelates in solubilizing Eu(III) under high base conditions. The number of hydroxide ions bound to the metal-ligand complex varies with the chelate type. The structural and energetic reasons for this variability in solution coorelate with density functional theory (DFT) calculations for gas phase La(III) clusters. In the case of EDTA solution complexes form with only one EDTA and one hydroxyl ions associated with the metal-chelate complex (i.e. $Eu(OH)EDTA^{2-}$) whereas in the case of NTA there are either one or two NTA molecules and at most only one hydroxyl associated with the metal-chelate complex (e.g. $EuOHNTA^-$ or $EuOH(NTA)_2^{4-}$). The first determination of the stability constants for two of these species ($Eu(OH)EDTA^{2-}$ and $EuOH(NTA)_2^{4-}$) as well as for the $Eu(OH)_4^-$ species are proposed. An aqueous thermodynamic model is presented which describes all of the available thermodynamic data for these chemical systems, to high ionic strength, and correlates with the DFT calculations. The implications of these results in regard to processing Department of Energy (DOE) high level wastes is also discussed.

Introduction

It has long been known that the organic chelates ethylenediaminetetraacetic acid (EDTA), and nitrilotriacetic acid (NTA) can form strong aqueous complexes with a variety of metal ions, including the rare-earths and trivalent actinides [Am(III), Cm(III), and Pu(III)] [1-3]. However, in the case of EDTA , the ionic size of the trivalent actinides and rare-earth ions makes it impossible for only simple sexidentate complexes to form and such solution species must involve additional coordinated water molecules[4,5]. These coordinated water molecules are then subject to displacement by other ligands in solution and this has led to the identification of several different types of mixed ligand complexes[6-8] including complexes where hydroxide ion has substituted for coordinated water under high pH conditions [1,3,9]. Similar mixed ligand complexes have also been identified for other metal ions including Cu(II), Ni(II), Co(II), and Al(III) [10-12]. In addition, the possibility exists, based upon analogy with Fe(III) complexes[13,14] that bridging dimeric complexes involving HEDTA, and possibly EDTA, can also form at high pH[15]. Formation of such mixed ligand complexes can have significant implications for the chemical speciation of trivalent actinides and rare-earths under basic conditions. Such highly basic condition are precisely the conditions encountered in high-level-waste (HLW) storage tanks at Hanford and other DOE sites which also can contain significant concentrations of chelating agents including EDTA and NTA [16-18]. The presence of these chelating agents along with trivalent actinides (Am(III), Cm(III)) can result in solubilizing the actinides from the sludge[19] and creating difficulties in subsequent actinide/radionuclide separation processes. As a result, developing a complete understanding of the effects of organic chelating agents on the solubility and aqueous speciation (including the identification of mixed chelate hydroxide complexes) of the trivalent actinides in HLW solutions is a key factor in developing effective waste processing strategies.

In this paper a combined experimental and molecular/thermodynamic modeling approach is used to study the hydrolysis and complexation of Eu(III) [a trivalent actinide analog] in NaOH solutions in both the presence and absence of the organic chelating agents EDTA and NTA. Eu(III) was selected for study since 1) it present in significant concentration in high level nuclear waste storage tanks at DOE sites which contain high base concentrations, 2) it is an excellent analog for trivalent actinides (e.g. Am(III), Cm(III), and Pu(III) which can also be present in tank solutions, 3) lanthanide species can be treated by high level molecular modeling simulations, 4) only limited experimental data on the hydrolysis and chelate complexation of lanthanides or trivalent actinides are available at high base (>1m) concentration, and 5) Eu(III) has excellent fluorescence properties for determining the molecular structure of the aqueous complexes. This combined approach consists of DFT calculations on gas phase clusters to help define the structure and relative stability of metal-chelate-hydroxide clusters, 2) use of Time Resolved Laser Luminescence Spectroscopy (TRLFS) to help determine the species present in solution, and 3) thermodynamic measurements and analysis of Eu(III)-NaOH-chelate solutions. The experimental thermodynamic data were obtained by solubility methods owing to the relatively low solubility of Eu(OH)$_3$ (c) under high base conditions.

The solubility studies were conducted over a broad range of NaOH concentrations in both the presence and absence of added EDTA and NTA.

Experimental Procedures

Eu(OH)$_3$(c) was prepared using the same procedure outlined by Rao et al.[20] for the preparation of crystalline Nd(OH)$_3$(c). The resulting material was examined by X-Ray Diffraction (XRD) and found to be fully crystalline Eu(OH)$_3$(c). This material was used in all subsequent studies.

The solubility studies using Eu(OH)$_3$(c) were conducted in a controlled atmosphere chamber filled with nitrogen gas to prevent the contact of the solutions with atmospheric CO$_2$. An upper limit of 0.01M in chelate concentration was selected since it is the approximate upper limit present in tank waste and at such chelate concentrations it is still possible to maintain a sufficient mass of solid Eu(OH)$_3$(c) in each tube. The NaOH concentrations ranged from 0.01M to 7.5M. Suspensions were sampled at various times to determine, or establish, equilibrium conditions. Sampling consisted of pH measurements, at NaOH concentrations lower than 0.1M, followed by centrifugation at 2000 g for 7 to 10 minutes. Aliquots of the supernate were filtered though Amicon-type F-25 Centriflo membrane cones with an approximate pore size of 0.0018 μ. Eu concentrations were determined by inductively coupled plasma mass spectrometry (ICP-MS). Total organic carbon, used to estimate dissolved chelate concentrations, was determined on selected samples using a Shimadzu TOC 5000A analyzer with a 680^{0}C furnace. Selected samples of the solid precipitates covering the entire range of NaOH concentration were analyzed by X-Ray diffraction analysis (XRD) after the end of the equilibration period. Fully crystalline Eu(OH)$_3$(c) was the only phase identified in the appropriate experiments. All studies were conducted at room temperature (22-23^{0}C).

The $^5D_0 \leftarrow {}^7F_0$ selective excitation spectra and fluorescence decay curves of Eu(III) were obtained through excitation of the Eu(III) samples with a Nd:YAG laser (Continuum) pumped dye laser (Lambda Physik) system and detection of Eu(III) fluorescence of the $^5D_0 \rightarrow {}^7F_2$ band at ~617 nm. To record the fluorescence decay curve, the sample was excited at the peak maxima of the $^5D_0 \rightarrow {}^7F_0$ selective spectra and detected at the emission maxima at ~617 nm. The data analysis was done with commercial software IGOR licensed by WaveMetrix.

Molecular Modeling Approach

We have previously shown that the geometries and frequencies of transition metal and actinide compounds can be predicted reliably at the local density functional theory level[21-26]. Thus we have used this approach to predict the structures of complexes of a Ln(III) metal with a variety of organic ligands of interest to this study. We initially attempted calculations on the Eu(III) complex with eight water molecules. Eu(III) nominally has the electronic structure f^6 which places 6 electrons in the 4f orbital. This makes convergence of the wavefunction extremely difficult and we were unable to get any of our calculations (using NWChem) [27-28] to converge even when starting from good estimates of the structure. We thus decided to use La(III) to model the Eu(III) species as the atomic radii for the ions are similar (1.28 Å vs. 0.98 Å) [29] and the fact that the La has no f electrons which minimizes convergence problems. All calculations were done with the density functional theory program DGauss[30-32] on SGI computer systems. The calculations were performed using a pseudopotential[33,34] for the Ln core electrons and the remaining electrons are treated with a polarized valence double zeta basis set (2s/1p/2d). The remaining atoms (C,N,O,H) were treated with the DZVP2 basis set[35]. All calculations were done at the local level with the potential fit of Vosko, Wilk and Nusair[36]. Geometries were optimized by using analytic gradients.

Thermodynamic Model

The aqueous thermodynamic model used in this study to interpret the solubility data is the ion-interaction model of Pitzer[37,38]. This model emphasizes a detailed description of the specific ion interactions in the solution. A detailed description of the exact form is given elsewhere[39-41]. The Pitzer thermodynamic model was used because it is applicable from zero to high concentration, and our solubility data extend to high ionic strength (I ~ 6m).

In these calculations the stability constants for EDTA, and NTA complexes with Eu(III) recommended by Martell and Smith[1], see Table 1, are included in all calculations.

Table 1. Equilibrium constants for Eu(III) aqueous complexes and solid phases used in this study. All values from Martell and Smith (1995) are for ionic strengths of zero or 0.1.

Reaction	LogK	Reference
$Eu^{3+} + EDTA^{4-} = EuEDTA^-$	17.29	(1)
$EuEDTA^- + OH^- = EuOHEDTA^{2-}$	4.87	This Study
$Eu^{3+} + HEDTA^{3-} = EuHEDTA^-$	15.45	(1)
$EuHEDTA + OH^- = Eu(OH)HEDTA^-$	4.03	(1)
$Eu^{3+} + NTA^{3-} = EuNTA$	11.32	(1)
$EuNTA(aq) + NTA^{3-} = Eu(NTA)_2^{3-}$	9.32	(1)
$EuNTA(aq) + OH^- = EuOHNTA^-$	6.84	(1)
$Eu(NTA)_2^{3-} + OH^- = EuOH(NTA)_2^{4-}$	3.95	This Study
$Eu^{3+} + 4OH^- = Eu(OH)_4^-$	18.0	This Study
$Eu(OH)_3(c) = Eu^{3+} + 3OH^-$	-25.6	(1)

Results and Discussion

In this section the experimental and modeling results for the chemical systems with and without added chelators are presented. A section comparing the stability constants calculated here with the molecular modeling results for gas phase clusters is also included.

EDTA

The solubility data for $Eu(OH)_3(c)$ in the presence and absence of different concentrations of EDTA, Figure 1, shows an extremely large increase in solubility as a result of EDTA complexation. In fact, this increase in solubility is far more than would be predicted by currently available thermodynamic models for this chemical system. As an example, the calculated solubility of $Eu(OH)_3(c)$ in these solutions using all of the stability constants recommended by Martell and Smith[1], Table 1, are several orders of magnitude lower than the experimental data. In these calculations the ion-interaction parameters for Na^+-$EDTA^{4-}$ were taken from the recent work of Pokrovsky et al.[42]. Clearly, there is a need to improve the thermodynamic models for this chemical system. These data also show a couple of other interesting features.

First, the solubility data at different equilibration times are similar, indicating that equilibrium, or at least steady-state, concentrations were reached fairly rapidly (i.e. <7 days). The only exception being the results at the lowest concentration of added NaOH and the highest EDTA concentration where a significant fraction of the $Eu(OH)_3(c)$ is dissolving increasing the time required to reach equilibrium.

Second, at high base concentration (i.e. 5M NaOH) the solubility data in the presence and absence of added EDTA are essentially identical indicating that inorganic hydrolysis species of Eu(III) are the dominant complexes. This result is similar to that found for the other chelators as will be described. Clearly, only very high base concentration is capable of completely displacing the Eu(III) from the EDTA chelate, and this effect requires over an order of magnitude more base concentration than current thermodynamic models, that do not include mixed metal-chelate-hydroxide complexes, would predict. The increase in $Eu(OH)_3(c)$ solubility at very high base concentration also means that the species present under these conditions most likely contain more than three hydroxides per Eu (i.e. $Eu(OH)_4^-$, ...), see Table 2.

Third, the slope of a reference line (~ -2) fit to the 0.01M EDTA data at 36 days of equilibration indicates, see Table 2, the formation of mixed metal-EDTA-hydroxide complexes.

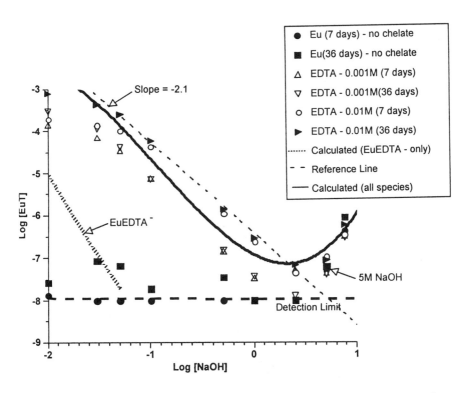

Figure 1. The solubility of Eu(OH)₃(c) as a function NaOH concentration in the presence and absence of EDTA.

Table 2. Possible chemical speciation reactions involving $Eu(OH)_3(c)$ and their dependence on base concentration. The indicated slopes are for a graph of the logarithm of the total Eu concentration (assuming the species formed is the dominant complex and the ionic strength is constant) versus the logarithm of the total base concentration.

Reaction	Slope
$Eu(OH)_3(c) + EDTA^{4-} = EuEDTA^- + 3OH^-$	-3
$Eu(OH)_3(c) + EDTA^{4-} = EuOHEDTA^{2-} + 2OH^-$	-2
$Eu(OH)_3(c) + EDTA^{4-} = Eu(OH)_2EDTA^{3-} + OH^-$	-1
$2Eu(OH)_3(c) + 2EDTA^{4-} = Eu_2O(EDTA)_2^{4-} + 4OH^- + H_2O$	-2
$Eu(OH)_3(c) + OH^- = Eu(OH)_4^-$	+1

Fourth, the soluble Eu concentration increases by approximately an order of magnitude when the EDTA concentration increases from 0.001M to 0.01M, Figure 1. This is a strong indication that the species present has a stoichiometry of one Eu per EDTA. This fact eliminates the possibility that the dominant aqueous complexes have more than one EDTA per Eu(III) but does not eliminate the dimeric species, see Table 2, which has been previously suggested as a possible species under slightly basic conditions (i.e. pH 10 to 12.5) [15].

In order to gain more insight into the species present in solution TRLFS measurements were performed on the samples with the highest dissolved Eu concentrations (i.e. $>10^{-5}M$). The restriction to $10^{-5}M$ was necessary owing to the detection limit of the TRLFS technique. The results of these measurements show that the emission spectral maxima of the Eu(III) does not significantly change in the higher base solutions, at least when contrasted with the spectra at lower base (pH 9) solutions which contain 1:1 Eu:EDTA complexes (Figure 2). However the relative intensity of the spectral bands, 579.5 nm vs. 580.0 nm, are different compared with spectra recorded at more acidic (pH 5) solutions (Figure 2). The emission peaks at 579.5 nm and 580.0 nm have been previously interpreted in terms of different numbers of coordinated water molecules to the EuEDTA⁻ complex. Specifically the peak at 579.5 nm was associated with 3 waters of hydration and 580.1 nm with 2[43]. Careful analysis of the peak at 580 nm indicated that there are two types of complexes with spectral maxima close to each other[43]. One has a Eu:EDTA ratio of 1:1 and the second with an Eu:EDTA ratio of 1:2. In the latter the second EDTA only weakly binds to Eu(III) so that the 1:1 and 1:2 complexes are in fast equilibrium on the time scale of the excited state lifetime and therefore these two species can not be time-resolved. It is interesting that it appears that the spectral maxima of the dominant complex(s) does not change in going from pH 9 to higher base. A broad new minor band does appears at ~ 578.8 nm and increases in predominace with base concentration. The broadness of the band suggests a high degree of structural heterogeneity of the complexes. However, this boadness does not suggest the formation of a dimer, since in a dimer the closeness of the coordination environments of the two cations will not display spectra with a large degree of broadness.

In terms of the measured fluorescence decay constants, for solutions having equal Eu(III) and EDTA concentrations the fluorescence decay constants are similar (3.1 ms-1) over a wide pH range from 3 to 10 in which only the EuEDTA- complex forms. But when EDTA is in large excess, formation of higher complexes occurs. At [EDTA]:[Eu(III)] ratios of 8 going from pH 3 to 11 fluorescence decay constants decrease from 3.1 ms-1 to 1.6 ms-1, indicating the partial binding of the second EDTA and removal of water molecules from the inner coordination sphere. The trend reversed when the solution pH is further increased. The fluorescence decay constant increased to 2.0 ms-1 at pH

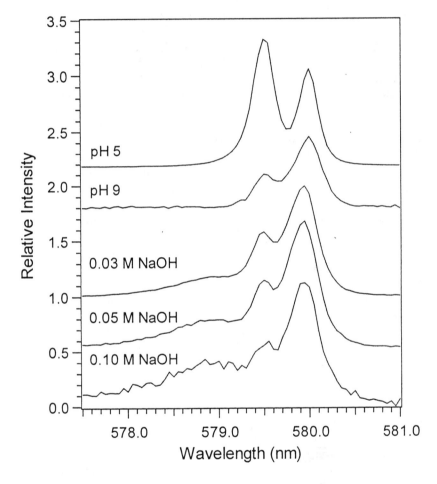

Figure 2. $^5D_0 \leftarrow {}^7F_0$ selective excitation spectra of Eu-EDTA solutions. For solutions at pH 5 and pH 9, [Eu(III)]:[EDTA] = 1:8. λ_{em} = 617 nm.

12 and 2.4 ms-1 in 0.05 M NaOH. The increased fluorescence decay constant at higher base concentration is an indication of the replacement of the second EDTA by hydroxide ions. The inner sphere waters can also be replaced by hydroxide ion. However such an exchange should lead to a decrease in the fluorescence decay constant since fluorescence quenching by water involves two –O-H vibrators. For iron (III) under basic conditions it has been pointed out that a dimeric complex which exhibits a bridging oxygen and a minimum or no waters of hydration ($Fe_2O(EDTA)_2^{2-}$) may have formed(13). Clearly such types of complexes did not form for europium-EDTA in sodium hydroxide solutions because a fluorescence decay constant of 2.4 ms-1 is far too large for a complex with no waters of hydration.

Further, if one assumes that the empirical correlation between the Eu(III) hydration number, q, and the fluorescence decay constants of Eu(III) measured in H_2O and D_2O proposed by Horrocks and Sudnick(44):

$$q = 1.05 \ (k_{H2O} - k_{D2O}) \qquad (1)$$

is valid for these solutions, then one arrives at an estimate of either 1 to 2 waters of hydration or one water of hydration and one to two associated hydroxyl ion for the complexes at higher base. We have already shown that a thermodynamic model that only considers a EuEDTA⁻ species cannot explain the observed solubilities. Further it is unlikely that a simple change in the number of waters of hydration for such a complex would increase the solubilities so significantly. Therefore, we conclude that the dominant complex at higher base concentration contains between one and two associated hydroxyl ions (i.e. $EuOHEDTA^{2-}$ or $Eu(OH)_2EDTA^{3-}$).

A thermodynamic analysis of the solubility data showed that the single best one species fit included the $EuOHEDTA^{2-}$ species. Including the $Eu(OH)_2EDTA^{3-}$ species did not significantly improve the fit, although this species cannot be completely ruled out given the large changes in ionic strength and the redundancies of the calculated standard state equilibrium constants and Pitzer ion interaction parameters at these high ionic strengths. The simplest model however includes only the $EuOHEDTA^{2-}$ species. The standard standard state equilibrium constant for the formation of an assumed $Eu(OH)_4^-$ species was also necessary to explain these data as well as the data in solutions without chelate. The values of these equilibrium constants (Table 1) do appear to be consistent with previously reported values for similar complexes. For example, the equilibrium constant for addition of a hydroxyl to EuEDTA⁻ to form $EuOHEDTA^{2-}$ (log K = 4.87) is consistent with a similar reaction for HEDTA to form Eu(OH)EDTA⁻ (Log K = 4.03). Also, the equilibrium constant for the formation of $Eu(OH)_4^-$ (Log K = 18.0) is consistent with the value of 18.6 for the corresponding $Nd(OH)_4^-$ formation reaction[1]. These results are also consistent with our molecular modeling results, which will be discussed in a later section.

NTA

The solubility data for $Eu(OH)_3(c)$ in the presence and absence of different concentrations of NTA, Figure 3, also show an extremely large increase in solubility as a result of NTA complexation. This increase in solubility is also far more than would be predicted by currently available thermodynamic models (Figure 3) for this chemical system. Interestingly, the existing thermodynamic data for NTA complexes also include stability constants for a 1:2 Eu:NTA complex (e.g. $Eu(NTA)_2^{3-}$). Our solubility data at high base also supports the formation of a 1:2 complex since the difference in solubility at long equilibration times between 0.002M and 0.01M NTA solutions is approximately a factor of ten, clearly greater than the factor of five difference expected for only a 1:1 Eu:NTA complex.

The TRLFS results (see Figure 4) show a single species is the predominant species in solution for all samples above the TRLFS detection limit. The fluorescence decay constants are also significantly less than those found for EDTA (NTA average is 1.5), indicating the NTA complexes are significantly less hydrated or hydroxylated than the EDTA complexes. Using Eq. (2) a ligand coordination number of 8 is calculated for the dominant species. A coordination number of eight is consistent with the formation of a 1:2 Eu:NTA complex (i.e. one nitrogen bond and three bonds to the carboxylate groups per NTA). Thus the dominant complex appears to be a 1:2 Eu:NTA complex with either one or two hydrated waters (i.e. just $Eu(NTA)_2^{3-}$) or a 1:2 complex with one hydrated water and one hydroxyl ion (i.e. $EuOH(NTA)_2^{4-}$). While it is possible to fit the solubility data in Figure 3 fairly well by adjusting only the formation constant for the $Eu(NTA)_2^{3-}$ species, the calculated value of the standard state equilibrium constant is quite high (log K = 11.5 for the reaction $EuNTA(aq) + NTA^{3-} = Eu(NTA)_2^{3-}$). This value for the addition of the second NTA^{3-} species is actually slightly greater than the established value for the addition of the first NTA^{3-} species to the Eu^{3+} ion (i.e. 11.32, Table 1). Clearly the value proposed by Martell and Smith[1], 9.32 would seem far more reasonable. In this analysis the ion-interaction parameters for Na^+-$HEDTA^{3-}$ (i.e. $HEDTA^{3-} = EDTA^{4-} + H^+$) reported by Pokrovsky et al.[42] were used as analogs for the unknown 1:3 electrolyte (i.e. Na^+-NTA^{3-}) values. This assumption influences the ionic strength dependence of the calculations more than the determination of the standard state equilibrium constants since the calculation of the standard state equilibrium constants tend to be anchored by the lower ionic strength data where the Pitzer ion-interaction parameters are less important. An example of this

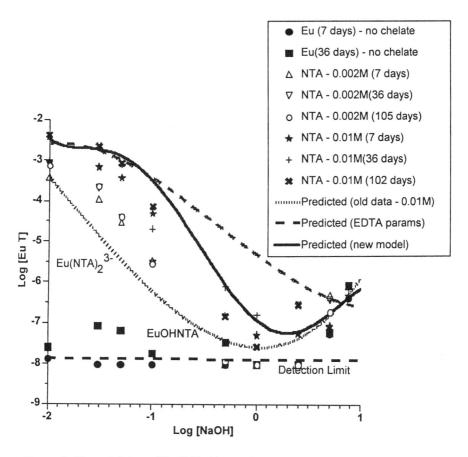

Figure 3. The solubility of Eu(OH)₃(c) as a function NaOH concentration in the presence and absence of NTA.

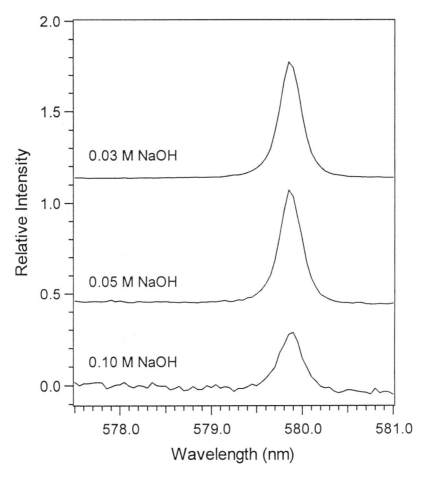

Figure 4. $^5D_0 \leftarrow {}^7F_0$ selective excitation spectra of Eu-NTA solutions. λ_{em} = 617 nm.

effect (see Figure 3, labeled EDTA params) is the calculation of the standard state equilibrium constant for the $EuOH(NTA)_2^{4-}$ species assuming the ion-interaction parameters for Na^+-$EDTA^{4-}$ apply to Na^+- $EuOH(NTA)_2^{4-}$. The calculated value of the standard state equilibrium constant involving this species differs by less than 0.5 log units from the value shown in Table 1, yet the ionic strength dependence of this model is clearly wrong. A determination of the ion-interaction parameters for the highly charged $EuOH(NTA)_2^{4-}$ species (i.e. Na^+-$EuOH(NTA)_2^{4-}$ parameters) is clearly required to explain these data. Therefore, we have fit the data in Figure 3 by adjusting the standard state equilibrium constant for the formation of the $EuOH(NTA)_2^{4-}$ species (Table 1) as well as the Pitzer ion-interaction parameters β^0 and β^1 for Na^+- $EuOH(NTA)_2^{4-}$. The calculated values of β^0 and β^1 for Na^+- $EuOH(NTA)_2^{4-}$ are somewhat high but reasonable for a 1:4 electrolyte (i.e. 2.0 and 18.0 for β^0 and β^1 Na^+-$EuOH(NTA)_2^{4-}$ compared to 1.016 and 11.6 for β^0 and β^1 Na^+- $EDTA^{4-}$). Further refinement of these values will not be possible without additional experimental data to determine the Na^+- NTA^{3-} values. The calculated stability constant for the addition of the hydroxyl ion to the $Eu(NTA)_2^{3-}$ complex also appears to be consistent with the DFT calculations as described below.

Finally it should be pointed out that the results for the NTA solutions can be explained without introducing a $EuOH(NTA)_2^{4-}$ complex if the solubility product for the $Eu(OH)_3(c)$ material is considerably (approximately two log units) higher than the recommended value in Martell and Smith[1]. At first thought such a large change in the solubility product would seem unlikely given that XRD analysis showed the material in our experiments was fully crystalline. However, the possibility still exists that small amounts of amorphous material could be present in the samples and this more soluble material could effectively control the observed solubilities. To investigate this possibility further, studies were conducted examining the solubility of the fully crystalline material in dilute (0.1M NaCl) solutions. The observed solubilities at 14 days of equilibration follow the theoretical line using Martell and Smith's[1] recommended solubility product quite closely, even at higher pH values where the solubilities are quite low and even traces of amorphous material would impact the solubilities. Clearly increasing the solubility product of the $Eu(OH)_3(c)$ by two orders of magnitude cannot be justified.

Electronic Structure Calculations

Calculations were done for a wide range of ligands bonded to La(III). Both hexavalent and octavalent aqueous isolated clusters of La(III) are predicted to be stable. We investigated the following complexes: $[La(EDTA)]^{-1}$, $[La(EDTA)(H_2O)]^{-1}$, $[La(EDTA)(H_2O)_2]^{-1}$, $[La(EDTA)(H_2O)_3]^{-1}$, $[La(EDTA)(OH)]^{-2}$,

[La(EDTA)(OH)$_2$]$^{-3}$,   [La(HEDTA)],   [La(HEDTA)   (H$_2$O)], [La(HEDTA)(H$_2$O)$_2$], [La(HEDTA)(OH)]$^{-1}$, [La(HEDTA)(OH)$_2$]$^{-2}$, [La(NTA)], [La(NTA)(H$_2$O)], [LaNTA(OH)]$^{-1}$, [La(NTA)$_2$]$^{-3}$, [La(NTA)$_2$(H$_2$O)]$^{-3}$, and [La(NTA)$_2$(OH)]$^{-4}$. The complexes with HEDTA were included since stability constants are available for mixed Eu-HEDTA-OH complexes (i.e. Eu(OH)HEDTA^{2-})[1]. These results demonstrate that for the free ion there are no significant steric hindrances to the formation of the types of new metal-hydroxyl-ligand complexes proposed in this paper. In addition, owing to the size of the trivalent rare-earth ions, the complexes with EDTA, HEDTA, and NTA can be quite asymmetric. This asymmetry creates binding locations for the associated waters of hydration or hydroxyls. This is true even for the structurally more complex EuOH(NTA)$_2$$^{4-}$ species. Even in this case, the two NTA ligands cannot completely surround the central trivalent rare-earth ion creating binding sites for the hydroxyl(s). The binding energies of all of these possible gas phase clusters also appear to correlate well with the calculated stability constants determined as part of this study (Figure 5). Although, there will certainly be differences between gas phase clusters and solution species, these calculations indicate that the binding energy from electronic structure interactions is consistent with the stability constants proposed here.

Summary and Implications for Tank Processing

In summary, EDTA and NTA both appear to form mixed metal-ligand-hydroxyl complexes under basic conditions. These complexes can significantly increase the solubility of trivalent rare-earths, and by analogy, trivalent actinides under basic conditions. A predictive thermoydnamic model has been developed that includes these complexes and is valid to high base concentration. The development of this model required the determination of the stability constants for three new metal-ligand-hydroxyl complexes.

The presence of these species can significantly impact waste tank processing strategies designed to either keep the trivalent actinides in the waste sludges or separate the actinides from the tank supernate. The solubility data themselves reveal that at chelate concentrations typical of tank solutions (i.e. ~0.01M) that basic solutions of 5M NaOH are required to remove the trivalent actinides from solution. This contrasts sharply with current model predictions that indicate that such removal should occur closer to 0.1M NaOH. The thermodynamic data developed here can be used in a wide range of applications including: evaluating the effects of competing metal ions on ligand displacement and solubility, evaluating the effects of waste mixing strategies from tank to tank, and determining the chemical forms and stability of complexes that can effect

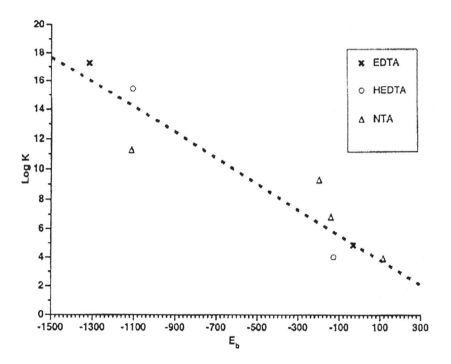

Figure 5. Relationship between gas phase binding energies calculated by DFT and solution phase equilibrium constants.

separation processes such as solvent extraction or ion exchange. Thermodynamic models capable of using these results are currently being used at Hanford and other sites[45-47]. Finally, it should be noted that a significant fraction of the organic carbon present in tank waste (23-61%) is comprised of low molecular weight organic acids, such as acetate, glycolate, and oxalate[18]. The results reported here indicate that relatively strong complexes will be required to solubilize trivalent actinides under basic conditions. Therefore, since most of the low molecular weight organic acids have relatively low binding constants with the trivalent actinides (Log K ~2-3) these ligands will probably not have much effect on tank processing strategies. Accurate thermodynamic models for these species are therefore not required. The major possible exceptions are HEDTA, oxalate (first association constant ~6.5) and citrate (first association constant ~9.5). These ligands can also form multi-ligand complexes. However, the complexes of oxalate and citrate are only predicted to significantly increase the solubility of trivalent actinides at high base if strong mixed metal-hydroxide-chelate complexes form. We recommend that thermodynamic data, or at least solubility studies, involving these ligands with trivalent actinides (or analogs) under highly basic conditions be evaluated.

Acknowledgements

This research was supported by the Department of Energy Environmental Management Sciences Program under the project entitled "Chemical Speciation of Strontium, Americium, and Curium in High-Level Waste: Predictive Modeling of Phase Partitioning During Tank Processing". We also would like to acknowledge the reviews of this manuscript by Dr. Dhanpat Rai at the Pacific Northwest National Laboratory and two anonymous reviewers.

References

1. Martell, R. E.; Smith,R. M. *Critically Selected Stability Constants of Metal Complexes Database, Version 2;* NIST Standard Reference Data program:Gaithersburg, MD 1995.
2. Anderegg, G.; *Critical Survey of Stability Constants for EDTA Complexes;* IUPAC Chemical Data Series No.14, 1977.
3. Anderegg, G.; *Pure and Appl. Chem.* 1982; 54, p 2693.
4. McConnell, A. A.; Nutall, R.H.; Stalker, D. M. *Talanta* 1978; 25, pp 425-434.
5. Nutall, R. H.; Stalker, D. M. *Talanta* 1977; 24, pp 355-360.

6. McChavan, M. C.; Palsokar, S.D.; Lokhande, N.R.; Deshpande, V.D. *Indian Journal of Chemistry* 1994; 33A, pp 343-345.

7. Kale, B. D.; Mhaske, T. H. *J. Indian Chem. Soc.* 1990; 67, pp 901-902.

8. Ternovaya, T. V.; Kostromina, N. A. *Russ. J. Inorg. Chem.* 1971; 16(11), pp 1580-1583.

9. Kostromina, N. A.; Ternovaya, T.V.; Yatsimirskii, K.B. *Russ. J. Inorg. Chem.* 1969; 14(1), pp 80-83.

10. Norkus, E.; Vaskelis, A. *Russ. J. Inorg. Chem.* 1987; 32(1), pp 130-131.

11. Bhat, T. R.; Krishnamurthy, M. *J. Inorg. Nucl. Chem.* 1963; 25, pp 1147-1154.

12. Ohman, L. O. *Polyhedron* 1990; 9(2/3), pp 199-205.

13. Lippard, S. J., Schugar, H.; Walling, C. *Inorg. Chem.* 1967; 6(10), pp 1825-1831.

14. Schugar, H. J; Hubbard, A. T.; Anson, F. C.; Gray, H. B. *J. Am. Chem. Soc.* 1969; 90, pp 71-77.

15. Spaulding, L.; Brittain, H. G. *Inorg. Chem.* 1983; 22, pp 3486-3488.

16. Campbell, J. A. *Organic Analysis Report FY 1997.* PNNL-11738, Pacific Northwest National Laboratory, Richland, WA 1997.

17. Grant, K. E.; Mong, G. M.; Luke, R. B.; Campbell, J. A. *J. Radioanalytical and Nuclear Chemistry Articles* 1996; 211(2), pp 383-402.

18. Campbell, J. A.; Stromatt, R. W.; Smith, M. R.; Koppenall, D. W.; Bean, R. M.; Jones, T. E.; Strachan, D. M.; Babad, H. *Anal. Chem.* 1994; 66(24), pp 1208A–1215A.

19. Delegard, C. H.; Gallagher, S. A. *RHO-RE-ST-3 P*, Rockwell Hanford Operations, Richland, WA 1983.

20. Rao, L.; Rai, D.; Felmy, A.R. *Radiochemica Acta.* 1996; 72, pp 151-155.

21. Parr, R. G.; Yang, W. *Density-Functional Theory of Atoms and Molecules;* Oxford University Press, New York 1989.

22. Sosa, C.; Andzelm, J.; Elkin, B. C.; Wimmer, E.; Dobbs, K. D.; Dixon, D. A. *J. Phys. Chem.* 1992; 96, p 6630.

23. Christe, K. O.; Dixon, D. A.; Mack, H. G.; Oberhammer, H.; Pagelot, A.; Sanders, J. C. P.; Schrobilgen G. J.. *J. Am. Chem. Soc.* 1993; 115, p 11279.

24. Casteel, Jr., W. J.; Dixon, D. A.; Mercier, H. P. A.; Schrobilgen, G. J. *Inorg. Chem.* 1996; 35, p 4310.

25. Casteel, Jr., W. J.; Dixon, D. A.; LeBlond, N.; Mercier, H. P. A.; Schrobilgen, G. J. *Inorg. Chem.* 1998; 37, p 340.

26. Nichols, J. A.; Dixon, D. A.; Harrison, R. A.; de Jong, W.; Windus, T.L. To be published 2000.

27. Bernholdt, D. E.; Apra, E., Fruechtl, H. A.; Guest, M. F.; Harrison, R. J.; Kendall, R. A.; Kutteh R. A.; Long, X.; Nicholas, J. B.; Nichols, J. A., Taylor, H. L.; Wong, A. T.; Fann, G. I.; Littlefield, R. J.; Niepolcha, J. *Int. J. Quantum Chem: Quantum Chem. Symp.* 1995; 29, p 475.

28. Guest, M. F.; Apra, E.; Bernholdt, D. E.; Fruechtl, H. A.; Harrison, R. J.; Kendall, R. A.; Kutteh, R. A.; Long X.; Nicholas, J. B.; Nichols, J. A.; Taylor, H. L.; Wong, A. T.; Fann, G. I.; Littlefield, R. J.; Niepolcha, J. *Future Generation Computer Systems* 1996; 12, p 273.

29. Emsley, J. *The Elements, 2nd Edition;* Clarendon Press, Oxford, UK 1991.

30. Andzelm, J.; Wimmer, E.; Salahub, D. R.; In *The Challenge of d and f Electrons: Theory and Computation.* Salahub, D. R.; Zerner, M. C., Eds.; ACS Symposium Series, No. 394, American Chemical Society: Washington D. C. 1989; p 228.

31. Andzelm, J. *In Density Functional Theory in Chemistry.* Labanowski, J., Andzelm, J., Eds.; Springer-Verlag: New York 1991; p 155.

32. Andzelm, J. W.; Wimmer, E. *J. Chem. Phys.* 1992; 96, p 1280.

33. Chen, H.; Kraskowski, M.; Fitzgerald, G. *J. Chem. Phys.* 1993; 98, p 8710.

34. Troullier, N.; Martins, J. L. *Phys. Rev. B* 1993; 1991, p 43.

35. Godbout, N.; Salahub, D. R.; Andzelm, J.; Wimmer, E. *Can. J. Chem.* 1992; 70, p 560.

36. Vosko, S. J.; Wilk, L.; Nusair, W. *Can. J. Phys.* 1980; 58, p 1200.

37. Pitzer, K. S. *J. Phys. Chem.* 1973; 77 (2), p 8.

38. Pitzer, K. S. *Activity Coefficients in Electrolyte Solutions, 2nd ed.* Boca Raton, Florida, CRC Press 1991.

39. Harvie, C. E.; Møller, N.; Weare , J. H. *Geochim. Cosmochim. Acta* 1984; 48, p 723.

40. Felmy, A. R.; Weare, J. H. *Geochim. Cosmochim. Acta* 1986; 50, p 2771.

41. Felmy, A. R.; Rai, D.; Schramke, J. A.; Ryan, J. L. *Radiochim. Acta* 1989; 48, p 29.

42. Pokrovsky, O. S.; Bronikowski, M. G.; Moore, R. C.; Choppin, G. R. *Radiochim. Acta* 1998; 80, pp 23-29.

43. Latva, M.; Kankare, J.; Haapakka, K.. *J. Coord. Chem.* 1996; 38, pp 85-99

44. Horrocks Jr., W. DeW.; Sunnick, D. R.. *J. Am. Chem. Soc.* 1979; 101, p 334.

45. MacLean, G. T. WHC-SD-WM-TA-184, Westinghouse Hanford Co., Richland, WA 1996.

46. MacLean, G. T. HNF-SD-TWR-PE-001, SGN Eurisys Services Corp., Richland, WA1997.

47. MacLean, G. T.; Eager, K. M. HNF-3257, COGEMA Engineering Corp., Richland, WA 1998.

Chapter 6

Eu^{3+} and UO$_2$$^{2+}$ Surface Complexes on SiO$_2$ and TiO$_2$: Models for Testing Mediated Electrochemical Methods for the Treatment of Actinide Processing Residues

Tammy A. Diaz[1], Deborah S. Ehler, Carol J. Burns, and Davis E. Morris[*]

Chemical and Environmental Research and Development Group, Chemical Science and Technology Division, and the G. T. Seaborg Institute for Transactinium Science, Los Alamos National Laboratory, Los Alamos, NM 87545
[1]Current address: N. M. Highlands University, Las Vegas, NM 87701

Batch sorption experiments and luminescence spectroscopic investigations of the resulting solid and supernate phases have been conducted for Eu^{3+} and UO$_2$$^{2+}$ on colloidal SiO$_2$ and TiO$_2$ phases from pH 2 to 6. Sorption data suggest that both sorbates form inner-sphere complexes on the metal oxide substrates, and the spectral data confirm significant structural perturbations for the surface complexes relative to the aqueous species from which they form. The perturbations are greatest for the species on TiO$_2$. Surface complexes change as a function of solution pH reflecting changes in both solution species contributing to sorption processes and availability of surface sites. Dehydration of the sorbate-bound solids at 200 °C results in a transition to a single surface species over the entire pH range for all systems except UO$_2$$^{2+}$ on TiO$_2$.

Introduction

The United States Department of Energy is faced with the stabilization and disposition of hundreds of metric tons of plutonium- and uranium-bearing residue materials resulting from 50+ years of nuclear weapons production activities (1). These materials are in storage throughout the weapons production complex. They pose an immediate potential threat due to instability and degradation, and a long-term problem because the volume of material to be disposed may exceed projected repository capacities. New and/or improved routes to stabilization and volume reduction of these residues are needed. Mediated electrochemical oxidation/reduction (MEO/R) processes for dissolution of the actinide material from these residues are one such approach (2, 3). They are low temperature, ambient pressure processes that operate in a non-corrosive environment, they can be designed to be highly selective for the actinides (i.e., no substrate degradation occurs), and they can be utilized for many categories of residue materials with little or no modification in hardware or operating conditions. However, some fundamental questions remain concerning the mechanisms through which these processes act, and how the processes might be optimized to maximize efficiency while minimizing secondary waste. In addition, further research is merited to extend the range of applicability of these electrochemical methods to other residue and waste streams.

One particularly important class of residues is incinerator ash. These are in general refractory materials comprised of often colloid-sized stoichiometric and sub-stoichiometric mixed-metal oxides having high surface area. The actinides can be found in the ash as discrete particulates, agglomerated with other metal oxides, and as surface and interstitial coatings (4). An important basis for the present studies is the recent advance made in understanding heterogeneous electron transfer processes in transition-metal and other semiconductor colloidal dispersions (5,6). For example, polarographic and voltammetric investigations of the reduction of SnO_2, TiO_2, and mixed $TiO_2/Fe(III)$ oxide colloids at mercury electrodes have been reported (7).

Our efforts have focused on model systems prepared to mimic the salient characteristics of incinerator ash (substrate composition, disposition of the actinides, etc.) while providing both electrochemical and spectroscopic properties to enable adequate characterization of local environments and redox properties. In particular, we have chosen colloidal SiO_2 and TiO_2 as the substrates since these have been identified as significant components in real ash samples (4) and yield stable aqueous suspensions at high solids concentrations. They have also been shown to sorb actinides from aqueous solutions. The sorbates include UO_2^{2+}, Pu^{4+}, and Eu^{3+} (as an actinide surrogate). These metals all possess one or more redox couples in aqueous solution and they all have spectroscopic properties that are sensitive to speciation, surface complexation, and changes therein. UO_2^{2+} and Eu^{3+}, in particular, have excellent luminescence characteristics that provide for exceptional sensitivity when probing speciation.

Before the electrochemical studies of heterogeneous electron transfer can be conducted and properly interpreted, it is necessary to first prepare and adequately characterize the actinide (or surrogate) - laden substrates. The present report is concerned with the preparation of UO_2^{2+} and Eu^{3+} sorption samples on SiO_2 and TiO_2 via conventional batch sorption reactions, and the characterization of the resulting surface complexes of these metals using luminescence spectroscopy. The sorption reactions have been carried out over the pH range 2 - 6 to vary the solution speciation and surface complexation with the aim of identifying surface species having differing redox energetics and/or kinetics. The ionic strength of the solution in the batch sorption reactions was also varied to assist in identifying the mechanism of the surface complexation process for these metals. There are numerous recent reports concerned with sorption and surface complexation of UO_2^{2+} and the trivalent lanthanides (Ln^{3+}) on metal oxides including silica and titania (8-15). This is the first report that includes both sorption data and spectroscopic characterization of the resulting surface speciation for these metals on SiO_2 and TiO_2.

Experimental

The metal oxide substrates were obtained from Degussa Corp. The SiO_2 is Aerosil 200. It is amorphous with a reported N_2 BET surface area of $\sim$ 200 m^2/g and an average particle size of $\sim$ 25 nm. The TiO_2 is Titandioxid P25. It is a mixed phase comprised of $\sim$ 70 % anatase and 30 % rutile with a BET surface area of $\sim$ 50 m^2/g and an average particle size of $\sim$ 20 nm. The pH at the point of zero surface charge (pH_{pzc}) for amorphous SiO_2 is typically 2-3 while that for TiO_2 is typically $\sim$ 6 (16). The Eu^{3+} source was $Eu(NO_3)_3 \cdot 6H_2O$ from Aldrich and the UO_2^{2+} source was $UO_2(NO_3)_2 \cdot 6H_2O$ from Strem. The substrates and the sorbates were used without further purification. All other reagents (HNO_3, $NaOH$, $NaNO_3$) were ACS reagent grade.

The batch sorption reactions were performed on solid suspensions that were pre-equilibrated for 24 - 48 hrs in deionized water or $\sim$ 0.1 M $NaNO_3$ solution at the native pH of the suspension. The solid concentration in the suspension was $\sim$ 12-24 g/L for the Eu^{3+} samples and $\sim$ 7-11 g/L for the UO_2^{2+} samples. The sorbate was introduced into the pre-equilibrated suspension as either a solid aliquot of the nitrate salt or a small volume of concentrated spike solution to give a final sorbate concentration of 3 - 8 mM for Eu^{3+} samples and 0.2 - 0.4 mM for UO_2^{2+} samples. The pH of the suspension was then adjusted to the target value between pH 2 and 6. The sorption reactions were allowed to equilibrate for 36 - 48 hrs while undergoing end-over-end rotation. The UO_2^{2+} / TiO_2 suspensions were covered with Al foil prior to introduction of the uranyl spike and for the duration of the equilibration time to prevent photochemical reduction/ surface precipitation of the uranium (9). The equilibrated suspensions were centrifuged to effect phase separation. The supernates were decanted and retained for analysis. Some of the solid phase was dehydrated in a convection oven at ~ 200 °C for 12-24 hrs to partially simulate the effects of residue incineration. The remainder of the solid phase was kept moist for spectroscopic interrogation and electrochemical studies (described elsewhere).

Quantitative evaluation of the uptake of sorbate onto the metal-oxide substrates was performed by inductively coupled plasma atomic emission spectroscopic analysis (Varian Model Liberty 220) of the supernates from the batch reactions under the assumption of mass balance between the solid and supernate phases (i.e., negligible container losses). This assumption has been verified in our labs as described elsewhere (17). The molecular luminescence investigations of the reaction supernates and the solids were carried out on a SPEX Industries Fluorolog 2 System consisting of a Model 1681 single-stage 0.22 m excitation monochromator and a Model 1680 two-stage 0.22 m emission monochromator with a thermoelectrically-cooled Hamamatsu Model R928 photomultiplier tube with photon-counting electronics. Continuous excitation was provided by a high-pressure 450 W Xe arc lamp and pulsed (time-resolved) excitation was provided by a Model 1934 Phosphorimeter attachment. Spectra reported here have not been corrected for monochromator or detector responses. Wet solids were contained in standard-bore borosilicate glass NMR tubes, dry solids were contained in borosilicate capillary tubes, and supernates were contained in standard fluorescence cuvettes for the spectral analyses. Most samples were examined at room temperature, but some uranyl samples were also run at ~ 77 °K in an insertion dewar.

Results and Discussion

Sorption Data

The strategies for the batch sorption experiments were based on published reports for UO_2^{2+} sorption on SiO_2 (11) and Co^{2+} sorption on TiO_2 (18). Specifically, these reports were used to guide in the selection of solid substrate concentration ranges (and therefore site densities) relative to the sorbate concentrations in solution. The oxide substrates used in these reports are not identical to those used here. However, these published reports were used to estimate surface sorption site densities for the materials used in this report as follows; 0.14 mmol sites /g solid for SiO_2 and 0.25 mmol sites / g solid for TiO_2. The aim was to have high sorbate concentrations on the solids at equilibrium to facilitate the subsequent electrochemical and spectroscopic investigations, yet not exceed the total sorption site density on the solids and risk introducing complicating sorption mechanisms like surface precipitation.

An additional important consideration was the solution speciation of the sorbate cations vs. pH in the sorption reactions. Uranyl ion is a much stronger

Lewis acid than Eu^{3+}. Thus, it is more susceptible to formation of insoluble hydroxide species at lower pH for any given concentration. For this reason, the total $[UO_2^{2+}]$ in the reactions was kept about a factor or 10 lower than the corresponding $[Eu^{3+}]$. Calculations using HYDRAQL (19) and the NEA/OECD database (20) for uranyl indicated that precipitation of uranyl hydroxide phases should not be important under the conditions applied here. Finally, sorption reactions were run with and without added $NaNO_3$ electrolyte (0.08 M) to assist in determining whether the sorption reactions were inner- vs. outer-sphere in nature. High electrolyte concentrations are known to reduce sorption by outer sphere mechanisms via mass-balance driven competition for these electrostatically governed sites (16).

Typical uptake curves for Eu^{3+} and UO_2^{2+} on the solid substrates are shown in Figure 1. All systems show the onset of a sorption "edge" typical of metal ion sorption onto oxide phases (16, 21). Note that UO_2^{2+} uptake is significantly greater on all solids than that for Eu^{3+}. In addition, the sorption edge occurs at much higher pH for the Eu^{3+} samples. This is consistent with the much higher metal ion concentration in these solutions than in the corresponding UO_2^{2+} systems, but it also likely reflects important differences in the strength of surface binding of Eu^{3+} vs. UO_2^{2+} as well as differences in the hydrolysis properties of these two cations. The edge for TiO_2 sorption occurs at lower pH than that for SiO_2 in all systems. This is consistent with results obtained for Co^{2+} sorption on these same substrates (21). Although the pH_{pzc} for SiO_2 is lower than that for TiO_2, the reversal in the pH of the sorption edge is interpreted to indicate

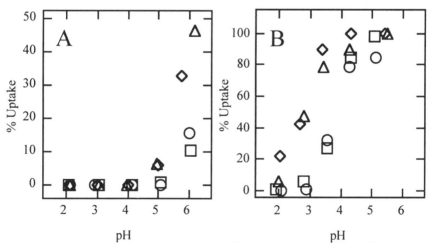

Figure 1. Sorption data for (A) ~ 3 mM Eu^{3+} and (B) ~ 0.2 mM UO_2^{2+} on oxide substrates. (O) SiO_2. (□) SiO_2 with 0.08 M $NaNO_3$. (Δ) TiO_2. (◊) TiO_2 with 0.08 M $NaNO_3$. Solid concentrations are ~ 12 g/L for Eu^{3+} and ~ 7.5 g/L for UO_2^{2+} samples.

stronger binding of the metal ion to titania than to silica (21). There is no discernible ionic-strength dependence in the UO_2^{2+} sorption data. This, coupled with the typical sorption-edge behavior indicates that the uranyl species are forming inner-sphere surface complexes on both substrates. The Eu^{3+} sorption data appear to have a very slight ionic-strength dependence, with the sorption suppressed somewhat at the higher ionic strength. However, the effect is minimal and only observed at the highest pH. This observation, coupled with the spectroscopic data described below, are more consistent with an inner-sphere description for the Eu^{3+} sorption complexes on both substrates as well.

Overall, these sorption data are consistent with those reported for related systems in other recent studies. For example, the sorption of Eu^{3+} and other trivalent lanthanides on silica (13) and the iron-oxide hematite (15) was interpreted to occur via an inner-sphere mechanism. Notably, in the SiO_2 sorption study the sorption edge at the highest Ln^{3+} concentration (0.1 mM) was at about the same pH as the edge observed in the present study (Figure 1A). Uranyl sorption has also been reported on SiO_2 (10-12) and TiO_2 (8), and in all cases the sorption mechanisms has been described as inner-sphere in nature. Thermodynamic modeling of sorption data has also been carried out in some of these previous studies, and surface binding constants have been published for some trivalent lanthanide ions on SiO_2 (13) and for UO_2^{2+} on SiO_2 (11).

Spectroscopic Data for Surface Complexes and Reaction Supernates

The luminescence characteristics of Eu^{3+} have been extensively exploited for investigations of the coordination environment surrounding this metal ion in chemical, biochemical, and, more recently geochemical systems (22). There have been surprisingly few reports of the use of Eu^{3+} luminescence to probe surface complexation processes. However, it is ideally suited for this purpose as well because it is highly sensitive in both an analytical sense and with respect to differentiating chemical environments. The most salient features of Eu^{3+} luminescence from which coordination environment information is deduced are the intensity in the "hypersensitive" transition ($^5D_0$ to $^7F_2$ at ~ 16,200 cm^{-1}) relative to that in the $^5D_0$ to $^7F_1$ transition at ~ 16,860 cm^{-1} and the luminescence lifetime. The former property is a reflection of the local symmetry around the ion (23) and the latter changes in response to changes in the number and type of coordinating molecules. Specifically, the lifetime has been shown to correlate with the number of coordinated water molecules (24), and a simple formula has been derived [$n_{H2O} = (1070/\tau_m)-0.62$ where τ_m is the measured lifetime in microseconds] to estimate this (25).

Representative emission spectral data for Eu^{3+} in the sorption reaction supernates and on the wet SiO_2 and TiO_2 solid phases are shown in Figure 2. These spectra have all been normalized to facilitate comparisons, but the unnormalized intensities track the concentrations determined by ICP/AES (Figure 1). Notably, although there is no analytically measurable uptake at the

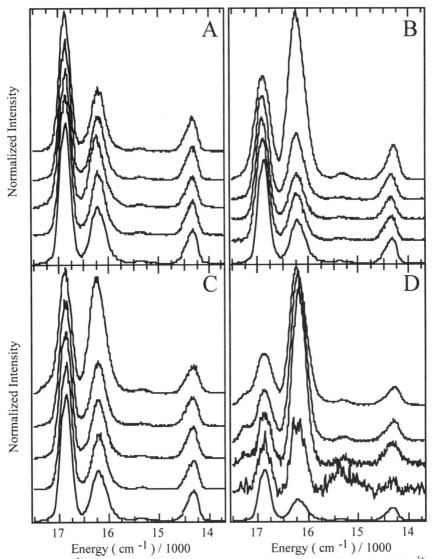

Figure 2. Eu³⁺ luminescence spectra from samples prepared with ~ 6 mM Eu³⁺ and 24 g/L of solid. (A) Supernate from SiO_2 sample. (B) Wet SiO_2 solid. (C) Supernate from TiO_2 sample. (D) Wet TiO_2 solid. All spectra have been normalized to the intensity in the band at 16,860 cm⁻¹. For all panels the order of the spectra from bottom to top is: 5 mM standard solution at pH 4, pH 3.0 sample, pH 4.0 sample, pH 5.0 sample, and pH 6.0 sample. Spectra were obtained with 393 nm pulsed excitation and gated detection using a 50 microsecond delay.

lower pH values, it is still possible to measure an Eu^{3+} luminescence spectrum on the solids for both SiO_2 and TiO_2 attesting to the sensitivity of this method to track surface complexation. (Samples prepared in other experiments at much lower initial $[Eu^{3+}]$ values and characterized by ICP/AES did show micromolar surface concentrations of Eu on both substrates consistent with the luminescence result here and previous batch sorption studies [13, 15].)

The spectra for the reaction supernate in the SiO_2 reactions are invariant over the entire pH range and look identical to that of the pH 4 standard solution (Figure 2A). This confirms that there is not a significant change in the solution speciation for Eu^{3+} over the entire pH range in these equilibrium solutions. This observation is consistent with the speciation data reported recently (albeit presumably at lower $[Eu^{3+}]_{tot}$) that indicates that the initial hydrolysis and carbonate complexes for Eu^{3+} solutions in equilibrium with atmospheric CO_2 (as in the present study) do not contribute significantly to the speciation below $\sim$ pH 6 (15). Thus, the fully aquated Eu^{3+} cation is the dominant species in these equilibrium supernates. A similar situation exists for the supernates from the TiO_2 reactions at all pH values except pH 6.0 (Figure 2C). For this latter supernate, the spectrum shows a much greater intensity contribution from the hypersensitive transition indicating some change in solution speciation. This change could reflect contributions from hydrolysis and/or carbonate species, but this seems unlikely since such species were not observed in the SiO_2 supernates, and the equilibrium Eu^{3+} concentration in the TiO_2 supernates is lower than in the SiO_2 supernates. In addition, the hypersensitive band does not have the general characteristics (splitting into two distinct bands and much greater intensity relative to the 16,860 cm^{-1} band) reported for carbonate species (26). The ICP analyses of these supernates did reveal some soluble Ti species from substrate dissolution; more so than Si species at comparable solid concentrations and pH. Perhaps the soluble Ti species are anionic titanates that can interact with the Eu^{3+} cation to lower the local symmetry and thereby give rise to increased intensity in the hypersensitive band.

The spectra for Eu^{3+} on the surface of SiO_2 (Figure 2B) are similar to those seen in the supernate for the lower pH values (2.0 to 4.0) suggesting that the structure of the surface complex is only minimally perturbed from the aquated cation (i.e., more like an outer-sphere complex). However, for the pH 5.0 and 6.0 samples the intensity in the hypersensitive transition increases steadily relative to that in the 16860 cm^{-1} band. The intensity ratio increases from ~ 0.5 at pH 2.0 - 4.0, to ~ 0.7 at pH 5.0, to ~ 1.7 at pH 6.0. This change reflects a much stronger interaction between the Eu^{3+} species and the surface that lowers the symmetry of the coordination environment around the Eu^{3+} cation relative to the aquated species, and likely signals the onset of a true inner-sphere complex. Note that this change takes place within the same pH interval that the uptake increases significantly (Figure 1A) and the system shows the appearance of a classical metal-oxide sorption edge. The emission lifetime data also support the conclusion of a change in surface speciation in the higher pH interval. Notably, the emission decay curves for all Eu^{3+} sorption samples were adequately analyzed using only a single exponential function. This indicates that a single surface species dominates the surface population at any given pH value. The measured lifetime for the wet SiO_2 solids remains constant from pH 2.0 through 5.0 ($\tau_{ave} = 111\pm1$ μs) but increases to 169 μs at pH 6.0. The former value agrees very well with the value measured for the 5 mM pH 4.0 standard (112 μs). Using the formula cited above (25), the short lifetime which dominates for the low pH surface species corresponds to 9.0 coordinated water molecules while the long lifetime seen in the pH 6.0 solid sample corresponds to 5.7 coordinated waters. Again, these data are consistent with a transition from a fully aquated outer-sphere complex to an inner-sphere complex having lower site symmetry and fewer coordinated waters. Interestingly, previous reports of Ln^{3+} sorption on SiO_2 (13) and iron oxides (15), while proposing inner-sphere complexation mechanisms, do not consider displacement of coordinated water in concert with this process. However, the loss of coordinated water (as demonstrated by the lifetime data reported here) must accompany the formation of inner-sphere bonds with the surface hydroxyl groups.

For the Eu^{3+} - laden TiO_2 wet solids the luminescence spectra show an even greater perturbation relative to the supernate spectral data. In this system, even at the lowest pH (2.0) the intensity in the hypersensitive transition exceeds that in the 16,860 cm^{-1} band (data at pH 2.0 are not included in Figure 2 for clarity sake). This intensity ratio increases with increasing pH to an approximately constant value in the pH 4.0 – 6.0 samples of 2.5. This perturbation in local symmetry clearly suggests that the surface interaction between Eu^{3+} and the TiO_2 surface sites is much greater, particularly at the lower pH values than is observed in the SiO_2 system. Other metal cation sorbates are also found to interact more strongly with transition-metal oxide substrates than with SiO_2 and its polymorphs (16). The emission lifetime data for the Eu^{3+} / TiO_2 samples are also consistent with a stronger surface interaction. The measured lifetimes range from 130 μs in the pH 4 solid (the emission was too weak to collect lifetime data for the lower pH samples), to 200 μs in the pH 5 solid, to 220 μs in the pH 6 solid. The calculated number of coordinated water molecules corresponding to these lifetimes are: 7.6 (pH 4), 4.7 (pH 5), and 4.2 (pH 6). Thus, the prediction is that more than half the original coordinated waters have been lost in forming the surface complex of Eu^{3+} with TiO_2 at the higher pH values.

For the Eu^{3+} - substrate samples that were dehydrated at 200 °C, the luminescence data (Figure 3) indicate that the heat treatment has a leveling effect on the surface speciation. That is, the variability observed in the spectra versus pH for the wet samples is essentially gone. The data for dehydrated samples do reflect significant perturbations in the surface complexation, even relative to that seen at the highest pH value for the wet solids. Note that the intensity ratio of the hypersensitive band to the 16,860 cm^{-1} band is greater still for these dehydrated samples than for the pH 6.0 wet solids for both SiO_2 and TiO_2. Consistent with observations from the wet solid data, this perturbation is also greater in the TiO_2 samples than the SiO_2 samples. As expected, the lifetime data from these dehydrated samples also reveal more substantial losses of coordinated H₂O than in the wet solids. However, it is interesting that for the dehydrated samples the Eu^{3+} appears to lose more water on dehydration on the SiO_2 surface than on the TiO_2 surface. The measured lifetimes range smoothly from 225 µs (4.1 waters) at pH 2.0 to 340 µs (2.5 waters) at pH 6.0 on SiO_2 whereas the range is only 220 µs (4.2 waters) to 280 µs (3.2 waters) on TiO_2.

The luminescence characteristics of the UO_2^{2+} ion have also made this an excellent chromophore for probing speciation in solution and on surfaces. There have been numerous recent reports on uranyl emission as a probe for speciation in groundwaters (27,28) and on environmentally relevant substrates (12,17,29,30). The way in which speciation information becomes encoded in the spectral response of the uranyl ion is much more complex than in the case of

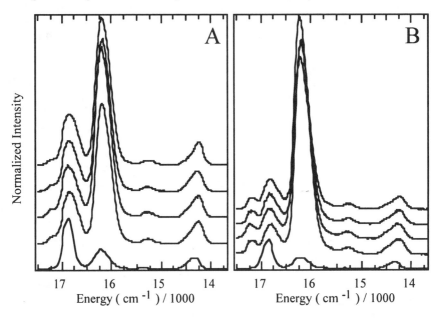

Figure 3. Eu^{3+} luminescence spectra from samples prepared with ~ 6 mM Eu^{3+} and 24 g/L of solid. (A) Dehydrated SiO_2 solid. (B) Dehydrated TiO_2 solid. The order of the spectra from bottom to top in both panels is: 5 mM standard solution at pH 4, pH 3.0 sample, pH 4.0 sample, pH 5.0 sample, and pH 6.0 sample. Spectra were obtained with 393 nm pulsed excitation and gated detection using a 50 microsecond delay. All spectra have been normalized to unit intensity in the band at 16,860 cm⁻¹.

Eu^{3+}. For systems containing the uranyl moiety the structural information can only be inferred based on trends in spectroscopic responses. Nonetheless, significant progress has been made in this area in relation to problems in environmental and surface complexation chemistries. With respect to the present study, uranyl poses several additional challenges. First, the onset of hydrolysis for uranyl occurs at low pH, even for the relatively dilute solutions used to prepare the sorption samples, and there are a large number of hydrolytic species possible in the pH range 2 to 6. This has the practical consequence of making it fruitless to compare in detail the supernate spectra from one pH to another given the changes in $[UO_2^{2+}]_{tot}$ in these solutions. Second, uranyl is photochemically very active and interacts with TiO_2 via energy and electron transfer mechanisms (9) that lead to substantially quenching of the luminescence signal. Thus, characterization of the uranyl surface complexes on TiO_2 by luminescence, if possible, may require additional steps to overcome the quenching (e.g., low temperatures).

Typical luminescence data obtained for UO_2^{2+} on SiO_2 and TiO_2 substrates are shown in Figure 4. These spectra have also been normalized to facilitate comparison. However, here too, the raw intensities track the concentration trends exhibited in the batch sorption data (Figure 1B). The spectra obtained for the wet SiO_2 solid samples (Figure 4A) show very little variation with pH. There is only a slight increase in the intensity of the shoulder at $\sim 18,300$ cm^{-1} with increasing pH which signifies a very minor contribution from an additional species that grows in at the higher pH. For comparison, the spectrum of the pH 2.1 supernate has been included. This solution is dominated by the fully aquated UO_2^{2+} monomer species, and the spectrum reflects this dominant population. The spectral benchmark for both hydrolysis (28) and inner-sphere surface complexation (17) in uranyl systems is a shift to lower energy (red shift) in the spectral band relative to that of the fully aquated monomer. This behavior is clearly observed for these wet SiO_2 samples indicating that the UO_2^{2+} is forming inner-sphere surface complexes. However, the essentially constant spectral behavior over the pH range investigated suggests that a single uranyl surface complex is dominating these samples. This observation is in agreement with recently published work on batch sorption and thermodynamic modeling of uranyl on SiO_2 which suggest a single dominant uranyl surface complex over the pH range investigated here (11). The luminescence lifetime data for these wet solid samples support the dominance of a single inner-sphere surface complex; only a single lifetime is observed in the decay data, and it remains constant within experimental error at 238 ± 14 μs. (Emission from the pH 2.1 sample was too weak to obtain a lifetime.)

The spectral data from the UO_2^{2+} / SiO_2 samples dehydrated at 200 °C are shown in Figure 4B. The constancy in the spectral band over the measured pH range is even more striking for these samples. In this respect, the dehydration has the same leveling effect on surface speciation as was observed in the Eu^{3+} samples (vide supra). Here, however, there is very little difference in the spectral band shape between the wet and the dehydrated samples. Thus, the surface complexes must not change significantly on dehydration in this uranyl system. The measured luminescence lifetimes for the dehydrated samples are

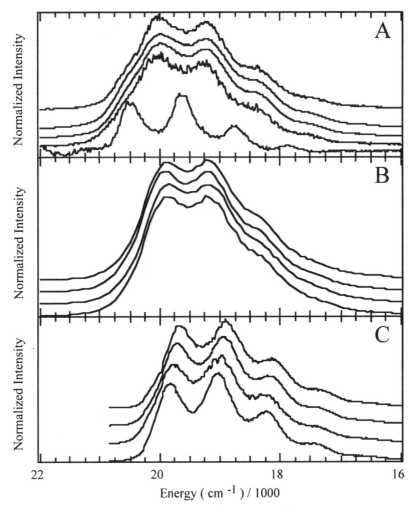

Figure 4. UO$_2^{2+}$ luminescence spectra for samples prepared with ~ 0.3 mM UO$_2^{2+}$ and ~ 12 g/L of solid. (A) Wet SiO$_2$ solid. (B) Dehydrated SiO$_2$ solid. (C) Dehydrated TiO$_2$ solid. The bottom spectrum in A is from the pH 2.1 supernate. The equilibrium pH values of the samples in A and B were (bottom to top): 2.1, 3.2, 3.5, and 4.0. The equilibrium pH values of the samples in C were (bottom to top): 2.1, 3.0, 3.2, and 3.6. All spectra were obtained using pulsed 400 nm excitation and gated detection with 50 us delay. Spectra in C were acquired at LN$_2$ temperature. All spectra are normalized at the point of maximum intensity.

also single-exponential denoting a single species, but the lifetime remains constant over the entire pH range here, too. It does, however, appear to increase slightly (258 ± 14 µs from all four samples) relative to the value for the wet samples consistent with the loss of water from the surface environment.

Unfortunately, it was not possible to obtain spectral data for the wet TiO_2 solids containing UO_2^{2+} at either room temperature or ~ 77 K. Even the supernate samples at low pH (those that contained an appreciable amount of unsorbed uranyl) gave only very weak luminescence suggesting that there is a strong quenching interaction between uranyl and dissolved Ti species. It was possible to obtain luminescence data for the dehydrated uranyl / TiO_2 samples. The room temperature data (not shown) are weak, broad, and poorly vibronically structured. However, on cooling these samples to ~ 77 K the spectral data shown in Figure 4C were obtained. There are several important features in this spectral series that distinguish it from the data obtained for the SiO_2 samples. First, the overall spectral band for the TiO_2 samples is shifted to lower energy by ~ 500 cm^{-1} relative to the SiO_2 spectra. This signifies an apparently greater extent of hydrolysis of the surface uranyl species in the TiO_2 samples compared to the SiO_2 samples. One possible mechanism for this is a stronger interaction with the surface titanium hydroxyl oxygens in which there is greater donation of electron density into the uranyl equatorial plane. Alternatively, the bonding interaction on the titanium surface could be of a bidentate nature (i.e., two surface oxygen atoms per uranyl). Either mechanism would be indicative of a stronger surface complex relative to what is observed in the SiO_2 system. The uranyl / TiO_2 spectra also reveal a slight yet monotonic red shift of the spectral bands with increasing pH. The shift amounts to ~ 150 cm^{-1} over the pH range examined and can be observed in both the first and second resolved vibronic bands. This is most likely indicative of a contribution from a second uranyl surface complex that increases in importance with increasing pH. The luminescence lifetime data for these samples support this assertion. The decay curves for all four dehydrated TiO_2 samples require two distinct lifetimes for adequate fits, and the relative contribution to the total intensity of each lifetime changes with pH. The average values for these two lifetimes are 74 ± 5 µs and 189 ± 6 µs. The observation of a second distinct surface complexation population for uranyl on TiO_2 (but not on SiO_2) probably indicates a difference in the surface itself that gives rise to multiple populations. Again, it is well established in other transition metal oxides (e.g., hematite and ferrihydrite) that both "strong" and "weak" sites exist for sorption of metal cations (31).

Conclusions

The surface complexation of Eu^{3+} and UO_2^{2+} on SiO_2 and TiO_2 appears to be dominated by inner-sphere processes under conditions of significant uptake of these metals. For the wet solids there is a pH dependence to the observed spectral signatures for these surface species suggesting that either the sorbing species, the characteristics of the surface, or both change as a function of the reaction pH and lead to structurally distinct surface complexes. The effect of dehydration of the solids in the presence of these surface complexes is to convert the various surface species into a single population (except for uranyl on TiO_2 where slight pH dependence is still observed). In general, the surface complexes of Eu^{3+} and uranyl interact more strongly with the TiO_2 surface than the SiO_2 surface.

Acknowledgment

The authors thank Dorsett and Jackson, Los Angeles, agents for Degussa Corporation, for proving samples of Titandioxid P25 for use in these studies. This work was supported by the United States Department of Energy, Offices of Environmental Management and Science, under the auspices of the Environmental Management Science Program. Additional support for TAD was provided by the Laboratory-Directed Research and Development program (Actinide Molecular Science) at Los Alamos.

References

1. U.S. DOE. Defense Nuclear Facilities Safety Board Recommendation 94-1 Implementation Plan, February 28, 1995.
2. Bray, L.A.; Ryan, J.L. In Actinide Recovery from Waste and Low-Grade Sources; Navratil, J.D.; Schulz, W.W., Eds.; Harwood Academic: London, 1982; pp 129-54.
3. Bourges, J.; Madic, C.; Koehly, G.; Lecomte, M. J. Less Common Metals 1986, 122, 303-11.
4. Behrens, R.G.; Buck, E.C.; Dietz, N.L.; Bates, J.K.; VanDeventer, E.; Chaiko, D.J. Characterization of Plutonium-Bearing Wastes by Chemical Analysis and Analytical Electron Microscopy. Argonne National Laboratory report ANL-95/35, September 1995.
5. Kamat, P.V. Prog. Reaction Kinetics 1994, 19, 277-316.
6. Mackay, R.A. Colloids Surfaces A. 1994, 82, 1-28.
7. Heyrovsky, M.; Jirkovsky, J.; Struplova-Bartackova, M. Langmuir 1995, 11, 4309-12.
8. Lieser, K.H.; Thybusch, B. Fresenius Z Anal. Chem. 1988, 332, 351-7.
9. Amadelli, R.; Maldotti, A.; Sostero, S.; Carassiti, V. J. Chem. Soc. Faraday Trans. 1991, 87, 3267-73.
10. Lieser, K.H.; Quandt-Klenk, S.; Thybusch, B. Radiochim. Acta 1992, 57, 45-50.

11. McKinley, J.P.; Zachara, J.M.; Smith, S.C.; Turner, G.D. Clays Clay Minerals 1995, 43, 586-98.
12. Glinka, Y.D.; Jaroniec, M.; Rozenbaum, V.M. J. Colloid Interface Sci. 1997, 194, 455-69.
13. Kosmulski, M. J. Colloid Interface Sci. 1997, 195, 395-403.
14. Hasany, S.M.; Shamsi, A.M.; Rauf, M.A. J. Radioanal. Nucl. Chem. 1997, 219, 51-4.
15. Rabung, T.; Geckeis, H.; Kim, J.-I.; Beck, H.P. J. Colloid Interace Sci. 1998, 208, 153-61.
16. Stumm, W. Chemistry of the Solid-Water Interface; John Wiley and Sons: New York, NY, 1992.
17. Morris, D.E.; Chisholm-Brause, C.J.; Barr, M.E.; Conradson, S.D.; Eller, P.G. Geochim. Cosmochim. Acta 1994, 58, 3613-23.
18. O'Day, P.A.; Chisholm-Brause, C.J.; Towle, S.N.; Parks, G.A.; Brown, Jr., G.E.. Geochim. Cosmochim. Acta 1996, 60, 2515-32.
19. Papelis, C., Hayes, K. F., and Leckie, J. O., "HYDRAQL: A program for the computation of chemical equilibrium composition of aqueous batch systems including surface-complexation modeling of ion adsorption at the oxide/solution interface", Stanford University Technical Report No. 306, 1988.
20. Grenthe, I. Chemical Thermodynamics of Uranium; North-Holland: New York, NY, 1992.
21. Brown, Jr., G. E.; Henrich, V.E.; Casey, W.H.; Clark, D.L.; Eggleston, C.; Felmy, A.; Goodman, D.W.; Gratzel, M.; Maciel, G.; McCarthy, M.I.; Nealson, K.H.; Sverjensky, D.A.; Toney, M.F.; Zachara, J.M. Chem. Rev. 1999, 99, 77-174.
22. Lanthanide Probes in Life, Chemical and Earth Sciences ; Bunzli, J.-C.G.; Choppin, G.R., Eds.; Elsevier: Amsterdam, 1989.
23. Bunzli, J.-C.G. In Lanthanide Probes in Life, Chemical and Earth Sciences; Bunzli, J.-C.G.; Choppin, G.R., Eds.; Elsevier: Amsterdam, 1989; pp 219-94.
24. Horrocks, W.D., Jr.; Sudnick, D.R. J. Am. Chem. Soc. 1979, 101, 334-40.
25. Kimura, T.; Choppin, G.R. J. Alloys Compounds 1994, 213/214, 313-17.
26. Moulin, C.; Wei, J.; Van Iseghem, P.; Laszak, I.; Plancque, G.; Moulin, V. Anal. Chim. Acta 1999, 396, 253-61.
27. Kato, Y.; Meinrath, G.; Kimura, T.; Yoshida, Z. Radiochim. Acta 1994, 64, 107-111.
28. Moulin, C.; Laszak, I.; Moulin, V.; Tondre, C. Appl. Spectros. 1998, 52, 528-35.
29. Morris, D.E.; Allen, P.G.; Berg, J.M.; Chisholm-Brause, C.J.; Conradson, S.D.; Donohoe, R.J.; Hess, N.J.; Musgrave, J.A.; Tait, C.D. Env. Sci. Tech. 1996, 30, 2322-31.
30. Glinka, Y.D.; Krak, T.B. Fresenius J. Anal. Chem. 1996, 355, 647-50.
31. Waite, T.D.; Davis, J.A.; Payne, T.E.; Waychunas, G.A.; Xu, N. Geochim. Cosmochim. Acta 1994, 58, 5465-78.

Chapter 7

Phase Chemistry and Radionuclide Retention of High-Level Radioactive Waste Tank Sludges

J. L. Krumhansl[1], P. V. Brady[1], P. C. Zhang[1], S. Arthur[1],
S. K. Hutcherson[1], J. Liu[2], M. Qian[2], and H. L. Anderson[1]

[1]Sandia National Laboratories, Albuquerque, NM 87185–0750
[2]Pacific Northwest National Laboratory, 3350 Q Street, Richland, WA 99352

The US Department of Energy (DOE) has millions of gallons of high
level nuclear waste stored in underground tanks at Hanford, Washington
and Savannah River, South Carolina. Decommissioning these tanks will
leave a residue of sludge adhering to the interior tank surfaces that may
contaminate groundwaters with radionuclides and RCRA metals.
Experimentation on such sludges is both dangerous and prohibitively
expensive so there is a great advantage to developing artificial sludges.
The US DOE Environmental Management Science Program (EMSP) has
funded a program to investigate the feasibility of developing non-
radioactive analogues for such materials. The following text reports on
the success of this program, and suggests that much of the radioisotope
inventory left in a tank will not move out into the surrounding
environment. Ultimately, such studies will play a role in developing safe
and cost effective tank closure strategies.

This work sponsored by Sandia National Laboratories, Albuquerque,
New Mexico, USA, 87185. Sandia is a multiprogram laboratory operated
by Sandia Corporation, a Lockheed Martin Company, for the U.S.
Department of Energy [DOE] under contract: DE-AC04-94AL8500.

INTRODUCTION

Cold War Pu production and purification activities produced many millions of gallons of high level nuclear waste (HLW). Most of this material is stored in underground tanks on the Hanford Reservation near Richland, Washington and at the Savannah River Site in South Carolina. Some tanks will eventually be emptied but routine-cleaning measures will leave significant amounts of sludge adhering to the interior tank surfaces. Because of this, the Environmental Management Science Program (EMSP) has funded research to assess whether artificial (nonradioactive) sludges can be prepared and used to assess rates at which sludges would release radionuclides to the environment adjacent to a decommissioned tank.

BACKGROUND

The radioisotope inventory of HLW storage tanks ultimately was derived from the irradiation of reactor fuel, but more than 99% of the contents consist of nonradioactive components from Pu purification and waste management practices. The greatest present-day hazard presented by HLW fluids arises from short-lived radioisotopes such as ^{154}Eu, ^{144}Ce, ^{137}Cs, ^{106}Ru, ^{90}Sr and ^{60}Co. Releases of these radioisotopes from a leaking tank may be of considerable importance in tracking short-term plume migration. Their association with sludge phases in the tanks may also have a large impact on waste removal and vitrification processes. However, they are not a long-term hazard to the environment since engineered barriers can be developed to isolate decommissioned tanks for several centuries (1). After 100 years the main fission products are still ^{137}Cs and ^{90}Sr, while after 1000 years the principal contributors are ^{99}Tc > ^{93}Zr > ^{94}Nb> ^{126}Sn > ^{79}Se ^{135},Cs> ^{151}Sm > ^{129}I, with a significant ^{59}Ni contribution from activation of stainless steel (2). After 10,000 years the order of the principal fission products is the same except that ^{151}Sm is no longer a significant contributor. Except for ^{99}Tc, all of these elements have non-radioactive isotopes that can be substituted for radioisotopes in order to study their affinity for artificial sludges. In oxidizing environments ReO_4^- is a reasonable substitute for TcO_4^-.

Actinides are more problematic. Actinides lighter than Bk all have at least one isotope that will persist for 10,000 years. Uranium (depleted) and thorium could be handled without specialized hot cell facilities, but a prohibitive administrative burden exists even though their heavy metal toxicity greatly outweighs any danger posed by radioactivity. Still, considerable insight into the behavior of Pu might eventually be gained using a combination of U(VI) and

Th(IV). The pentavalent actinides [(Pu(V) and Np(V)] may be important in some settings but good nonradioactive analogues do not exist. However, americium and curium (both dominantly +3 valence) have good non-radioactive surrogates in the rare earth elements.

Finally, HLW fluids may contain a number of non-radioactive metals that are of concern for their chemical toxicity – notably Pb, Cd, and Cr. Chromium was derived from two sources, corrosion of metal components, for which appropriate amounts of Cr^{+3} were added to the basic sludge recipes (Table 1). However, chromium, as $CrO_4^=$, was also added during some reprocessing operations so hexavalent chromium was included in the list of surrogates.

Many waste streams ultimately contributed to the contents of the HLW tanks at Hanford and Savannah River. This study is focused on the waste stream for each main Pu purification process that contained the greatest amount of radioactivity (Table 1). This excluded consideration of cladding wastes except when, as with the early $BiPO_4$ process, they were mixed with wastes from fuel reprocessing. For the $BiPO_4$ process the "1C/CW" waste was selected based on tabulations in Kupfer, Table C-5 (3). REDOX process waste streams were approximated based on the CWR1 and CWR2 waste streams from Table D2-3 in the same source. Waste solution chemistry from U-recovery operations employing the TBP process was developed from Kupfer (3) using metal wastes from the $BiPO_4$ processes (also Table C-5) variously supplemented by input from Agnew (4). PUREX wastes were derived principally from Agnew (4) with supplemental data from other sources (5, 6, 7). Early studies also used a simplified neutralized current acid waste ("NCAW") mix to simulate late-stage PUREX processes wastes (8).

Prior to being placed in tanks the acid wastes were rendered strongly caustic with NaOH and the NO_3^-/NO_2^- ratio was adjusted to minimize corrosion of the mild steel tank liners. *It is the premise of this research that the phase chemistry of the sludges was fixed at the time the wastes were neutralized.*

EXPERIMENTAL METHODS AND RESULTS

Synthesizing artificial sludges involved mixing solutions with the compositions presented in Table 1 using principally nitrate salts of the major metals. Where nitric acid concentrations were tabulated (3-8) appropriate amounts were also added. If data was unavailable 10 ml of concentrated nitric acid was added to 70 ml of starting fluid. Radionuclide and RCRA metal surrogates (Pb, Cd, Cs, Ba (for Ra), Sr, Nd (for Am, Sm, and Cm), $CrO_4^=$, Se, ReO_4^- (for TcO_4^-) and Co) were also added to the acidified solutions. Surrogate concentrations (Table 3 about 40 ppm, and Table 4 about 80 ppm) are higher than occur in actual wastes. However, this was necessary so that they would be detectable in various post-test analyses. Sodium hydroxide pellets were slowly added slowly so the fluids did not boil, though occasionally the bottles became too hot to be handled without an insulating glove. Typically, about half an hour was needed to bring the strongly acidic solutions to a pH between 12 and 13. Finally, they were allowed to cool before splits were taken for various tests.

Table 1 - Molar concentrations of significant sludge components in artificial acid wastes.

	BiPO4	TBP – U recovery	REDOX	PUREX	PUREX	NCAW - PUREX
	Al Cladding	Al Cladding	Al>>Zr Cladding	Al Cladding	Zr Cladding	Al/Zr Clad
Al	8E-2	0	1.1	8E-1	0	6E-1
Fe	3E-2	5E-2	5E-2	1E-1	4E-2	1E-1
Al/Fe ratio	2.6	0	22	6.7	0	0
Cr	3E-3	3E-3	7E-2	8E-3	3E-3	2E-2
Ni	2E-3	2E-3	4E-3	1E-2	1E-3	6E-2
Zr	3E-4	0	0	0	1E-1	4E-3
Bi	1E-2	0	0	0	0	0
Ca	2E-2	2E-2	0	6E-2	2E-2	0
Si	6E-2	4E-3	4E-2	5E-2	0	8E-3
F	2E-1	0	0	0	8E-1	1E-1
P	2E-2	1E-1	0	2E-2	0	0
Pb	4E-4	1E-4	1E-2	1E-3	0	0
Mn	7E-3	0	0	3E-2	0	0
$SO_4^=$	5E-2	2E-1	2E-2	2E-2	2E-2	1E-1
Cd	1E-5	8E-6	0	1E-3	1E-3	0

SLUDGE PHASE CHEMISTRY

Major phase identification was done using powder X-ray diffraction (XRD). Transmission electron microscopy (TEM) supplemented by energy dispersive (EDS) analysis confirmed the XRD determinations and detected a number of minor phases (Table 2). Unfortunately, a comparable suite of EDS data did not accompany the archived images of real sludges (Figures 1,2 and 4) so it is not possible to compare partitioning of minor elements between real and artificial sludges.

Table 2 - Phases identified in artificial sludges

Data	Waste Stream	Detected by XRD	TEM detected phases with elements by EDS
23° C 169 Days	BiPO$_4$ Process	Apatite 2-line Ferrihydrite	Apatite, Ca>Si>P>Al,Fe>Bi Bi-Fe-oxide, Fe,Ca,Si>Bi>P>Al>>Mn>Cr Zeolite, Si>Al>Ca>Na>>Fe>Cd
90° C 169 Days	BiPO$_4$ Process	Apatite Zeolite	Apatite, Ca>P>>Na>Bi>Pb, Trace Mn,Fe Zeolite, Al, Si>>Na, Bi, Ca, Fe>Mn, P, > Cr, Trace Cr, Ni, Pb Bi metal, Bi>>Na, Trace P Fe-Bi oxide, Fe,Bi>>Mn,Si, >Al, P, Ca, Na, Trace Pb CaSO$_4$, Ca, S>>Na ZrO$_2$, Zr>>Na Goethite, Fe>>Na Hematite, Fe>>Na
23° C 169 Days	TBP-U Recovery	Apatite 2-line Ferrihydrite	Apatite, Ca,P>>Fe>>Pb, Cr, Na FeOOH, Fe>>Al, Si, P, Pb, Ca,Cr
90° C 169 Days	TBP-U Recovery	Apatite 2-line Ferrihydrite	FeOOH, Fe>>Ca>Si>Ni, Pb Apatite, Ca, P > Fe, Na PbO, Pb>>Fe>Ca
23° C 169 Days	REDOX	Boehmite, Gibbsite Hematite (trace)	Boehmite, Al>Si>Cr, Fe >Ca Gibbsite, Na>>Al (Fe,Cr,Al,Si)OOH, Al>Fe,Cr>Si>Ca Trace Pb, Ni Portlandite Ca>>Na Hematite Fe>>Si, Ca>Pb, Na>Ni
90° C 169 Days	REDOX	Zeolite, Boehmite, Hematite (trace), Goethite (trace)	Zeolite, Al>Si>>Ca, Fe, Si>>Cr boehmite, Al>>Si, Fe, Cr> Ca>Na> FeOOH, Al,Fe>Cr>Na, Pb, Si, P> Ni
90° C 169 Days	PUREX – Al Cladding	Boehmite, Zeolite, Hematite	Boehmite, No EDS taken Zeolite, Al>Si>Na>>Fe, Ca Apatite, No EDS tanken FeOOH, Fe>>Al>Bi>Mn

23° C 169 Days	PUREX – Zr Cladding	Na_3ZrF_7, $NaFe_3(SO_4)_2$ $NaNO_3$ $Ca_5(SiO_4)_2$ $(OH,F)_2$	Fluorite, Ca, F>> Na, Zr, Fe >Cr, Si $NaFeO_2$, Fe, Na>>Zr , Ca, Cr, Ni Na_2FeF_6, Na, F, Fe>Ca>Cr, Zr Na_3ZrF_7, Na, Zr, F >>Ca, Cr, Fe, Si
90° C 169 Days	PUREX – Zr Cladding	Mostly Unknown, 2-line Ferri-hydrite	Fluorite, Ca, F, Trace Na, P, Pb FeOOOH, Fe > Ca, F, P , Si, Al, Na>Pb, Ni, Zn, Cs ZrO_2, Zr>>Ca, Fe, Si, Na > Pb
90° C 266 Days	NCAW: generic late stage PUREX	Boehmite Fe-Ni-Hydro-talcite	Boehmite, Al>>Fe>>Ni>Cr,Ca, P Fe-Ni-Hydrotalcite, Al>Ni>Fe>>Cr, Ca Portlandite,, Ca

Most sludges are comprised primarily of hydrous iron and aluminum oxides. Thus, the comparison between artificial and actual sludge phases containing these components is of particular importance. Figure 1 demonstrates that for hydrous iron oxides there is a good match between real and artificial sludges. The long-term lack of crystallinity is noteworthy since in other systems ferrihydrite often alters to goethite (or other crystalline phases) in a matter of days. However, Al (and other impurities) is known to have a large effect in retarding the crystallization of such materials (9).

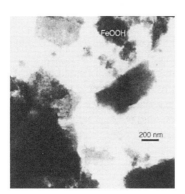 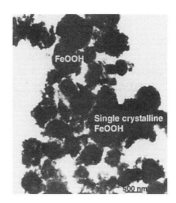

Figure 1. Comparison of hydrous iron oxides from synthetic sludges aged at 90° C for 169 days (left) and actual sludge from tank C-107 (right).

The correspondence with the Al-bearing compounds is not as good (Figure 2). Both real and artificial sludges are rich in boehmite (AlO(OH)), but the artificial material is not as well crystallized. Zeolites are another Al-containing phases that may be abundant in sludges (10). In this case, both real and artificial sludges are typically too fine grained to exhibit much crystal structure though X-ray patterns are typically quite sharp.

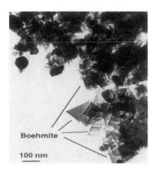

Figure 2. Boehmite formed in tanks (left) and artificial sludges (right)

SLUDGE-SUPERNATE INTERACTIONS

Both short-term (Table 3) and long-term (Tables 4,5) experiments were carried to assess which elements would partitioned into the sludge solids (and could be released later). In the short-term experiments the synthesis fluid was filtered and analyzed by direct current plasma (DCP) emission spectroscopy about an hour after the sludges were precipitated. This confirmed that that those surrogates expected to have very low solubilities at a high pH (11) were quantitatively scavenged from solution during sludge synthesis. Surprisingly, small decreases were also observed for ReO_4^- and $SeO_4^=$. Neither anion usually sorbs onto hydrous metal oxides from high pH solutions. Further, these losses were only observed in high aluminum REDOX and Al-PUREX sludges. Later, tests using TcO_4^- rather than ReO_4^- confirmed this pattern (Table 3, analytic precision about 10% in all cases). However, unlike ReO_4^-, more than half of the TcO_4^- was released by heating the sludge to 90° C overnight.

Table 3 - Decreases in anions during preparation of artificial sludges

Sludge Type	Re	Tc	Se
Al-PUREX	40 to 29 ppm	1 to 0.6 ppm	40 to 26 ppm
REDOX	40 to 27 ppm	1 to 0.8 ppm	40 to 31 ppm

Detailed studies involving ReO_4^- sorption onto boehmite (AlOOH) in low ionic strength solutions (Figure 3) indicate that ReO_4^- sorption should not be significant at the final pH of the synthesis solutions (pH 11-12). However, most of the boehmite would have precipitated between pH 3 and pH 8. In this pH interval ReO_4^- (and other anions) would sorb on the freshly forming surfaces and then become occluded by further growth of the boehmite crystals. Anions sorbed in this manner would not be available to desorb until the material recrystallized. Heating the samples may have initiated Ostwald ripening, thus, causing TcO_4^- to be released.

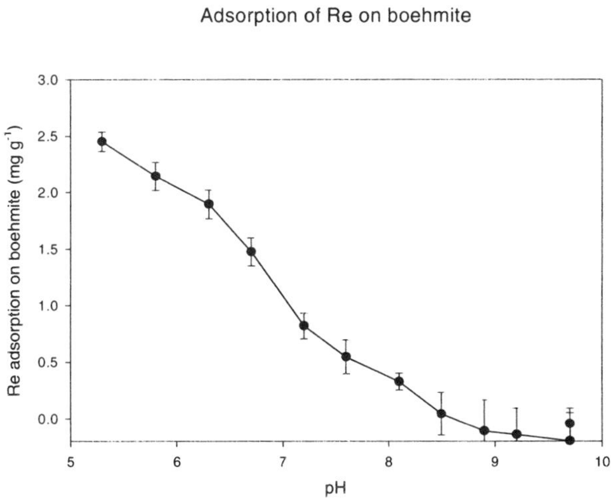

Adsorption of Re on boehmite

Figure 3. ReO_4^- sorption onto boehmite with pH in 0.0005 M $NaNO_3$. . Boehmite was added as a powder, not precipitated in contact with ReO_4

The release of Tc during heating suggests that recrystallization might impact surrogate retention. Thus, longer two-week experiments were carried out at 25° and 90° C, and, this time, the synthesis fluids were analyzed by ICP-MS (Table 4). The elevated temperature was employed to accelerate aging reactions and simulate the fact that many of the tanks were near the boiling point for long periods of time. The "Al:Fe-Molar" ratio (Table 4 and also 5) are given for each sludge immediately below the temperature. These values reflect post-test analyses of the sludge solids after the leaching and washing procedures were completed. These ratios differ from values shown in Table 1 since much of the Al initially present remains dissolved in the highly caustic synthesis fluids.

Table 4 - Analysis of post-precipitation fluids (PPM) after two weeks aging relative to solid Al/Fe ratios

Element	$BiPO_4$	TBP	REDOX	PUREX-Al	PUREX-Zr
25° C Expts					
Al/Fe-Molar	1.1	0	15.7	12.2	0
Surrogate Initial Conc.	87 ppm	85 ppm	84 ppm	79 ppm	88 ppm
Cr	77	73	108*	9.9	79
Co	<0.40	<0.40	<0.30	<0.30	<0.20
Sr	0.74	1.050	<0.20	<0.20	0.22
Cd	<0.40	<0.40	<0.20	<0.20	<0.20
Ba	7.94	4.67	6.04	4.36	5.69
Nd	<0.40	<0.40	<0.20	<0.20	<0.20
Pb	<0.30	0.57	0.65	<0.20	1.18
Se	73 (-16%)	65 (-15%)	66 (-21%)	53 (-33%)	71(−19%)
Re	81 (-7%)	70 (-16%)	71 (-16%)	67 (-15)	74 (-16%)
Cs	75 (14%)	67 (-15%)	97 *	72 (-99%)	79 (-10%)
90° C Expts					
Al/Fe-Molar	2.0	0	9.1	2.9	0
Cr	179*	141*	691*	157*	163*
Co	0.63	<0.40	<0.40	<0.30	<0.30
Sr	0.88	0.86	<0.40	<0.20	<0.20
Cd	<0.40	<0.40	<0.40	<0.20	<0.20
Ba	8.2	3.2	7.1	5.5	<0.20
Nd	<0.40	<0.40	<0.40	<0.20	<0.20
Pb	<0.30	0.80	1.88	<0.20	<0.20
Se	72 (-17%)	64 (-25%)	65 (-23%)	58 (-27%)	72 (-18%)
Re	83 (-5%)	72 (-15%)	77 (-9%)	69 (-13%)	80 (-9%)
Cs	82 (-6%)	77 (-9%)	75 (-11%)	72 (-9%)	89*

"*" Indicates concentrations that exceed amount added initially.

( ) Shows % decrease in aqueous concentration after sludge synthesis.

Aside from the molar Al/Fe ratios the reminder of the values in Table 4 are analyzed dissolved concentrations (in ppm). Chromate is often higher than what was initially added as a surrogate. This arises from the highly oxidizing properties of the synthesis fluids that transformed Cr^{+3} in the sludge recipe to soluble chromate, a "surrogate". The remaining two anomalous high values probably represent analytic errors. In contrast, Co, Sr, Cd, and Nd were universally scavenged from solution while the scavenging of Sr was spotty and that of Ba relatively poor. Some scavenging of Re and Se also occurred, but not to the extent documented in Table 3. Further, this time no preference was evident for the high-Al sludges

The absence of a clear correlation between Al content and anion sorption suggests that the ubiquitous hydrous iron oxides dominated sorption-desorption process. The shift in behavior can be understood in light of the fact that the surface properties of both Fe and Al oxides are quite variable. Depending on the particular oxide species the zero point of charge (e.g., the pH below which a surface is positively charged and hence sorbs anions) for Al oxides and hydroxides ranges from pH 9 to pH 5. For Fe oxides and hydroxides the range is pH 8.5 to pH 6.7 (12). Thus, the characteristics of the oxides that happen to form or evolve in a particular experiment will determine which (if any) hydrous oxide plays a dominant role in anion sorption.

The final stage of the laboratory study was directed at observing surrogate releases as the pH and ionic strength fell as would occur if groundwater gained access to a decommissioned tank. This was accomplished by first dialyzing the sludge suspension to remove the supernate solution and then suspending the sludge directly in deionized water and titrating the pH downward with dilute nitric acid. This process was developmental and the resulting pH values were generally lower than what might be expected within a decommissioned tank. However, both cation desorption and metal hydroxide solubility would be expected to increase at lower pH values. Thus, the residual sludge compositions (Table 5) represent a somewhat pessimistic assessment of what would remain fixed in the sludge under actual in-tank conditions.

An added complication in obtaining the data in Table 5 was that the one-piece filter construction made it impractical to recover and directly weigh the small amounts of trapped sludge solids. Thus, rather than reporting weight percentages of the various components, the sludge compositions are reported relative to the amount of iron in the analysis. From the sludge recipes (Table 1), an initial surrogate: Fe ratio was calculated and the analysis of the samples caught on the filters yielded a post leach surrogate : Fe ratio. Dividing the post-leach surrogate : Fe ratio by the initial surrogate : Fe ratio gives a measure of the relative enrichment (values greater than 1) or depletion (values below 1) of the surrogate due to the washing process (Table 5).

Table 5 - Post-leach solid analyses: post-leach surrogate: Fe ratios vs. "as mixed" surrogate : Fe ratios

Element	BiPO$_4$	TBP	REDOX	PUREX-Al	PUREX-Zr
25° C Expts					
Final pH	5.9	5.7	5.6	5.6	5.7
Al/Fe-Molar	1.1	0	15.7	12.2	0
Ba	*1.62*	*3.00*	0.29	0.71	0.67
Sr	0.07	0.71	0.06	0.16	0.07
Pb	0.74	*1.43*	1.10	1.03	*1.49*
Cd	0.12	*3.31*	0.65	0.90	1.08
Co	0.10	*1.42*	0.96	*1.76*	1.10
Cr	0.07	0.89	1.04	*1.33*	0.61
Nd	*1.52*	*2.74*	0.99	*2.14*	*1.49*
Re	9E-4*	2E-3	3E-3	8E-3	3E-3
Cs	0.01	6E-3	1E-3	4E-3	6E-6
90° C Expts					
Final pH	6.2	5.1	4.2	4.8	5.5
Al/Fe-Molar	2.0	0	9.1	2.9	0
Ba	*1.37*	0.85	0.49	*1.74*	*1.30*
Sr	0.98	0.13	0.08	1.05	0.20
Pb	0.76	1.03	*1.49*	0.88	*1.74*
Cd	0.97	1.12	0.61	1.12	0.41
Co	0.83	0.84	0.75	0.95	0.88
Cr	0.14	0.28	0.45	0.51	0.37
Nd	1.01	*2.39*	0.52	*1.98*	*1.32*
Re	5E-3	4E-3	1E-2	4E-3	7E-3
Cs	0.01	0.02	5E-3	9E-3	4E-3

Notation such as 9E-4 denotes 0.0009

Values greater than roughly 1.15 are problematic and cannot be ascribed to analytic errors. Nor is it likely that they reflect the systematic loss of iron from a sample since, in such instances, all the ratios for a particular sample would be skewed toward high values. Rather, they probably represent a lack of sample homogeneity; a nugget effect in which a phase enriched in a particular element was preferentially incorporated into the sample. Most nuggets seem to be associated with Ba, Pb, Nd, and 25° C TBP sludges. On the surface, the existence of Ba-rich nuggets and the absence of similar Sr-rich materials seems problematic. However, this can be explained noting that hydrous iron oxides would be expected to scavenge Sr better than Ba at high pH values (13, 14).

Thus, Ba would have been more available than Sr to participate in precipitation reactions occurring later in the titration

A possible suspect would be the high pH nucleation of apatite grains. The existence of clumps of apatite crystals has been verified for both artificial and actual sludges (Figure 4). This may also explain why so many of the nuggets seem to be associated with the TBP sludge. This sludge (Table 1) contains almost an order of magnitude more phosphate than the other mixes, and comparable amounts of calcium. In contrast, the REDOX waste contains no phosphate, cannot form apatite and has only one Pb-rich nugget. Neodymium nuggets may arise from formation of either insoluble hydroxides or phosphates.

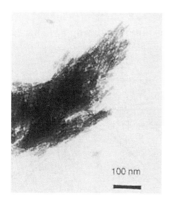 

Figure 4. Clumps of apatite crystals from artificial sludge (left) and actual sludge from Tank BY-108 (right).

The sludge analyses reveal almost no Re and Cs, and no detectable Se. Analysis of fluids from the last dialysis wash and the acidification step also failed to detect significant Re and Se, but did detect Cs. Thus, anion sorption was readily reversible but some phase apparently retained Cs. Cs releases were also greatest in experiments associated with Al-bearing sludges. Cancrinite zeolites occur in sludges (10) and seem to be a logical choice as a Cs sink. However, other studies suggest that freshly precipitated hydrous Al-hydroxides are more likely to retain Cs than cancrinite zeolites (15).

Finally, where Fe–hydrous oxides dominate, it is possible to employ theoretical geochemical models (e.g., a code named REACT) to predict the sorption of different components (Figure 5). Results of this modeling reflect several trends observed in Tables 4 and 5. At higher pH value it is predicted that Ba should, and did, remain in solution to a greater degree that Sr. However, raising the carbonate concentration above 10^{-4} molar reverses the predicted order due to the lower solubility of $BaCO_3$ relative to $SrCO_3$. As the pH falls it is also predicted that more Sr than Ba should be released. Where phosphate does not interfere (e.g., the REDOX and PUREX-Zr experiments) the sludges exposed to lower pH values do contain less Sr than Ba (Table 5). Lead and Eu (for Nd) are strongly removed from solution. All of the recipes contain at least some hydrous iron oxide and Pb is predicted to strongly sorb onto this component over the entire pH range as well as form insoluble $PbCO_3$ above about a pH of 8. Thus, it is unclear why the TBP and REDOX fluids from the lab tests still have detectable Pb levels, when the others do not. Theoretically, Co should be relatively immobile over the entire pH range. Low Co concentrations in the synthesis fluids (Table 4) and generally high Co retention in the acid-treated sludges (Table 5) supports these predictions. Finally, the code also predicts that $SeO_4^=$ would not be strongly sorbed over the upper half of the pH range examined. Thus, it is not surprising that little was removed during synthesis, and that which was scavenged was readily removed by later washes.

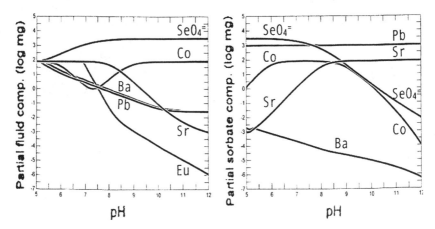

Figure 5. Theoretical model of surrogate precipitation and sorption on hydrous iron computed using the REACT code (13). Computed based on 80 ppm Se, Co, Pb, Ba, Sr, and Eu (for Nd), in a matrix solution containing 0.3 molal sodium and nitrate and 0.001 molal carbonate. 1000 grams of solution are mixed with 100 grams of freshly precipitated FeO(OH). Changes in dissolved components (left) reflect both sorption and precipitation reactions. The amounts that are actually sorbed appear on the right. Note: No Eu entry appears on the right panel since the sorption database contains no data on Eu sorption.

CONCLUSIONS

This study was undertaken to address two issues: is it possible to develop non-radioactive artificial sludges that resembling those found in high HLW storage tanks and, what fraction of the radioisotope inventory left in tanks would be mobilized if groundwater gained access to the sludge. At least for the major sludge phases, it is apparent that one can mimic what forms in HLW tanks if one is willing to let the samples age for about a month. The one significant deficiency seems to be the size (but not the phase) of the boehmite crystals, which are significantly larger in real sludges. This is unlikely to change the chemistry of sludges but could impact rheologic test results. Although much remains to be learned regarding potential sludge-groundwater interactions, the existing data clearly demonstrates that the complete radionuclide inventory of a decommissioned tank should not be regarded as mobile and, hence, a threat to the surrounding environment. Instead, it appears that a combination of modeling based on well described properties of hydrous iron oxide, solubility limits derived from thermochemical databases, and some experiments similar to what were described here would provide reasonable guidelines in developing performance assessment source terms for a variety of tank decommissioning scenarios. These, in turn, should help significantly in choosing the safest and most cost effective tank closure options.

REFERENCES

1. Westrich, H.R.; Krumhansl, J.L.; Zhang, P.-C; Anderson, H.L.; Molecke, M.A.; Ho, C.; Dwyer, B.P.; and McKeen, G. *Stabilization of in-tank residual wastes and external-tank soil contamination for the Hanford Tank Closure Program: Applications to the AX Tank Farm*; SAND98-2445, U.S. DOE; 1998, 112 p.
2. Johnson, L.H.; and Shoesmith, D.W. *Radioactive waste forms for the future;* Lutze, W.; Ewing, R.C., Eds.; Elsevier Science Publishing Co: New York, NY, 1988; table 1, p 61.
3. Kupfer, M.J. et al. (14 others) *Standard inventories of chemicals and radionuclides in Hanford Site tank Waste*; HNF-SD-WM-TI-740, U.S. DOE; 1997.
4. Agnew, S.F. *Hanford tank chemical and radionuclide inventories: HDW Model Rev. 4*; LA-UR-96-3860, U.S. DOE; 1996.
5. Colton, N.G. *Sludge pretreatment chemistry evaluation: Enhanced sludge washing separation factors*; TWRSPP-94-053, U.S. DOE; 1994.
6. Hsu, C.L.W.; Ritter, J.A. *Nuclear Technology* **1996**, *116*, p 198, Table I.

7. Fowler, J.R. *American Nuclear Society Transactions* **1982**, *41*, 159-160.
8. Norton, M.V.; and Torres-Ayala, F. *Summary letter report: Laboratory testing in-tank sludge washing,* US DOE, PNL-10153, 1994, p 3-4.
9. Cornell, R.M., and Schwertmann, U. *The iron oxides: structure, properties, reactions, occurrence, and uses*; VCH Verlagsgesellschaft, mbH Waldhiem, New York, NY ,1996, p.313-374.
10. Fowler, J.R.; Wallice, M. *Memorandum: CRC Zeolite in SRP Waste, Technical Division, Savannah River Laboratory*; DPST-80-488, U.S. DOE; Dec.2, 1980.
11. Baes, C.F.; Mesmer, R.F. *The hydrolysis of cations*; John Wiley & Sons: New York, NY, 1976.
12. Silva, R.J.; Nitsche, H. *Radiochimica Acta* **1995**, 70/71, 377-396.
13. Bethke, C.M. *Geochemical reaction modeling, concepts and applications*; Oxford University Press, New York, N.Y., 1966.
14. Dzombak, D.A.; Morel, F.M.M. *Surface complexation modeling – hydrous ferric oxide*; John Wiley and Sons: New York, NY, 1990, 299-306.
15. Nyman, M.; Krumhansl, J.L.; Zhang P.-C.; Anderson, H.L.; and Nenoff, T.M. *Chemical evolution of leaked high-level liquid wastes in Hanford soil - DSSF interactions,* Scientific Basis for Nuclear Waste Management XXIII, Materials Research Society (in press), 1999.

Separations and Treatment Chemistry

Chapter 8

Use of Cage-Functionalized Macrocycles and Fluorinated Alcohols in the Liquid–Liquid Extraction of NaOH and Other Sodium Salts

Strategies Toward Waste-Volume Reduction

Bruce A. Moyer[1,*], Peter V. Bonnesen[1], C. Kevin Chambliss[1], Tamara J. Haverlock[1], Alan P. Marchand[2,*], Hyun-Soon Chong[2], Artie S. McKim[2], Kasireddy Krishnudu[2], K. S. Ravikumar[2], V. Satish Kumar[2], and Mohamed Takhi[2]

[1]Chemical and Analytical Sciences Division, Oak Ridge National Laboratory, Bethel Valley Road, Oak Ridge, TN 37830–6119
[2]Department of Chemistry, University of North Texas, NT Station, Avenue C at Sycamore Street, Denton, TX 76203–5070

Concepts for the selective separation of sodium hydroxide and other sodium salts from alkaline high-level wastes are described together with initial results. Eight extraction mechanisms may be envisaged for transferring NaOH equivalents to an organic solvent by liquid-liquid extraction. Selectivity derives from principles of solvation, host-guest chemistry, and cation exchange. Initial results are presented on the synthesis and properties of new cage-functionalized macrocyclic hosts and fluorinated alcohol cation exchangers. Such compounds show promise toward reducing the overall waste volume by removal of bulk quantities of sodium salts.

The goal of the research described in this paper is to acquire fundamental knowledge regarding the separation of sodium hydroxide and other predominant sodium salts from alkaline nuclear waste stored at various United States Department of Energy sites. The principal environmental benefit of this research is the potentially major reduction in the volume of the low-level waste stream that remains after the predominant radionuclides have been separated. The need to remove bulk constituents that would otherwise have to be vitrified in the residual low-level waste stream has recently been identified (*1*). Owing to the large volume of alkaline waste (55 million gallons at the Hanford site alone) and the high cost of vitrification, large cost savings could be realized. Excluding water, the composition of such waste is dominated by sodium hydroxide, nitrate, nitrite, carbonate, and aluminate salts (*2*). A successful technical approach to volume reduction via sodium nitrate crystallization from the waste has recently been demonstrated (*3*). A case for sodium hydroxide removal based on electrochemical salt splitting has also been proposed (*4,5*). It has been estimated that the recovery of sodium hydroxide could reduce the overall volume of low-activity waste by as much as 32% (*4*). Moreover, recovered sodium hydroxide could be reused for neutralization of newly generated waste, corrosion inhibition, or dissolution of alumina from sludge. Without such recycle, fresh sodium hydroxide would have to be added to the waste stream, ultimately increasing the waste volume and worsening the overall problem.

It was our thought that the technological options for removal of sodium hydroxide from the waste could be further enhanced by exploiting the high-throughput potential of liquid-liquid extraction. As befits the Environmental Management Science Program, the task of identifying optimal extractants and solvent components entails addressing some exciting forefront questions in chemical science. Perhaps the most fascinating question deals with the design of extractants that confer the needed selectivity for the simultaneous separation of a target cation, sodium, and a target anion, hydroxide. In this review, we present our multi-faceted approach to this problem, illustrating selected concepts with initial results. The following section outlines the array of extraction mechanisms that could be considered and their expected characteristics. Next we report our progress toward developing novel cage-functionalized macrocycles for the liquid-liquid extraction of various sodium salts, particularly sodium hydroxide. Finally, we report extraction results using a novel approach based on weakly acidic fluorinated alcohols. It will be seen that progress has been made toward both a practical separation process and an understanding of the fundamental chemistry of the synthesis and properties of applicable extractants.

Eight Basic Extractive Approaches to NaOH Separation

Since the chemical literature offers little information on the liquid-liquid extraction of NaOH, we make at the outset no presumption as to the best practical method but rather consider here a heirarchy of applicable fundamental chemical processes. These rely on principles of solvation, acid-base reactions, and host-guest chemistry. Table I lists eight basic approaches that one might take. The first five approaches entail ion-pair extraction processes in which the extracted cation M^+ and anion X^- may either be solvated or complexed. The receptor in each case is indicated by a circle without implying any particular topology. The last three approaches entail an acid-base reaction to transfer a hydroxide *equivalent* to the solvent with or without a receptor for the cation M^+. The cation exchanger is depicted with an alkyl tail to indicate lipophilicity. As written for all cases except 5 and 8, the product cation and anion species given in each case are dissociated in the solvent, implying *ideally* little or no influence of the cation and anion upon one another. However, secondary effects such as ion pairing or aggregation can be expected to influence ion selectivity.

The simplest approach in concept entails choice of a water-immiscible solvent that by itself effects the extraction of NaOH from an aqueous mixture of salts (Table I, Case 1). The solvent molecules must therefore completely accommodate the Na^+ cation and OH^- anion. For Na^+, this means supplying electron-pair donor (EPD) groups for coordination (6). Likewise for OH^-, hydrogen-bond donor (HBD) groups are needed (7). Since the thermochemical radii of both ions are small $[r_{Na^+} = 0.102$ nm and $r_{OH^-} = 0.133$ nm (8)], the EPD and HBD groups must be significantly stronger than the H_2O molecules in the source phase for efficient ion partitioning. This is difficult to achieve in a water-immiscible liquid, and indeed, positive Gibbs energies of ion transfer (8) lead one to expect weak extraction.

Extraction can be enhanced by use of a cation receptor (Case 2), anion receptor (Case 3), both cation and anion receptors in synergistic combination (Case 4), or a ditopic ion-pair receptor (Case 5). For present purposes, a receptor may be defined as a molecule having multiple EPD or HBD groups for

Table I. Fundamental Approaches Applicable to NaOH Separation using Host-Guest and Liquid-Liquid Extraction Principles

Case	System	Organic-phase species
1	No receptors	$M^+ + X^-$
2	Cation receptor	$\langle M^+ \rangle + X^-$
3	Anion receptor	$M^+ + \langle X^- \rangle$
4	Cation receptor + Anion receptor	$\langle M^+ \rangle + \langle X^- \rangle$
5	Ditopic ion-pair receptor	$\langle M^+ X^- \rangle$
6	Cation exchanger	$\sim\!\!\sim\!\!\sim A^- + M^+$
7	Cation exchanger + Cation receptor	$\sim\!\!\sim\!\!\sim A^- + \langle M^+ \rangle$
8	Ditopic cation exchanger-receptor	$\langle M^+ \rangle\!-\!A^-$

partial or complete encapsulation of guest ions. With use of any receptor, the Gibbs energy of complexation augments the Gibbs energy of ion transfer, thereby boosting overall extraction strength (9). The enhancement can also be made selective. Receptors for Na^+ and other alkali cations are plentiful among crown ethers, cryptands, and calixarenes, for example (10). Receptors for anions are less plentiful, and OH^- ion has yet to be considered (11). Naturally, strong HBD groups directed at the oxygen atom would be favorable, possibly together with an EPD group directed to interact with the hydrogen atom. Ion-pair receptors are still rare, but a few designed examples have been reported (12).

As ion-pair extraction processes, Cases 1 to 5 in Table I can be adapted well to cyclic sodium hydroxide extraction and stripping. Taking Case 2 as an example, the aqueous cation (M^+) and anion (X^-) are transferred to a solvent phase, where the cation is then complexed with a crown ether:

$$Na^+ (aq) + X^- (aq) + Crown (org) \rightleftharpoons [NaCrown]^+ (org) + X^- (org) \quad (1)$$

It may be expected from eq 1 that high Na^+ concentration drives extraction, and extraction may be subsequently reversed by contacting the solvent with a low-salt aqueous solution, ideally water. Such a cycle is ideal for treatment of high-salt wastes, such as alkaline high-level tank waste, and use of water for stripping introduces no new chemicals or dissolved solids to the process (13).

It is likely that a successful approach employing ion-pair extraction would require use of an anion receptor to obtain sufficient selectivity for OH^- ion. When the anion is solvated as in Cases 1 and 2 in Table I, one generally observes Hofmeister-type selectivity (14). That is, extraction strength is biased in favor of larger, more charge-diffuse anions (7). Thus, the abundant anion nitrate would be preferentially extracted. In the event that anti-Hofmeister behavior could be demonstrated, fluoride extraction would compete (7). Although this would be desirable in a scheme to separate nitrate or possibly fluoride salts, a bias-type selectivity would not provide OH^- selectivity. As mentioned above, a recognition approach would entail building a molecule that directs appropriate HBD and EPD groups in a geometry complementary to OH^-.

Ironically, cation exchange provides an alternative approach for an effective extraction of OH^- ion. Possessing exchangeable acidic protons, such extractants (HA) have many variants, but all function according to a common exchange process, which (neglecting complications due to aggregation and ion pairing) may be written most simply as eq 2 or its equivalent in terms of OH^- (eq 3):

$$Na^+ (aq) + HA (org) \rightleftharpoons H^+ (aq) + Na^+ (org) + A^- (org) \quad (2)$$

$$Na^+ (aq) + OH^- (aq) + HA (org) \rightleftharpoons H_2O (aq) + Na^+ (org) + A^- (org) \quad (3)$$

The reverse reaction affords recovery of sodium hydroxide upon stripping with water, whereby the alcohol returns to its protonated form in the organic phase. When used in tandem, the forward and reverse steps constitute a cyclic process affording the transfer of alkali metal hydroxide from an aqueous mixture into water. To function efficiently for hydroxide recovery, HA must possess weak acidity (pK_a *ca.* 9-14) so that contact of the loaded solvent with water readily regenerates the protonated form of the extractant. Surprisingly, a single study involving phenols represents the only citation of such a process in the literature (*15,16*). As before, a cation receptor may be added to the solvent as a synergist (*17*) for the cation exchanger (Case 7), and one may envision that the cation receptor could also contain the cation-exchange functionality in the same molecule (Case 8). Whereas the desired extraction-stripping cycle is again possible using water for stripping, the major advantage of any of the cation-exchange approaches is the potentially high selectivity for OH⁻ ion. Only highly basic anions can undergo the acid-base process given in eq. 3.

Synthesis of Novel, Cage-Functionalized Host Systems

In relation to the problem of separating sodium salts from high-salt wastes, the structure of macrocyclic cation receptors allows control of the cation selectivity. Investigators at the University of North Texas have synthesized several cage-functionalized crown ethers and related aza-crown systems, each of which contains a 4-oxahexacyclo[$5.4.1.0^{2,6}.0^{3,10}.0^{5,9}.0^{8,11}$]dodecane moiety. In several instances, incorporation of this cage unit into these systems has produced a dramatic effect upon their ability to function as ionophores. Here, the cage unit functions as a lipophilic "spacer" that also serves to increase the rigidity of the crown system in the resulting macrocycle relative to the corresponding non-cage-containing analog, thereby influencing the overall conformational mobility of the host. Incorporation of the cage unit also affects the shape and size of the cavity in the host system. In addition, the *furano* oxygen atom in the 4-oxahexacyclo[$5.4.1.0^{2,6}.0^{3,10}.0^{5,9}.0^{8,11}$]dodecane cage moiety potentially can participate along with the remaining Lewis base atoms in the macrocycle during formation of an eventual host-guest complex.

Our first attempts to incorporate the 4-oxahexacyclo[$5.4.1.0^{2,6}.0^{3,10}.0^{5,9}.0^{8,11}$]dodecane cage moiety into crown ethers resulted in the synthesis of two novel host systems, **5** and **6**. The methodology that was employed successfully for this purpose is summarized in Scheme 1 (*18*).

Thus, reaction of pentacyclo[$5.4.1.0^{2,6}.0^{3,10}.0^{5,9}$]undecane-8,11-dione ("PCU-8,11-dione", **1**) with excess vinylmagnesium bromide afforded the corresponding diol, **2**, in 60% yield. Dehydration of diol **2** produced 3,5-

Scheme 1

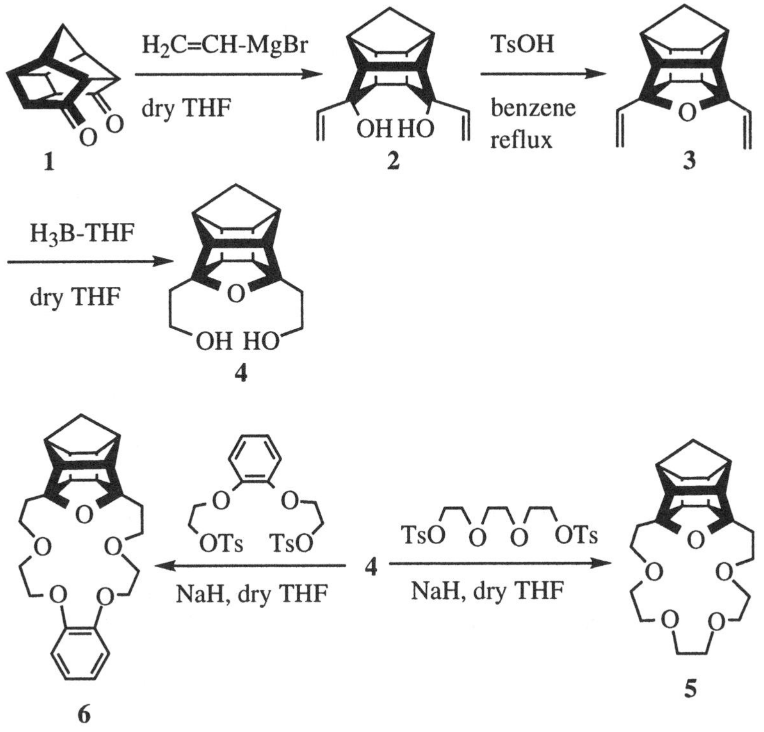

divinyl-4-oxahexacyclo[5.4.1.0^{2,6}.0^{3,10}.0^{5,9}]dodecane, **3** (77% yield). Subsequent hydroboration-oxidation of the carbon-carbon double bonds in **3** afforded the corresponding cage diol, **4**, in 85% yield. Diol **4** thereby prepared proved to be a key starting material for the preparation of several of the new cage-functionalized crown ethers and cryptands reported herein.

In an effort to probe the importance of proximity effects (i.e., "host preorganization") on host-guest complexation properties, a series of 3,5-difunctionalized 4-oxahexacyclo[5.4.1.0^{2,6}.0^{3,10}.0^{5,9}.0^{8,11}]dodecanes has been prepared. These compounds serve as templates to develop a series of novel complexing agents that can be used for metal ion separation and transport. Thus, for example, 1-aza(12-crown-4) (i.e., 1,4,7-trioxa-10-azacyclododecane) and several substituted 1,4-diaza(12-crown-4) derivatives have been prepared, and these species subsequently have been affixed as pendant "arms" to the polycyclic template by using the synthetic strategy shown in Scheme 2 (*19*).

Whereas the pendant (12-crown-4) moieties in **8**, **9**, and **10** are

Scheme 2

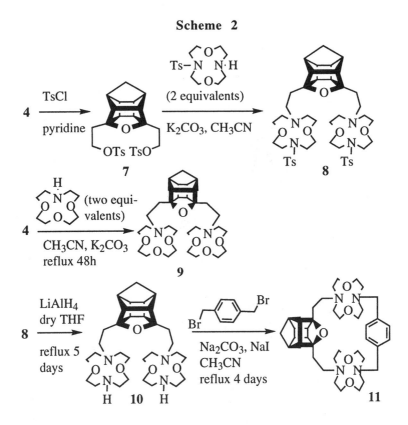

conformationally mobile, the situation is quite different in **11**. Thus, in **11**, the use of a *p*-xylyl unit as a covalent linking agent results in the formation of a "molecular box" (*20*) with concomitant "forced cooperativity" between the otherwise distant 1,4-diaza(12-crown-4) moieties. Thus, it was of interest to determine how the anticipated increase in host preorganization brought about through "forced proximity" between the opposing 1,4-diaza(12-crown-4) moieties might influence the avidity and/or selectivity of **11** *vis-à-vis* **8-10** toward, for example, complexation of alkali metal picrates.

In addition, several new cage-functionalized crown ethers and cryptands have been prepared as part of this study (see Scheme 3). Thus, **12-14** contain one or more 2,6-pyridiyl moieties in addition to the 4-oxahexacyclo[5.4.1-.0^{2,6}.0^{3,10}.0^{5,9}.0^{8,11}]dodecane cage (*20*). Another series of novel host molecules, **15-20**, is comprised of cage-functionalized diaza(17-crown) ethers (*21-23*). Finally, two additional cage-functionalized host systems, **21** (*24*) and **22** (*25*), have been synthesized. Based upon literature precedent (*26*), it was

Scheme 3

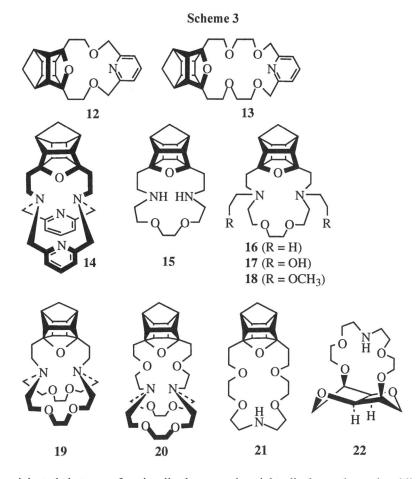

anticipated that cage-functionalized cryptands might display enhanced avidity and/or selectivity toward complex formation with alkali metal cations due to the increased preorganization inherent in their structural frameworks.

Liquid-Liquid Extraction

Survey of the Cation Selectivity of Macrocycles by Picrate Extraction

Cation selectivity was assessed qualitatively by extraction of individual alkali metal picrate salts from neutral solution and by subsequent measurement of the spectrophotometric absorbance of the picrate anion in the organic phase.

Mechanistically, the test formally reflects the selectivity of simple ion-pair extraction in Case 2 (Table I), but by extension it also applies to Cases 4 and 7. The procedures employed to prepare alkali metal picrates (*18, 27*) and also the relevant alkali metal picrate extraction techniques (*28, 29*) used in these studies have been described previously. Non-competitive (i.e., single alkali metal picrate) cation extraction experiments were performed by using 5 mM solutions of individual receptors in $CHCl_3$ as the host systems. Extraction data were obtained, and averages and standard deviations for the extraction data were calculated for a minimum of three and a maximum of five independent extraction experiments in each case. It should be pointed out that this test does not definitively measure metal cation extraction, since only the anion is monitored. Allowance should be made for the possibility that hydrogen ion can also be extracted, which for the more basic aza compounds would likely give a non-zero baseline picrate extraction. This effect is being investigated.

In order to gain insight into the effect of cage annulation upon the ability of crown ethers and cryptands to perform as hosts for extraction of alkali metal picrates, the relative extraction efficiencies of novel, cage-functionalized hosts toward alkali metal picrates are compared with those of corresponding, non-cage-functionalized model compounds, in each case. In Table II, the results of alkali metal picrate extractions performed by using host systems **5** and **6** (Scheme 1) are compared with the corresponding results obtained by using 15-crown-5 and benzo(15-crown-5) as model systems. We note that cage-functionalized crown

Table II. Results of Alkali Metal Extraction Experiments

Host Molecule	Percent of Picrate Extracted (%)				
	Li^+	Na^+	K^+	Rb^+	Cs^+
15-crown-5	2.3 ± 0.2	13.9 ± 1.3	14.3 ± 0.3	9.6 ± 0.8	BLD[a]
5	2.8 ± 0.3	18.5 ± 1.0	29.0 ± 0.2	8.4 ± 1.3	BLD[a]
benzo-(15-crown-5)	BLD[a]	11.8 ± 1.8	19.3 ± 1.9	5.9 ± 1.2	4.1 ± 1.2
6	BLD[a]	10.6 ± 0.6	10.0 ± 1.8	5.7 ± 1.5	2.3 ± 1.2
18-crown-6	1.9 ± 0.8	4.5 ± 0.6	68.2 ± 0.6	56.6 ± 1.3	30.3 ± 0.9

[a]BLD = Below limit of detection.

ether **5** displays somewhat enhanced avidity and selectivity toward extraction of Na^+ and K^+ picrates *vis-à-vis* 15-crown-5. However, addition of a benzo group to give **6** weakens extraction *vis-à-vis* that of **5** and benzo-15-crown-5 overall but improves selectivity for Na^+ vs. K^+. The reduced efficiency of **6** vs. **5** as a Na^+ and/or K^+ picrate extraction may be due to conformational and ring-size differences between these two host molecules. Alternatively, this difference in behavior might result simply from the reduced Lewis basicity of the two phenolic oxygen atoms in **6**.

The results of alkali metal picrate extractions performed by using host systems **8-11** (Scheme 2) are presented in Table III (*19*). Hosts **8** and **10** are poor performers. However, **9** displays improved avidity toward extraction of alkali metal picrates *vis-à-vis* **8** and **10**. Further improvement in this regard is achieved by **11**, wherein the linking *p*-xylyl moiety constrains the movement of opposing 1,4-diaza(12-crown-4), thereby placing them in mutual proximity. The improved performance of **11** as an alkali metal extractant *vis-à-vis* **8-10** provides a clear demonstration of the advantages conferred by increased host preorganization concomitant with its "molecular box" structure.

Cryptands **14** (*21*), **19** (*24*), and **20** (*22*) (Scheme 3) constitute a series of highly preorganized host systems of varying cavity dimensions. Relevant alkali metal extraction data for these three cryptands and related model crown ethers are given in Table IV. Here, it can be seen that all three cryptands display significantly higher avidities toward all five alkali metal picrates studied when compared in each case with the less highly preorganized corresponding model crown ether.

Table III.   Results of Alkali Metal Extraction Experiments

Host Molecule	Percent of Picrate Extracted (%)				
	Li^+	Na^+	K^+	Rb^+	Cs^+
8	7.9 ± 0.5	1.9 ± 0.5	6.1 ± 0.1	6.0 ± 0.7	6.5 ± 0.5
9	41.7 ± 0.6	55.6 ± 0.8	44.3 ± 0.5	34.8 ± 0.1	31.6 ± 0.5
10	10.6 ± 0.6	8.6 ± 0.3	8.8 ± 0.5	7.4 ± 0.9	8.9 ± 0.7
11	66.8 ± 0.2	86.0 ± 0.3	58.0 ± 0.6	62.4 ± 0.4	65.3 ± 0.1

Table IV. Results of Alkali Metal Extraction Experiments

Percent of Picrate Extracted (%)

Host Molecule	Li$^+$	Na$^+$	K$^+$	Rb$^+$	Cs$^+$
4,13-diaza-(18-crown-6)	11.0 ± 0.7	10.4 ± 1.3	13.0 ± 2.0	9.1 ± 0.6	10.1 ± 0.3
	31.1 ± 0.6	30.9 ± 1.3	30.3 ± 0.5	33.0 ± 0.9	29.8 ± 1.4
14	81.9 ± 0.7	70.7 ± 0.8	40.3 ± 0.5	46.0 ± 0.6	40.9 ± 0.9
N,N'-diethyl-4,13-diaza-(18-crown-6)	3.3 ± 0.5	22.4 ± 0.6	34.8 ± 0.8	21.2 ± 0.7	16.0 ± 0.7
19	39.8 ± 0.8	95.1 ± 0.8	89.7 ± 1.2	56.1 ± 0.1	33.7 ± 0.7
N,N'-bis-(2-methoxyethyl)-4,13-diaza-(18-crown-6)	17.5 ± 0.5	36.0 ± 0.7	46.3 ± 0.8	34.6 ± 0.9	21.3 ± 0.5
20	39.3 ± 0.6	50.0 ± 0.7	74.4 ± 0.8	68.4 ± 1.0	60.9 ± 0.8

Cryptand **19** (*24*) functions as a selective and highly avid Na$^+$ and K$^+$ picrate extractant, whereas **14** (*21*), whose cavity dimensions are significantly more restrictive than those of **19**, displays high avidity and selectivity toward extraction of Li$^+$ picrate. Finally, as expected on the basis of the size-match ("optimal spatial fit") principle (*30*), **20** displays particularly high avidity toward K$^+$ and Rb$^+$ picrates (*22*).

Extraction of Sodium Salts by Selected Macrocyclic Hosts

Although it was expected that the cage-annulated macrocycles would mainly function as cation receptors, it was of interest to address the question as to their effect, if any, on anion selectivity. In particular, the aza-macrocycles containing >N-H linkages possess some hydrogen-bond donating ability, which would offer a site for anion interaction that crown ethers with all oxa donor groups lack. Cage-annulated aza-macrocycles **15**, **21**, and **22** were investigated in non-competitive extraction studies of a series of sodium salts of varying anion radius:

NaF, NaCl, NaBr, NaOH, NaNO$_3$, and NaClO$_4$. Nitrobenzene was selected as the diluent in these extraction studies, since it possesses a high dielectric constant [ε = 34.8, (*31*)], allowing one to compare the relative extractability of each anion under conditions where ordinary ion-pairing effects are weak. In addition, its lack of H-bond donor ability (*31*) confers minimal specific effect on anion selectivity. Extractions were performed by contacting equal volumes of the organic phase containing 0.05 M macrocycle with the aqueous phase by gentle end-over-end rotation for 30 minutes at 25 ± 0.2 °C. The aqueous phases consisted of 1 M solutions of the individual sodium salt, spiked with ^{22}Na radiotracer as ^{22}NaCl in water (at *ca.* 10-15 nM). Duplicate samples were evaluated for each data point. After centrifugation, aliquots of each phase were removed for ^{22}Na gamma activity analysis. Sodium distribution ratios (D_{Na}), reproducible to within ± 5%, were calculated as the ratio of the ^{22}Na activity in the organic phase to the ^{22}Na activity in the aqueous phase at equilibrium. It should be pointed out that the anion extraction here is inferred from charge balance and remains to be confirmed by direct measurements of anion distribution. Further, since reagent-grade nitrobenzene was used, the issue of impurity effects due to phenols remains to be addressed. Work in progress shows that purified nitrobenzene yields somewhat lower D_{Na} values, though the qualitative trends described below are preserved.

The sodium extraction results for each anion by macrocycles **15**, **21**, and **22**, plotted as the log of the sodium distribution ratio (average of replicates) vs. the reciprocal of the thermochemical ionic radius (*r*, in nm) for each anion (*8*), are shown respectively in Figure 1. Each plot exhibits three regions indicated by the straight line segments. The top-most horizontal segment simply indicates saturation level of the macrocycle by Na$^+$ cation, whence a maximum D_{Na} value of 0.05 is expected if a 1:1 macrocycle:metal stoichiometry is obtained. The steeply descending straight line segment indicates the natural bias (*7*) expected for extraction of large anions, which may be termed the Hofmeister effect (*14*). Roughly linear dependence on $1/r$ with negative slope is expected based on the form of expressions for Gibbs energy of ion transfer derived from the Born charging equation (*7*). Hence, such a plot provides a convenient and qualitatively meaningful vehicle for presenting the data on the basis of ion size. Given the limited data at present, the line segments corresponding to ClO$_4^-$, NO$_3^-$, Br$^-$, and Cl$^-$ ions were drawn arbitrarily, with comparable decreasing dependencies on $1/r$. Perchlorate is strongly extracted and exhibits significant saturation in each case. Since perchlorate is large and practically unhydrated in the organic phase, it is most likely to conform to the ideal ion-pair model given by eq 1 (Table I, Case 2), in which the macrocycle binds the cation and the anion is dissociated. On this basis, the inherent macrocycle extraction strength toward Na$^+$ ion may be given as **21** > **15** > **22**.

126

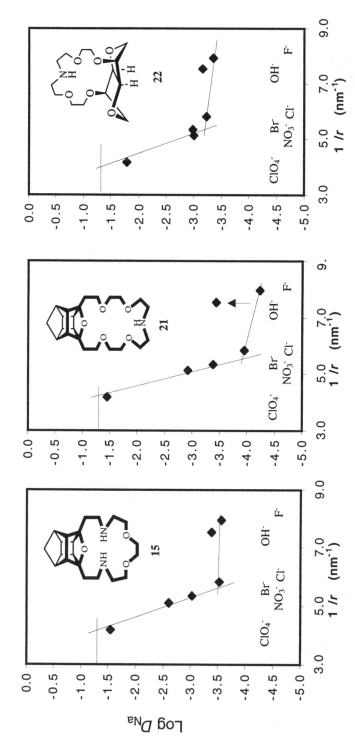

Figure 1. Plots of the logarithm of sodium distribution ratios vs. reciprocal thermochemical anion radii for three aza macrocycles. Conditions: 0.05 M macrocycle in nitrobenzene, 1 M aqueous sodium salt, 1:1 phase ratio, and 25 °C.

Although this ordering seems readily defensible, it persists for none of the other anions. Thus, anion selectivity appears to depend upon the structure of the macrocycle, and the mechanism (eq 1) cannot be generalized to the other anions.

Although extrapolation of the electrostatic arguments based on the radii of bare ions would predict that extraction of NaOH and NaF by macrocycles **15**, **21**, and **22** would be exceedingly low, Figure 1 shows that the D_{Na} values for these anions do not continue to follow the steeply decreasing trend. Indicated arbitrarily by a straight line segment joining the points for Cl⁻ and F⁻, the third region of the plots in fact exhibits no obvious dependence upon ion radius. A straightforward explanation under investigation lies in the fact that the small anions are highly hydrated in organic solvents equilibrated with water. For example, the ions Br⁻ and Cl⁻ have respectively an average of 1.8 and 3.3 water molecules per ion in water-saturated nitrobenzene at 23 °C (*32*). Although data are lacking, OH⁻ and F⁻ ions would be expected to have even greater hydration. The effect of ion hydration in the organic phase is to make the Gibbs energy of ion partitioning more favorable (*7*), compensating for the decreased ion radius. One could say simply that the highly hydrated anions behave as if they have effectively comparable solvation in the organic phase.

The D_{Na} values corresponding to the anions Cl⁻, OH⁻, and F⁻ in Figure 1 vary depending upon the macrocycle employed. The values follow the order **21** < **15** < **22**, the reverse order observed for ClO_4^- ion. Logically, the macrocycle may exert an effect on anion selectivity either if the anion is ion paired with the macrocycle-Na⁺ complex or if the macrocycle interacts directly with the dissociated anion. The latter possibly could occur via hydrogen bonding mediated by water molecules, but there is no prior evidence in the literature by which the likelihood of this occurrence can be judged. The former possibility entailing ion pairing may occur, since ion pairing is favored by small ion radius and incomplete encapsulation of the metal cation by the macrocycle (*33*).

Cation-Exchange Approach to Sodium Hydroxide Separation

Whereas the macrocyclic cation receptors presented above rely on ion-pair extraction to effect an actual transfer of both Na⁺ and OH⁻ ions to the solvent phase, Cases 6-8 in Table I offer a powerful alternative based on a principle of cation exchange. A pseudo sodium hydroxide extraction, Case 6, has been demonstrated (*34,35*) using 1H,1H,9H-hexadecafluorononanol (HDFN) in 1-octanol (Figure 2). This and related fluorinated alcohols are expected to exhibit requisite weak acidity and good stability. For example, the pK_a values of 2,2,2-trifluoroethanol and 2,2,2-trifluoro-1-(4-methylphenyl)-ethanol are 12.37 (*36*) and 12.04 (*37*), respectively. 1-Octanol was chosen as the diluent so as to provide good solvation for the Na⁺ cation and putative alkoxide anion (*6,7,31*). The ability to extract NaOH is clearly enhanced by the addition of 0.2 M HDFN

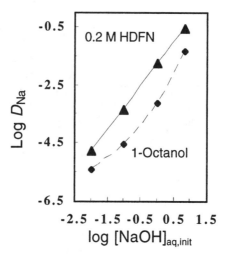

Figure 2. HDFN at 0.2 M in 1-octanol extracts Na⁺ ion significantly more than the blank (1-octanol alone). Conditions: 1:1 phase ratio, 25 °C.

relative to the baseline behavior observed for the 1-octanol diluent alone. This result demonstrates a significant advantage of Case 6 over Case 1 (Table I).

As expected from eq 3, NaOH could be recovered from the solvent following extraction. On contact with 7 M aqueous NaOH, the organic phase was found to contain 0.29 M Na^+ ion by ^{22}Na gamma tracer radiometry, somewhat higher than can be accounted for by stoichiometric deprotonation of HDFN (0.2 M) and the background extraction by 1-octanol (0.050 M). On contact of the loaded HDFN organic phase with water, essentially all of the extracted Na^+ cation was back-extracted from the 1-octanol phase within experimental error (±2%). By titration, the aqueous OH^- ion concentration upon back-extraction was shown to be equivalent to that of Na^+ ion.

The selectivity and recyclability of 0.2 M HDFN in 1-octanol were found to be excellent (*34,35*). Selectivity was judged under competitive conditions using a simulant (*38*) of a Hanford alkaline tank waste high in K^+ (0.945 M), NO_3^- (3.52 M), Cl^- (0.102 M), $Al(OH)_4^-$ (0.721 M), and free OH^- (1.75 M) ions. Equilibrium data were collected by ^{22}Na radiometry, acid-base titrimetry, inductively coupled plasma atomic emission spectrophotometry, and ion chromatography. A convenient measure of selectivity is the separation factor (α_{OH^-/X^-}), defined by the ratio $\{[OH^-]_{org}/[OH^-]_{aq}\}/\{[X^-]_{org}/[X^-]_{aq}\}$. Here, this quantity indicates the relative distribution strength of OH^- equivalents vs. that of other anions X^-. From the analytical data, α values were found as follows:

$\alpha_{OH^-/NO_3^-} = 35$, $\alpha_{OH^-/Cl^-} = 2.5$, $\alpha_{OH^-/Al(OH)_4^-} = 280$. The extraction was also selective for sodium, by analogy expressed as $\alpha_{Na^+/K^+} = 3.5$. The loaded organic phase Na^+ concentration was 0.12 M, and contact with water resulted in near-quantitative recovery of the extracted NaOH equivalents. Four cycles of the same solvent gave identical results, demonstrating that HDFN does not significantly partition to the aqueous phase and that the solvent is recyclable.

Although the selectivity data point to cation exchange via eq 3 (Case 6) as the predominant extraction mechanism, no definitive conclusion is implied at present. The question arises as to whether HDFN facilitates extraction of NaOH via cation exchange or by ion-pair extraction (Case 1), which are formally indistinguishable by mass-action behavior. That is, comparing the simplest mass-action models, a hydroxide equivalent in the organic-phase could exist as the alkoxide anion or as a solvate of hydroxide ion with a molecule of HDFN. Experiments are under way to answer this question via spectroscopy. In the meantime, the fact that the enhancement of Na^+ extraction by HDFN takes place significantly only when the anion is OH^- (*34,39*) suggests a process of cation exchange is taking place, since enhancement with all anions capable of receiving H-bonds would be expected if the mechanism were solvation (*7*).

By exploiting the strong inductive effect of fluoro- or fluorine-containing substituents in sufficiently hydrophobic alcohols, the effective reversible extraction of sodium hydroxide from waste appears feasible. We show elsewhere (*34*) that the efficiency of the process is consistent with the expected acidity of ionizable protons in the structures of tested extractants.

Conclusions

Eight fundamental approaches have been identified for the separation of NaOH from high-salt wastes by liquid-liquid extraction. Data show that one approach involving a putative cation-exchange process using weakly acidic alcohols appears especially promising (*34,35*). In fact, this study represents the first example of the use of fluorinated alcohols as cation exchangers in liquid-liquid extraction. This study also demonstrates a repeatable cyclic process for the selective recovery of hydroxide ion from aqueous salt mixtures. Investigations on such systems will continue, with a view both toward practical applications and toward understanding the underlying principles.

In developing other approaches involving ion receptors, synthetic methodology has been successfully devised for the synthesis of a series of novel cage-annulated oxa- and aza-crown ethers. These crowns vary in size and type of donor atom and accordingly vary in strength and selectivity toward alkali metal cations in picrate extraction surveys. It was possible to show that in the extraction of a series of sodium salts by three candidate macrocycles, the selectivity toward different anions also varies according to macrocycle structure.

It was further shown that anion radius has relatively little effect on extraction of sodium salts of very small anions, and a rationale based on anion hydration and ion pairing was proposed.

Acknowledgments

Research at Oak Ridge National Laboratory was sponsored by the Environmental Management Science Program of the Offices of Science and Environmental Management, U. S. Department of Energy, under contract number DE-AC05-96OR22464 with Oak Ridge National Laboratory, managed by Lockheed Martin Energy Research Corp. A. P. M thanks the Robert A. Welch Foundation (Grant B-963) and the Environmental Science Program of the U. S. Department of Energy (Grant DE-FG07-98ER14936) for financial support. The participation of C. K. C. was made possible by an appointment to the Oak Ridge National Laboratory Postgraduate Program administered by the Oak Ridge Associated Universities.

References

1. *Tanks Focus Area Annual Report: 1997*; Report DOE/EM-0360, U.S. Department of Energy, 1997.
2. Bunker, B.; Virden, J.; Kuhn, B.; Quinn, R. Nuclear Materials, Radioactive Tank Wastes. In *Encyclopedia of Energy Technology and the Environment,* Bisio, A.; Boots, S., Eds.; John Wiley & Sons, Inc., New York, 1995; pp 2023-2032.
3. Herting, D. L. "Clean Salt Disposition Options"; Report WHC-SD-WM-ES--333, Westinghouse Hanford Co., Richland, WA, 1995.
4. Kurath, D. E.; Brooks, K. P.; Hollenberg, G. W.; Sutija, D. P.; Landro, T.; Balagopal, S. *Sep. Purif. Technol.* **1997**, *11*, 185-198.
5. Kurath, D. E.; Brooks, K. P.; Jue, J.; Smith, J.; Virkar, A. V.; Balagopal, S.; Sutija, D. P. *Sep. Sci. Technol.* **1997**, *32*, 1-4.
6. Moyer, B. A.; Sun, Y. In *Ion Exchange and Solvent Extraction*; Marcus, Y., Marinsky, J. A., Eds.; Marcel Dekker: New York, 1997; Chap. 6, pp 295-391.
7. Moyer, B. A.; Bonnesen, P. V. In *The Supramolecular Chemistry of Anions*; Bianchi, A. Bowman-James, K. Garcia-Espana., E., Eds.; VCH: Weinheim, 1997.
8. Marcus, Y. *Ion Properties*; Marcel Dekker: New York, 1997.
9. Cox, B. G.; Schneider, H. *Coordination and Transport Properties of Macrocyclic Compounds in Solution*; Elsevier: New York, 1992.

10. Izatt, R. M.; Pawlak, K.; Bradshaw, J. S.; Bruening, R. L. *Chem. Rev.* **1995**, *95*, 2529.
11. *The Supramolecular Chemistry of Anions*; Bianchi, A. Bowman-James, K. Garcia-Espana., E., Eds.; VCH: Weinheim, 1997.
12. Reetz, M. T., in *Comprehensive Supramolecular Chemistry*, F. Vögtle, Ed.; Pergamon: Oxford, 1996; Vol. 2, pp 553-562.
13. Leonard, R. A.; Conner, C.; Liberatore, M. W.; Bonnesen, P. V.; Moyer, B. A.; Presley, D. J.; Lumetta, G. J.; *Sep. Sci. Technol.* 34, 1043-1068 (1999).
14. Hofmeister, F. *Ach. Exptl. Pathol. Pharmakol.* **1888**, *24*, 247.
15. Grinstead, R. R. U.S. Patent 3,598,547, Aug. 10, 1971.
16. Grinstead, R. R. U.S. Patent 3,598,548, Aug. 10, 1971.
17. McDowell, W. J. *Sep. Sci. Technol.* **1988**, *23*, 1251-1268.
18. Marchand, A. P.; Kumar, K. A.; McKim, A. S.; Milnaric-Majerski, K.; Kragol, G. *Tetrahedron.* **1997**, *53*, 3467.
19. Marchand, A. P.; McKim, A. S.; Kumar, K. A. *Tetrahedron*, **1998**, *54*, 13421.
20. Castro, R.; Davidov, P. D.; Kumar, K. A.; Marchand, A. P.; Evanseck, J. D.; Kaifer, A. E. *J. Phys. Org. Chem.* **1997**, *10*, 369.
21. Marchand, A. P.; Chong, H.-S.; Alihodzic, S.; Watson, W. H.; Bodige, S. G. *Tetrahedron.* **1999**, *55*, 9687.
22. Marchand, A. P.; Chong, H.-S.; *Tetrahedron.* **1999**, *55*, 9697.
23. Marchand, A. P.; Alihodzic, S.; McKim, A. S.; Kumar, K. A.; Milnaric-Majerski, K.; Kragol, G. *Tetrahedron Lett.* **1998**, *39*, 1861.
24. Marchand, A. P.; McKim, A. S.; Krishnudu, K. Unpublished results.
25. Marchand, A. P.; Ravikumar, K. S. Unpublished results.
26. Lamb, J. D.; Izatt, R. M.; Christensen, J. J. In: Izatt, R. M.; Christensen, J. J., Eds. *Progress in Macrocyclic Chemistry,* Wiley-Interscience: New York, Vol. 2, 1981, pp 41-90.
27. Wong, K. H.; Ng, H. L. *J. Coord. Chem.,* **1981**, *11*, 49.
28. Ouchi, M.; Inoue, Y.; Wada, K.; Iketani, S.; Hakushi, T.; Weber, E. *J. Org. Chem.* **1987**, *52*, 2420.
29. Bartsch, R. A.; Eley, M. D.; Marchand, A. P.; Shukla, R.; Kumar, K. A.; Reddy, G. M. *Tetrahedron* **1996**, *52*, 8979.
30. Weber, E.; Vögtle, F. In: Vögtle, F.; Weber, E., Eds., *Host Guest Complex Chemistry. Macrocycles. Synthesis, Structures, Applications*, Springer-Verlag: New York, 1985, pp 1-41.
31. Marcus, Y. Principles of Solubility and Solutions. *In Principles and Practices of Solvent Extraction*; Rydberg, J., Musikas, C., Choppin, G., Eds.; Marcel Dekker: New York, 1992, p.24.
32. Kenjo, T.; Ito, T. *Bull. Chem. Soc. Jpn.,* **1968**, *41*, 1757-1760.

33. Moyer, B. A. In *Molecular Recognition: Receptors for Cationic Guests*; Gokel, G. W., Ed.; *Comprehensive Supramolecular Chemistry*; Atwood, J. L.; Davies, J. E. D.; MacNicol, D. D.; Vögtle, F.; Lehn, J.-M., Eds.; Pergamon: Oxford, 1996, Vol. 1; pp 377-416.

34. Chambliss, C. K.; Haverlock, T. J.; Bonnesen, P. V.; Engle, N. L.; Moyer, B. A. *Environ. Sci. Technol.* (In preparation).

35. Moyer, B. A.; Chambliss, C. K.; Bonnesen, P. V.; Keever, T. J. "Solvent and Process for Recovery of Hydroxide from Aqueous Mixtures," U.S. Patent Application Ser. No. 09/404,104, Sept. 23, 1999.

36. Ballinger, P. *J. Am. Chem. Soc.* **1959**, *81*, 1050.

37. Stewart, R. *Can. J. Chem.* **1960**, *38*, 399.

38. Haverlock, T. J.; Bonnesen, P. V.; Sachleben, R. A.; B. A. Moyer *Radiochim. Acta.* **1997**, *76*, 103-108.

39. B. A. Moyer, C. K. Chambliss, P. V. Bonnesen, J. C. Bryan, A. P. Marchand, H.-S. Chong, A. S. McKim, K. Krishnudu, K. S. Ravikumar, V. S. Kumar, and M. Takhi, 218th Nat. Mtg. of the American Chemical Society, Aug. 22-26, 1999 (Paper No. NUCL-108).

Chapter 9

The Use of Synthetic Inorganic Ion Exchangers in the Removal of Cesium and Strontium Ions from Nuclear Waste Solutions

Paul Sylvester and Abraham Clearfield[*]

Department of Chemistry, Texas A&M University, College Station, TX 77843

Three pillared clays were prepared along with a sodium titanate and a sodium mica. These exchangers were examined for removal of Cs^+ and Sr^{2+} in two types of groundwater and a highly alkaline tank waste (NCAW). Two zeolites were included for benchmark comparisons. The pillared clays showed a strong affinity for Cs^+ in the groundwaters but not for Sr^{2+}. Other synthesized exchangers exhibited high Sr^{2+} ion Kd values. The great variety, stability and specificity of inorganic ion exchangers requires that they be seriously considered for nuclear waste remediation.

One of the most compelling environmental problems facing our nation is the remediation of enormous stocks of nuclear waste that exists throughout the land. The most critical waste is that accumulated as a result of our nuclear weapons program and is termed high level waste (HLW). It contains varying quantities of ^{137}Cs and ^{90}Sr and smaller amounts of actinides in the fluid part of the stored waste. This waste arose mainly as byproducts of the processes utilized for the separation of uranium and plutonium and originally was stored in steel tanks underground. To prevent corrosion of the tanks the acid was treated with sodium hydroxide. This procedure precipitated the insoluble hydroxides but held in solution large quantities of alumina and silica. In most cases excess base was added so that the fluid portion existing above the solid "sludge" is 1-3 M in NaOH and 5-7 M in $NaNO_3$. Over time, evaporation has resulted in the formation of a salt cake at the top of the fluid portion.

Most of this HLW is stored at the Hanford, Washington site and in Savannah River, South Carolina with lesser amounts at Oak Ridge and the Idaho National Engineering and Environmental Laboratory (INEEL). Current thinking in terms of tank waste treatment is to remove the Cs^+ and Sr^{2+} from the fluid portion with possible dissolution of the salt cake portion for similar treatment. These elements would then be encased in a special boron glass. The remaining fluid waste could then be treated as low level waste (LLW) and made into a cement or grout that could be stored above ground. The glass containing HLW would be stored underground in steel containers (1).

The composition of the tanks can vary greatly from tank to tank. For example, some of the tanks contain considerable amounts of complexing agents and others do not. Potassium ion interferes with the removal of Cs^+ for most ion exchange processes so that adjustments to the procedures need to be based on these factors. In addition there are groundwaters and contaminated soils adjacent to the tanks where leakage or intentional spillage has taken place. Thus, it appears to us that no one process will be able to solve the diverse problems of remediation and that a veritable arsenal of ion exchange materials, solvent extraction and other separation procedures are required. Apparently, with this in mind the Department of Energy has supported a wide-ranging research program. Much of the results of this research program are published in the open literature or in reports generated at several national laboratories. However, it does not follow, that those directly responsible for the actual remediation work, are aware of the many potentially useful results generated by the research program. A very useful activity would be the compilation of abstracts of the research with some insight on how they can be applied to present remediation problems.

This current article describes some of our results with inorganic ion exchange materials that may be applicable to groundwater and soil remediation. Groundwaters vary considerably in composition and thus exchanger performance is likely to vary significantly from groundwater to groundwater. The main ions present in groundwaters are Ca^{2+}, Mg^{2+}, Na^+, and K^+ which will compete with ^{90}Sr and ^{137}Cs for the available ion-exchange sites in any ion-exchange materials. Strontium is typically present at concentrations of a few tenths of a part per million (ppm), but cesium is present at much lower levels, typically parts per billion (ppb) or less. Consequently, ion exchangers will need to exhibit high selectivities for cesium and strontium relative to Ca^{2+}, Mg^{2+}, Na^+, and K^+, which are present at concentrations several orders of magnitude greater than the target elements.

Smectite or swelling clays are layered materials that consist of aluminum or magnesium-oxygen octahedra sandwiched between layers of silica-oxygen tetrahedra. Montmorillonite is a member of the smectite clay family in which the octahedral positions are occupied by aluminum ions and is termed a dioctahedral clay (2). Substitution of aluminum by divalent ions, usually magnesium, occurs in nature and results in a net negative charge on the clay lattice which is counterbalanced by the presence of cations, usually Na^+ and Ca^{2+} between the clay layers. These cations are readily exchangeable and can be replaced by large polymeric cations such as the Al_{13} Keggin ion, $[Al_{13}O(OH)_{24}(H_2O)_{12}]^{7+}$ or cationic zirconium species based upon the tetrameric unit $[Zr_4(OH)_8(H_2O)_{16}]^{8+}$. After calcination, the clay lamellae are permanently propped open by pillars of alumina or zirconia, leading to microporosity (3-9). A simplified schematic showing the pillaring of a clay by alumina is illustrated in Fig. 1. These materials have relatively high surface areas, uniformly sized

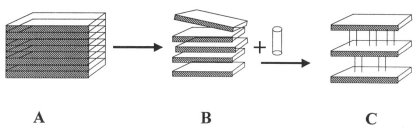

A   B   C

Figure 1. Cartoon depiction of formation of a pillared clay. (A) unswollen clay, (B) swelling of clay and (C) insertion of inorganic polymer as pillar.

pores, and no longer swell or disperse in aqueous media. A preliminary study of the ion-exchange properties of pillared clays was undertaken by Dyer and Gallardo (10) who showed that the ion-exchange behavior of pillared clays was largely associated with the pillar rather than the host clay, and that pillaring actually increased the ion-exchange capacity of the clays. However, no ion selectivity data were presented, and their results did not take into account that solubility of the pillar may occur at low or high pHs. Subsequently, Molinard et al. (11) showed that some alumina does dissolve in both acid and base, and if not taken into account yields misleading results in the ion-exchange titration data.

The present study investigated the Cs^+ and Sr^{2+} ion selectivity of silica, alumina and zirconia-pillared montmorillonites and compared the selectivity to that of the parent clay. In addition, certain other exchangers to be described later and layered sodium nonatitanate were included as well as chabazite and clinoptilolite for baseline comparisons.

Experimental

Materials. All chemicals used to prepare simulants were Fisher Analytical grade. [137]Cs was obtained from Amersham Life Science and had a specific activity of 2.5×10^4 Ci/g total Cs. ^{89}Sr was purchased from Isotope Products and had a specific activity of 69.3 mCi/g total Sr. Liquid scintillation counting was performed on a Wallac 1410 counter using Fisher Scientific Scintisafe Plus 50% scintillation cocktail.

Preparation of Ion Exchangers. A sample of montmorillonite, SWy-1 (Crooks County, Wyoming) was obtained from Source Clays, University of Missouri, Columbia. This raw mixture was purified by wet sedimentation and converted to the sodium form. Approximately 3 g (dry weight) of purified SWy-1 was suspended in 200 ml of deionized water and heated to $100°$ C. 3-aminopropyltrimethoxy silane (10 g), $NH_2(CH_2)_3Si(OCH_3)_3$, was refluxed in 300 ml water for 20 h and added to the clay dispersion. This mixture was kept at $100°C$ for 24 h during which time an additional 100 ml of water was added. The solid product was recovered by centrifugation, washed and dried at $65°$ C. It was then heated to $500°$ C for 19 h. This sample is designated PS-II 55(12). The alumina-pillared clay was obtained from Laporte Absorbents and was prepared by exchanging the Al_{13} Keggin ion, $[Al_{13}O_4(OH)_{24}(H_2O)_{12}]^{7+}$ into a Los Trancos montmorillonite followed by calcination at $550°$ C for 4 h. It is designated Al-PILC.

Zirconia Pillared Clay. The zirconia-pillared clay (Zr-PILC), was synthesized according to the method described by Dyer et al. (13). The clay (33 g) was stirred for 24 hours at room temperature with 1000 mL of a 0.1 M $ZrOCl_2$ solution. The clay was then centrifuged and washed repeatedly with deionized water until the washings were free of chloride ions and did not cause a precipitate with silver nitrate solution. The product was then calcined at $300°C$ for 24 hours and characterized using surface area analysis and X-ray powder diffraction.

Sodium Mica. A phlogopite mica of ideal formula $KAl_3Si_3AlO_{10}(OH)_2$ was treated with 2M NaCl at $180°C$ to convert it to the sodium ion form. A complete description of the preparation of sodium micas by hydrothermal treatment was provided previously (14). The sample used in this study was obtained from AlliedSignal Corp.

Sodium Nonatitanate, $Na_4Ti_9O_{20} \bullet H_2O$. 8.4 g of a 50% NaOH solution was added dropwise to 7.7 g of titanium isopropoxide (Aldrich, 97%) in a Teflon round bottom flask with continuous stirring. The mix was then heated to $110°C$ for 4 h and then transferred to a Teflon lined pressure vessel using 13 ml of deionized water in the process. The vessel was kept at $200°C$ for 20 h and then quickly cooled under running tap water. The solid was filtered off and washed with 200 ml of ethanol and dried at $85°C$ for two days (15).

Kd Determinations. The selectivities of the exchangers for ^{137}Cs and ^{89}Sr were investigated using simple batch studies. Approximately 0.05 g of material was contacted with 10 mL of solution (giving a V:m ratio of 200) spiked with ^{137}Cs or ^{89}Sr, as appropriate, for 20 hours with constant rotary mixing. The solution was then filtered through a Whatman No. 42 filter paper (which had previously been shown to be sufficient to exclude all of the clay fines from the supernate), and the activity in the aqueous phase measured using liquid scintillation counting. Distribution coefficients (K_ds) were then calculated according to:

$$K_d = [(C_i - C_f/C_f]V/m \qquad (1)$$

Where C_i = initial activity of solution

$\quad$ C_f = final activity of solution

$\quad$ V = volume of solution (mL)

$\quad$ m = mass of exchanger (g)

The radiotracers added resulted in a total cesium concentration in the N- Springs groundwater of approximately 2 ppb and a total strontium concentration of approximately 0.1 ppm in the NCAW. The contributions of the radiotracers to the total strontium in the groundwater (1.48×10^{-6} M) and the total cesium in the NCAW (5.0×10^{-4} M) were negligible. All Kd determinations were performed in duplicate with variations greater than 10% repeated until consistent results were obtained.

Instrumentation

$\quad$ X-ray powder diffraction patterns were obtained on a Scintag PAD V diffractometer using nickel-filtered CuKα radiation. Surface area measurements were obtained on a Quantachrome Autosorb 6 unit using nitrogen adsorption-desorption at 77 K.

Preparation of Simulants. Two groundwater simulants were prepared according to information supplied by the Pacific Northwest National Laboratory (PNNL) (16) and INEEL (15), and spiked with ^{137}Cs to give a total cesium concentration of approximately 2.5 ppb and 0.13 ppm ^{89}Sr. The composition of these simulants is given in Table 1. N-Springs represents an unusually alkaline groundwater found in the Hanford 100 area and contains relatively high concentrations of magnesium and calcium but no potassium. Since Mg^{2+} and Ca^{2+} primarily compete with Sr^{2+} while K^+ competes with Cs^+, most ion exchangers would be expected to exhibit relatively high K_ds for Cs^+ because of the lack of K^+ in the N-Springs simulant. The TAN (Test Area North) simulant was developed in-house at Texas A&M University and contains the major cations found in a sample of groundwater taken from the TSF-05 injection well at INEEL (17). Contaminants of concern in this groundwater include chlorinated hydrocarbons, ^{90}Sr, and ^{137}Cs. This groundwater is less alkaline, has a greater total ionic strength than the N-Springs, and also contains significant potassium (2.45 ppm). NCAW (neutralized current acid waste) represents Tank

102AZ diluted to 5 M Na^+ and is thought to contain the highest levels of Cs^+ in the Hanford site.

Kinetic Experiments. The rate of uptake of ^{137}Cs from 0.1 M $NaNO_3$ solution was investigated for the Los Trancos clay, the Al-PILC, and the Zr-PILC. Material (0.05 g) was weighted into a scintillation vial and 10 mL of $NaNO_3$ solution, spiked with ^{137}Cs, added. The vial was then capped and placed on a rotary shaker. After a specific period of time the vial was removed and the aqueous phase sampled immediately using a syringe equipped with a 0.2 µm filter. The aqueous phase was then analyzed using liquid scintillation counting and the percentage removal of cesium form the aqueous phase calculated.

Table 1
The composition of NCAW (Neutralized Current Acid Waste) and N-Springs and TAN Groundwater Simulants

Compound	NCAW (mol/L)	N-Springs (mol/L)	TAN (mol/L)
$NaNO_3$	0.258	—	—
KNO_3	0.120	—	6.27×10^{-5}
Na_2SO_4	0.150	—	—
Na_2HPO_4	0.025	—	—
$NaOH$	3.40	—	—
$Al(NO_3)_3$	0.430	—	—
Na_2CO_3	0.230	1.25×10^{-4}	1.93×10^{-3}
NaF	0.089	1.05×10^{-5}	—
$NaNO_2$	0.430	—	—
$RbNO_3$	5.00×10^{-5}	—	—
$CsNO_3$	5.00×10^{-4}	—	2.46×10^{-5}
$Ca(NO_3)_2$	—	9.55×10^{-5}	1.36×10^{-3}
$CaCl_2$	—	2.26×10^{-5}	—
$Ca(OH)_2$	—	6.09×10^{-4}	—
$Ba(OH)_2$	—	1.12×10^{-7}	—
$Ba(NO_3)_2$	—	—	1.01×10^{-6}
$MgSO_4$	—	2.16×10^{-4}	—
$Mg(NO_3)_2$	—	—	0.811×10^{-3}
$Sr(NO_3)_2$	—	1.48×10^{-6}	1.1×10^{-5}
$Zn(NO_3)_2$	—	—	$6.11 \times 10-6$
$Fe(NO_3)_3$	—	—	3.13×10^{-5}
pH	>14	11.2^a	$7.5 - 9.5^b$

[a]Theoretical pH, [b]pH range of actual groundwater

Results

The physical characteristics of the Los Trancos and Swy-1 clays and the PILCs prepared from them are shown in Table 2. X-ray powder diffraction showed that the Los Trancos clay was relatively pure with the only obvious impurity being a minor amount of quartz. Earlier work (18) had shown that this clay has a cation exchange capacity of approximately 1 meq/g. The major exchangeable cation is Ca^{2+} together with minor amounts of Na^+ and K^+. Both the alumina and zirconia pillared clays were microporous but the silica pillared clay (SWy-1 PILC) contained substantial (~44%) mesopores. The initial d-spacing in the X-ray powder patterns (Table 2) correspond to the interlayer

Table 2
Physical Characteristics of the Los Trancos and Swy-1 Clays and the Zr-PILC and the Al-PILC and Si-PILC

Clay	Pillar	Calcination Temperature (°C)	Surface Area (m²/g)	Micro-porosity (%)	d-Spac-ing (Å)
Los Trancos	None	65	73.7	45	15.1
Al-PILC	Alumina	550	264	82	18.3
Zr-PILC	Zirconia	300	171	84	Not seen
SWy-1	None	—	41.8	38.8	9.65
SWy-1-Si	Silica	500	170	56	17.6

Spacings of the clays and their pillared products. The low interlayer spacing of 9.65 Å for the SWy-1 clay shows that it was not swollen, whereas the layer spacing and surface area for the Los Trancos clay indicates that it is in the swollen condition. The increased spacings obtained for the heat treated Al-PILC and SWy-1-Si sample, as well as the porosity indicate that the clays have been successfully pillared instead of delaminated. The Zr-PILC did not yield a well defined peak corresponding to the interlayer spacing after calcination. However, the increase in surface area and the high level of microporosity correlate well with the results reported by Dyer et al (13). Pore size measurements gave a mean micropore diameter of approximately 6.2 Å for the Al-PILC and 5.6 Å for the Zr-PILC.

The Kd values for removal of Cs^+ from the N-Springs simulant is shown in Table 3. Both the Al-PILC and the Zr-PILC gave excellent extractions of Cs^+ both at the V:m ratio of 200 and 1000. These results indicate a truly high selectivity of the PILC's since less selective materials usually show a decrease in Kd at high V:m ratios. This was shown to be the case for the commercially available sodium silicotitanate IE-911 which decreased from 900,000 ml/g to 88,6000 ml/g under similar conditions (19).

It is evident from the results in Table 4 that the pillared clays have very little affinity for Sr^{2+} and probably for alkaline earths in general. This lack is in contrast to the relatively high preference for alkali metals and suggests their use in separation of these two groups of metals. The best exchanger by far is the powdered pure form of the sodium titanium silicate. We have previously described the synthesis, structure and ion exchange behavior of this compound (20, 21). The sodium mica is interesting. We have found that as the sodium ion is replaced with larger ions the layers come together trapping the ions permanently. Therefore, this exchanger may be useful to entrap Cs^+ and Sr^{2+} radioisotopes in contaminated soils and keep them from migrating into the environment. The decrease in Kd for the sodium mica may indicate that this exchanger removes some K^+, Cs^+ and Ca^{2+} from the groundwater as well as Sr^{2+}. Consequently, for low ionic strength solutions such as the N-Springs groundwater, the exchange sites become filled by the more abundant competing ions and the Sr^{2+} uptake is reduced. Further work in column flow studies are necessary to determine whether this hypothesis is correct.

Kinetic Data. The rate of uptake of Cs^{2+} and Sr^{2+} ion by selected exchangers is provided in Tables 5 and 6, respectively. The cesium ion data were obtained in 0.1 M $NaNO_3$ solution. The rate of cesium ion uptake by the pillared clays was found to be initially rapid followed by a gradual increase in Cs^+ removal. By contrast, Cs uptake by the unpillared Los Trancos clay was variable with no trend with increased time observed.

Table 3. Kd values and % removal of Cs^+ from the N-Springs simulant

Sample	V:m = 200		V:m = 1000	
	Kd(ml/g)	% Cs^+ Removal	Kd(ml/g)	% Cs^+ Removal
Clinoptilolite	16,400	98.78	39,400	97.62
Na-Mica	11,400	98.26	3,300	77.65
Al-PILC	115,000	99.83	246,000	99.60
Zr-PILC	850,000	99.98	863,000	99.89
Si-PILC	23,000	99.14	—	—
PS-II 55	23,300	99.15	—	—
SWy-1	560	73.7	—	—
Los Trancos Clay	1880	90.34	1650	61.2

Table 4. Kd values and % removal of Sr^{2+} from N-Springs simulant.

Sample	V:m = 200		V:m =1000	
	Kd (ml/g)	% Sr^{2+} Removal	Kd (ml/g)	% Sr^{2+} Removal
AW 500	30,600	99.35	26,700	96.41
Clinoptilolite	27,200	99.34	7,700	88.17
Na-Mica	240,000	99.92	14,200	93.93
Al-PILC	440	68.71	1050	51.5
Si-PILC	513	71.95	–	–
$Na_4Ti_9O_{20} \bullet 2H_2O$	25,300	99.21	322,000	99.69
IE-911	25,800	99.23	73,600	98.68
$Na_2Ti_2OSiO_4 \bullet 2H_2O$	326,000	99.94	863,000	99.89

Table 5. Uptake of Cs^+ from 0.1 M $NaNO_3$ solution by selected pillared clays as a function of time.

Sample	5 min	10 min	20 min	30 min	1 h	2h	5h	24h
Clay[a]	75.81	76.75	77.24	77.96	77.97	76.01	74.89	82.09
Zr-PILC	96.02	97.06	97.91	98.18	98.38	99.09	99.30	99.57
Al-PILC	90.00	92.66	94.68	94.89	96.23	97.14	97.81	97.87

% Cs removal at

[a]Los Trancos

Table 6. Uptake of Sr^{2+} from N-Springs simulant by selected exchangers as a function of time.

Percent Strontium Removal

Exchanger	5min	10 min	20 min	30 min	60 min	24 h
AW 500	96.56	97.69	98.23	98.44	98.84	99+
$Na_4Ti_9O_{20} \bullet 2H_2O$	99.95	99.95	99.95	99.95	99.95	99.90
Al-PILC	29.5	30.8	41.6	42.7	48.4	51.5

Table 7. K_ds and % Cs Removal from the TAN Groundwater Simulant

Sample	V:m = 200		V:m = 1000	
	K_d(mL/g)	% Cs Removed	K_d(mL/g)	% Cs Removed
AW500	3.68×10^4	99.46	4.37×10^4	97.82
IE-96	5.32×10^4	99.63	5.86×10^4	98.32
Clinoptilolite	1.63×10^4	98.79	2.04×10^4	95.27
Los Trancos Clay	1.58×10^3	88.91	9.94×10^2	50.47
Zr-PILC	5.45×10^4	99.64	3.88×10^4	97.44
Al-PILC	1.87×10^4	98.94	2.54×10^4	96.18

From the data in Table 6 it is evident that the uptake of Sr^{2+} by the sodium nonatitanate is very rapid being complete in 5 min. The uptake by the zeolite, AW 500 is also rapid but followed by a slow approach to equilibrium. This type of behavior also characterizes the Al-PILC but at a much slower rate of uptake.

Tan Groundwater. The equilibrium uptake of Cs^+ from TAN groundwater is presented in Table 7. All of the exchangers exhibited high uptakes except the Los Trancos clay. There was a slight decrease in the percent uptake at the higher V:m ratio and this is probably due to the high content of Na^+ and K^+ in the TAN groundwater. The effect of electrolyte concentration on uptake of Cs^+ by the Al and Zr PILCS had been reported earlier (18). Potassium ion was the most effective followed by Na^+ in reducing the exchange of Cs^+. As already shown, alkaline earths are not readily taken up by the pillared clays and hence only a slight reduction in Kd values for Cs^+ were observed even in 0.1 M supporting electrolyte solutions.

NCAW. The Cs^+ uptake of the pillared clays in the NCAW simulant were extremely low (12) due to the high level of Na^+ and K^+ in the simulant. The results for removal of Sr^{2+} are given in Table 8. Sodium titanate and the sodium titanium silicate gave excellent results. For the latter compound the Kd value decreased by 16% but the sodium titanate exhibited a much larger decrease. Nevertheless, this exchanger was able to remove Sr^{2+} from 1000 bed volumes before breakthrough and nearly 4000 bed volumes to 50% BT (15). The Kd values for the two PILCs were unexpectedly high. However, it is quite likely that the high pH for the NCAW solution dissolved some silica that in turn precipitated strontium silicate. In alkaline solutions not only are clays unstable but the alumina pillars are also soluble (11). Long exposure to strong alkali solutions would then result in destruction of the PILCs.

Table 8. Kd values and percent uptake of Sr^{2+} from NCAW simulant.

Exchanger	V:m = 200		V:m = 1000	
	Kd(ml/g)	% Sr^{2+} Removed	Kd(ml/g)	% Sr^{2+} Removed
AW 500	260	56.5	—	—
Clinoptilolite	48	19.3	—	—
Swy-1	274	57.8	—	—
Si-PILC	6620	97.1	—	—
Al-PILC	3300	94.3	—	—
$Na_4Ti_9O_{20} \bullet 2H_2O$	235,000	99.92	39,600	97.52
$Na_2Ti_2O_3SiO_4 \bullet 2H_2O$	269,500	99.93	226,000	99.56

Conclusions

This study has shown that both alumina and zirconia pillared clays have a high affinity for Cs^+. Since they are easy to prepare and relatively inexpensive, they may be useful for removal of ^{137}Cs from ground waters, certain process waters and for immobilization of Cs^+ in soils. It should also be remembered that the sodium mica has a high affinity for both Cs^+ and Sr^{2+}, with the difference that the mica layers retract on continued loading, trapping the ions permanently. These altered clays and micas may also prove useful as barrier materials surrounding leaking tanks where both ^{137}Cs and ^{90}Sr migration in the soil need to be halted.

Clay type materials are not useful at pH values below 2 and above 11 because both acid and alkali can remove alumina from the clays altering their ion exchange behavior. The same is true for zeolites. However, the sodium nonatitanate has been found to be stable in 3-4 M NaOH and remove Sr^{2+} from such high alkaline waste solutions.

Our laboratory has prepared dozens of inorganic ion exchangers ranging from amorphous beads to fine crystals (22). Some can be used in acid solutions and others in neutral and basic solutions. The fine powders can be fashioned into spheres of cylinders with or without inorganic binders. In the present report we have shown that pillared clays exhibit interesting ion exchange behavior. Many other layered exchangers and those with tunnel or cavity structures have been prepared and show high selectivities for specific ions under different solution conditions. The point is that the exchangers can be prepared to fit the needs of the different solutions and conditions that are encountered in the overall nuclear waste remediation program.

Acknowledgement

We thank Garrett N. Brown, formerly of PNNL and Terry A. Todd of INEEL for supplying data on the groundwater compositions. We also thank Dr. Suheel Amin of BNFL Plc for donating the sample of clinoptilolite, Dr. Malcolm Buck of Laporte for donating the sample of alumina-pillared montmorillonite and the Los Trancos clay, and Dr. Boris Shpeizer for performing the surface area measurements. We gratefully acknowledge financial support from Lockheed-Martin Energy Systems, Inc. through Grant 9800585, Oak Ridge National Laboratory, and the DOE, Basic Energy Sciences, Grant 434741-00001, funded by the Office of Environmental Management (EMSP).

References

1. Science and Technology for Disposal of Radioactive Tank Waste, Plenum Press, New York, 1998.
2. C.J.B. Mott, Catal. Today 2, 19 (1998).
3. D.E. W. Vaughan, in Perspectives in Molecular Sieve Science (W.H. Flank and Y.E. Whyte, Eds., American Chemical Society Symposium Series, vol. 38, American Chemical Society, Washington, DC, 1988, p. 308.
4. T.J. Pinnavaia, Science, 220, 365 (1983).
5. R. Burch (Ed.) , Catalysis Today, Special Issue: "Pillared Clays," Elsevier, Amsterdam, 1988.
6. A. Clearfield, in Advanced Catalysts and Nanostructured Materials, Academic Press, New York, NY, 1996, pp. 345-394.
7. G.J.J. Bartley and R. Burch, Appl. Catal., 19, 175 (1985).
8. E.M. Farfan-Torres, E. Sham and P. Grange, Catal. Today 15, 515 (1992).
9. J.W. Johnson and J.F. Brody, US Patent 5,248,644 (1993).
10. A. Dyer and T. Gallardo, in Recent Developments in Ion Exchange 2 (P.A. Williams and M.J. Hudson, Eds.), Elsevier Applied Science, London, 1990, pp. 75-84.
11. A. Molinard, A. Clearfield, H.Y. Zhu, and E.F. Vansant, Microporous Mater., 3, 109 (1994).
12. P. Sylvester and A. Clearfield, Sep. Sci. Tech. 33, 1605 (1998).
13. A. Dyer, T. Gallardo and C.W. Roberts in Zeolites: Facts, Figures, Future, P.A. Jacobs and R.A. van Santen, Eds.: Elsevier Sci., Amsterdam, 1989, p. 389.
14. L.N. Bortun, A.I. Bortun and A. Clearfield, in Ion Exchange Developments and Applications, Proc. of IEX '96, Royal Soc. Chem. 182, 313 (1996).
15. E.A. Behrens, P. Sylvester, G. Graziano and A. Clearfield in Science and Technology for Disposal of Radioactive Tank Wastes, W.W. Schulz and N.J. Lombardo, Eds. Plenum Press, New York, 1998. pp. 287-300.
16. G.N. Brown, PNNL, private communication.

17. T.A. Todd, Lockheed-Martin, private communication.
18. P. Sylvester, A. Clearfield and R.J. Diaz, Sep. Sci. Tech. 34, 2293 (1999).
19. P. Sylvester and A. Clearfield, Solv. Extr. Ion Exch. 16, 1527 (1998).
20. D.M. Poojary, R.A. Cahill and A. Clearfield, Chem. Mater. 6, 2364 (1994).
21. A.I. Bortun, L.N. Bortun and A. Clearfield, Solv. Extr. Ion Exch. 14, 341 (1996).
22. A.I. Bortun, L.N. Bortun and A. Clearfield, Solv. Extr. Ion Exch. 15, 909 (1997).

Chapter 10

Stepwise Assembly of Surface Imprint Sites on MCM-41 for Selective Metal Ion Separations

M. C. Burleigh[1,2], Sheng Dai[1,*], E. W. Hagaman[2], C. E. Barnes[3],
and Z. L. Xue[3]

[1]Chemical Technology and [2]Chemical and Analytical Sciences Divisions,
Oak Ridge National Laboratory, Bethel Valley Road, Oak Ridge, TN 37831
[3]Department of Chemistry, University of Tennessee, Knoxville, TN 37996

A new stepwise surface functionalization methodology has been developed to synthesize mesoporous sol-gel sorbents containing the ethylenediamine functionality. The N-[3-(trimethoxysilyl)-propyl]ethylenediamine (TMSen) ligand is first coated on MCM-41, followed by sorption of Cu(II). A second exposure to TMSen results in the formation of a 2:1 TMSen:Cu(II) complex on the surface of the mesopores. Acid washing protonates the amino groups which release the copper ions. This results in the formation of binding sites that are uniquely designed with the coordination environment Cu(II) prefers. The mesoporous MCM-41 used for coating was synthesized under basic conditions using cetyltrimethylammonium bromide (CTAB) as the structure directing agent. Mesopores are formed by the calcination of the surfactant micelles giving this material relatively large surface areas and good mass transfer. A copper(II) imprinted gel has shown significant enhancement of copper(II) uptake capacities at various concentrations relative to a non-imprinted gel made without the metal ion template. The imprinted material also exhibits improved selectivity for removal of copper(II) ions from Cu(II)/Zn(II) aqueous solutions. This material has copper distribution coefficients $(K_d) > 100,000$ at concentrations less than 20 ppm. Separation factors, $k > 100$ w.r.t. copper(II), were achieved with an aqueous Cu(II)/Zn(II) system at pH 5.0.

Introduction

The discovery of a new family of mesoporous silicon oxides (M41S) by scientists at Mobil Oil Research and Development (*1,2,3,4,5*) has led to great interest in this area of materials science. These materials exhibit large internal surface areas and narrow pore size distributions similar to those found in microporous zeolites. Unlike zeolite materials, with maximum pore dimensions <20 Å, the pores of M41S can be engineered with diameters from 15-100 Å. The relatively large pore sizes available make these materials very attractive for applications such as catalysts, catalytic supports, and sorbents, especially when the molecules involved are too large to fit in the microporous channels of conventional zeolites.

The syntheses of various M41S materials is based on the ability of surfactant molecules to form supramolecular assemblies in solution. These self-assembled structures act as templates for the formation of the metal oxide materials. Removal of the organic template via calcination or extraction results in the mesoporous metal oxide. The final structure of the M41S material depends on many factors such as surfactant chain length and charge, solution pH, temperature, and ratios of the various precursors. A unique member of the M41S family, MCM-41, has a hexagonal arrangement of cylindrical mesopores. Huo *et al.* (*6*) have reported the synthesis of mesoporous silicas exhibiting a variety of phases (hexagonal, cubic, lamellar *etc.*) by using tetraethoxysilane (TEOS) as the silicon-containing precursor and quaternary ammonium surfactants (various chain lengths, head group sizes) as the structure directing agents. Pinnavaia and co-workers (*7,8,9*) have used nonionic surfactants to synthesize mesoporous silica and alumina with wormhole motifs based on hydrogen bonding interacions at the organic/inorganic interface. More recently, Zhao *et al.* (*10*) also reported the synthesis of a wide variety of silica mesophases using nonionic poly(alkylene oxide) surfactants and block copolymers as the organic templates.

One potential application of mesoporous silica is as a support for metal ion sorbents. The removal of toxic metal ions from aqueous solutions is a high priority for environmental cleanup and proper treatment of industrial and municipal effluents. The relatively large pore size and hydrophilic character of mesoporous silica results in good accessibility of small molecules and ions to the large surface areas of its inner pores. This provides the material with the potential for fast kinetics and high sorbent capacities. The presence of reactive silanol groups on the pore surfaces of MCM-41 make it ideal for the grafting of organofunctional ligands with a high affinity for metal ions. Liu and Co-workers (*11,12*) have grafted ligands containing a thiol functionality to mesoporous silica to produce a sorbent for the removal of mercury and other heavy metals from contaminated solutions. Mercier and Pinnavaia (*13*) have shown that functionalized mesoporous silica, with its uniform pore diameters, has a higher metal ion uptake than a sorbent with similar pore size and surface area made with amorphous silica. The functionalization of the amorphous material may cause a blockage of the bottleneck pores by ligands, making many of the functionalized sites inaccessible, even to the relatively small metal ions.

The technique of molecular imprinting involves the incorporation of a template into a host matrix by combining it with host monomers which polymerize, surrounding the template. Subsequent removal of the template results in a material that contains imprints with the ideal size, shape, and chemical environment to rebind the template. The imprinting approach based on organic polymer hosts was first developed by Wulff and Sarhan (*14*) who used this technique to produce polymers for the resolution of racemic mixtures. It has since been utilized in the devlopement of artificial enzymes (*15*) and antibodies (*16,17*), chromatographic resins (*18*), and metal ion sorbents (*19*).

We have been interested in harnessing the unique properties of molecularly imprinted materials for the selective removal of metal ions from mixed solutions (*20,21,22,23*). By combining the best attributes of molecular imprinting, organic functionalization of silica, and the surfactant template approach to mesoporous silica, we have recently developed an imprint coating methodology for the synthesis of sorbents that display not only high capacities but also selectivities for the metal ion of interest. The concept behind the imprint coating methodology (*20*) is to construct binding sites on the surface of the sorbent support that are tailored in order to have a high affinity for the

Scheme 1

Schematic Illustration of Imprint-Coating Process Involving the Introduction of
Complexes Between Target Metal Ions and Bifuctional Ligands, Followed by
the Hydrolysis and Condensation Reactions of the Siloxane Groups in the
Bifunctional Ligands

Complexes in solution   Complexes coated on mesopores

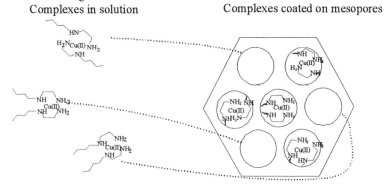

metal ion template. A metal-ligand complex is formed in solution and then
grafted to the surface of mesoporous silica (**Scheme 1**). The ligands are
arranged in a favorable configuration so that following the removal of the metal
ion, the cavities that remain have a higher affinity for the metal ion template
than those created by conventional surface coating. For comparison, non-
imprinted sorbents are made using an identical mesoporous support and
reaction conditions, except that no metal ion complex is formed prior to the
grafting process. We have found that the overall ligand coverage of the
imprinted sorbents can be considerably lower than that of the non-imprinted
samples. This can be attributed to the increased steric hindrance experienced
by the relatively large metal ion complexes when trying to incorporate them
into the inner pore spaces of MCM-41 as compared to the smaller single
ligands. In spite of this, the imprint coated materials show better selectivity for
the metal ion template when in the presence of competitor species than
conventionally coated sorbents. This drawback in our original method has
prompted us to seek an alternative approach to the imprint coating. In order to
minimize the discrepancy in overall ligand coverage between the imprinted and
non-imprinted sorbents, we have developed a stepwise assembly technique.
Herein, we describe the preparation of a mesoporous sol-gel sorbent containing
the ethylenediamine functional group and imprinted with Cu(II) cations via the
stepwise assembly approach.

Experimental Section

Materials. The N-[3-(trimethoxysilyl)propyl]ethylenediamine (TMSen) ligand was obtained from Gelest, Inc. All other chemicals were obtained from Aldrich. All chemicals were used as received.

Synthesis of the Mesoporous Support. A procedure similar to that used by Huo and co-workers (6) was employed here. Cetyltrimethylammonium bromide (CTAB) was used as the surfactant in order to engineer mesoporosity via the S^+I^- assembly pathway. CTAB was added to deionized water and stirred to form a micellar solution. The basic catalyst, sodium hydroxide, is then added. The sol-gel precursor tetraethylorthosilicate (TEOS) is then added, while stirring. A snow white precipitate soon appears. The mixture is then stirred at 100° C for 16 hours. The molar ratios used were 1.0 TEOS: 0.28 CTAB: 0.29 NaOH: 80 H_2O. The hydrothermal precipitate is recovered by vacuum filtration and washed with plenty of deionized water and ethanol. The as-synthesized material is then placed in a furnace and ramped at 1° C/minute to 550° C, at which temperature it is calcined for three hours to remove all surfactant (**Scheme 2**).

Functionalization and Imprinting. One gram of calcined MCM-41 is placed in 100 ml toluene and 4.58 mmol TMSen is added. The mixture is stirred and refluxed for 5 hours. The gel is recovered by filtration and washed with ethanol and air dried. The gel is then placed in 100 ml 10^{-2} M Cu(NO$_3$)$_2$ 3H$_2$O solution and sonicated for 15 minutes, recovered by filtration, washed with deionized water, and placed under vacuum at 90° C for one hour. The material is then placed in 100 ml toluene with 4.58 mmol TMSen and refluxed for another 5 hours. The resulting gel is recovered by filtration, washed with ethanol, air dried, and placed under vacuum at 90° C for one hour. A non-imprinted control sample was generated by a similar protocol, without the metal salt solution step.

Post-synthetic Treatment. The metal ion template is then removed by mixing the sorbent in 100 ml 8M HNO$_3$ and gently heating for one hour. The gel is recovered by filtration and placed in 100 ml deionized water and titrated with 2N NaOH to pH 7.0. The neutralized material is filtered, washed with plenty of deionized water, and placed under vacuum at 90° C for one hour. The non-imprinted control sample is treated in an identical manner to ensure experimental consistency.

Batch Procedures. All metal ion solutions were buffered to a specific pH with sodium acetate/acetic acid (0.05M). In a typical run, 0.1 grams sorbent and 10 ml metal ion solution were placed in a capped plastic vials and sonicated for 30 minutes. The resulting mixture was filtered and the filtrate was analyzed via ICP/AE spectroscopy in order to measure the metal ion

Scheme 2

Micelle Formation

CTAB, H₂O, NaOH

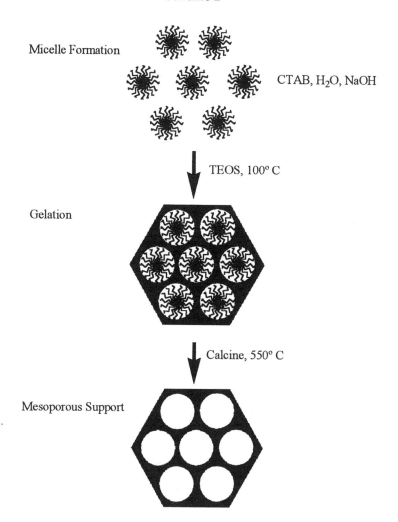

TEOS, 100° C

Gelation

Calcine, 550° C

Mesoporous Support

concentration. The overall capacity of the sorbent for a given metal ion was then calculated by the change in concentration between the filtrate and the original metal ion solution.

Analysis. All metal ion concentration analyses were performed with a Thermo Jarrel Ash Iris inductively coupled plasma spectrometer. Surface areas and pore volumes were measured on a Micromeritics Gemini 2375 surface area analyzer (Micromeritics Corp.). Nitrogen was used as the adsorbent. Adsorption isotherms were measured at 77° K after degassing the samples at

80° C for three hours. Small angle X-ray scattering of the mesoporous samples was conducted at the SAXS user facility of the Oak Ridge National Laboratory.

Results and Discussion.

MCM-41 Characterization. Nitrogen adsorption analysis was performed on the mesoporous support following the calcination step. A B.E.T. (*24*) surface area of 1030 m^2/g, a total pore volume of 1.0 cc/g, and a narrow B.J.H. (*25*) pore size distribution with a peak at 28 $\mathring{A}$ were calculated from the isotherm. Small angle X-ray scattering data shows a peak at $2\theta = 2$ degrees, characteristic of hexagonally packed mesoporous materials. The d-spacing value is 35 $\mathring{A}$. Subtracting the average pore diameter from the B.J.H. data from the d-spacing gives an approximate wall thickness of 10 $\mathring{A}$.

Stepwise Methodology. The stepwise assembly approach involves three principal steps (**Scheme 3**). In the primary step, the mesoporous support is coated with the bifunctional ligand by refluxing in toluene. During this step, the ligand concentration and reflux time are used to control the ligand coverage. Secondly, the sorbent is placed in an aqueous solution of the metal salt and allowed to absorb the metal ion template. The relatively high Cu(II) concentration and 100 ml volume of the metal salt solution ensures that there are enough copper ions to form a 1:1 complex with all of the ligands in the one gram sample. In the final step, the sorbent is again coated with the bifunctional ligand, which first forms a 1:2 metal:ligand complex and then grafts to a nearby silanol group on the silica surface. The Cu(II) acts as a catalyst for this step due to the high affinity it has for the ethylenediamine functional group on the ligand molecules. The solubility characteristics of toluene make it an appropriate solvent since the ligand molecules are quite soluble in toluene while copper and the copper:ligand complexes involved here are not. The condensation reaction between the ligand and the surface silanol groups produces methanol. The high solubility of methanol in toluene may help to drive this reaction which is required to graft the ligand to the silica support. A non-imprinted control sample is synthesized by excluding the second step in this process. Therefore, it is important not to attain too high a coverage during the primary step, so that silanol sites remain for the final step and ligand coverages for the copper-imprinted and non-imprinted materials are similar. With a similar number of surface grafted ligands, the metal ion sorption behavior of these two materials can be compared to see the effect of imprinting.

Scheme 3: Stepwise Assembly Mechanism

I. Reflux calcined MCM-41 in toluene/TMSen

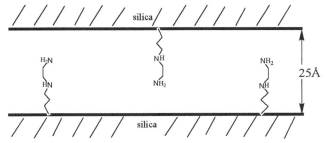

II. Place sorbent in aqueous Cu(II) solution

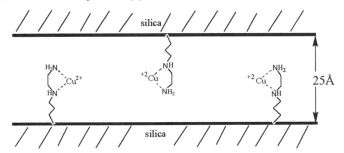

III. Reflux in toluene/TMSen

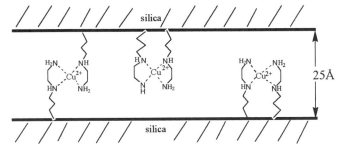

Cu(II) Uptake Capacity. The copper imprinted sol-gel material showed an enhanced capacity relative to the nonimprinted gel for the absorption of copper ions from aqueous solutions. The ability of a sorbent to remove a given ion from solution can be expressed in terms of a distribution constant (K_d),

$$K_d = \{C_0 - C_f)/C_f\} \bullet \{\text{solution volume(ml)}/\{\text{mass sorbent(g)}\}. \tag{1}$$

where C_0 and C_f are the initial and final concentrations of the given ion in solution. **Table I** summarizes the copper uptake of the sorbents from solutions ranging from 6.35-63.55 ppm Cu(II). A direct comparison shows the higher affinity of the imprinted gel for the Cu(II) ion at concentrations up to 31.77 ppm. The non-imprinted sample absorbed more copper at the higher concentrations studied. A simple depiction of the probable arrangement of ligands in the mesoporous channels of these sorbents is shown in **Figure 1**. On the non-imprinted sorbent the ligands are homogeneously distributed throughout the channels. Due to the grafting locations of the ligands, the formation of a 1:2 metal to ligand complex is unlikely. In contrast to this, the ligands on the copper-imprinted sorbent are "paired up", making a 1:2 metal to ligand complex the most likely configuration upon metal ion sorption. The relatively large stability constant associated with the addition of a second diamine moiety to metal ions (*26*) gives the imprinted material a much higher affinity for Cu(II) ions. This can explain the enhancement of Cu(II) capacity shown by the imprinted material at the lower concentrations. A consequence of this arrangement of ligands is that although the non-imprinted sites may not have the great affinity for the metal ion that the imprinted sites have, there are approximately twice as many of them. This may explain the larger distribution coefficients for the non-imprinted materials for Cu(II) at the higher solution concentrations.

Table I. Copper(II) Adsorption

Sorbent	C_0 Cu^{2+} (ppm)	C_f Cu^{2+} (ppb)	Cu^{2+} K_d (ml/g)
Non-imprinted	6.35	166	3,730
Cu-imprinted	6.35	6	109,000
Non-imprinted	15.89	280	5,580
Cu-imprinted	15.89	14	115,000
Non-imprinted	31.77	585	5,340
Cu-imprinted	31.77	127	25,000
Non-imprinted	47.66	505	9,340
Cu-imprinted	47.66	4220	1,030
Non-imprinted	63.55	769	8,170
Cu-imprinted	63.55	7000	808

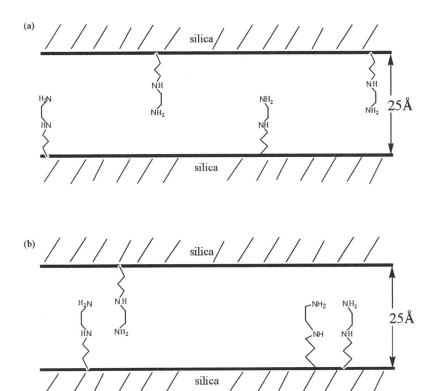

Figure 1: Structures of Imprinted and Non-imprinted TMSen Sorbents
(a) MCM-41 Functionalized with TMSen (Non-Imprinted): Ligands are
homogeneously distributed throughout the mesoporous channels. Due to the
grafting locations a 1:2 metal : ligand complex is not likely to form.
(b) MCM-41 Functionalized with TMSen (Cu-Imprinted): Ligands are
distributedin pairs throughout the mesoporous channels. Due to the grafting
locations a 1:2 metal : ligand complex is likely to form.

Metal Ion Selectivity. Competitive ion binding studies were run with
Cu(II) and Zn(II) ions in order to measure the selectivity of the imprinted
material. The Zn(II) ion was chosen as the competitor species since it has the
same charge, nearly identical size, and also binds well with the diamine ligand.
Zinc is also usually present in industrial effluents containing copper, such as
electroplating waste and acid mine drainage (*27*). A selectivity coefficient, *k*,
for the binding of a specific metal ion in the presence of competitor species can
be obtained from equilibrium binding data (*19*) according to equation (3):

$$M_1(\text{solution}) + M_2(\text{sorbent}) \rightleftharpoons M_2(\text{solution}) + M_1(\text{sorbent}) \qquad (2)$$

$$k = \{[M_2]_{\text{solution}}[M_1]_{\text{sorbent}}\} / \{[M_1]_{\text{solution}}[M_2]_{\text{sorbent}}\} = K_d(\text{Cu})/K_d(\text{Zn}) \qquad (3)$$

Comparison of the k values for the imprinted and control blank gels can show the effect that imprinting has on the metal ion selectivity for a given material. A measure of the increase in selectivity due to molecular imprinting can be defined by the ratio of the selectivity coefficients of the imprinted and non-imprinted materials:

$$k' = k_{\text{imprinted}}/k_{\text{non-imprinted}} \qquad (4)$$

The results of competitive ion binding batch tests at pH 5.0 are summarized in **Table II**. The copper-imprinted sorbents absorbed more Cu(II) and less Zn(II) than the non-imprinted material at all concentrations studied. The separation factors (k) of ~ 150-400 and the selectivity coefficients (k') of ~ 850-2,500 for our copper-imprinted materials synthesized via stepwise assembly show a considerable improvement over the values of k (0.5-220) and k'(1.7-40) achieved previously with imprint coated sorbents under similar experimental conditions (20).

Table II. Sorbent Metal Ion Selectivity (Cu^{2+} Vs. Zn^{2+})

Sorbent	C_0 Cu^{2+} ppm	C_f Cu^{2+} ppb	Cu^{2+} K_d ml/g	C_0 Zn^{2+} ppm	C_f Zn^{2+} ppb	Zn^{2+} K_d ml/g	k	k'
Non-imp.	6.35	125	4,980	6.54	16	42,200	0.118	
Cu-imp.	6.35	15	42,100	6.54	1770	269	156	1,330
Non-imp.	15.89	210	7,470	16.35	101	16,000	0.466	
Cu-imp.	15.89	32	49,100	16.35	7300	124	396	850
Non-imp.	6.35	269	2,260	65.39	229	28,500	0.0791	
Cu-imp.	6.35	10	67,100	65.39	15,100	334	201	2,540

Conclusions

A new stepwise assembly approach has been used to synthesize mesoporous sol-gel sorbents that exhibit enhanced ionic recognition. Base-catalyzed MCM-41 was used as the sorbent support, which was coated with the surface imprints made with the bifunctional ligand, TMSen, and Cu(II) ions. Acid washing removes the metal ions and creates binding sites with the size, shape, and coordination environment that the metal ion prefers. Detailed characterization of these materials is underway in order to elucidate the molecular imprinting mechanism.

Acknowledgment

This work was conducted at the Oak Ridge National Laboratory and supported by the Division of Chemical Sciences, Office of Basic Energy Sciences, U.S. Department of Energy, under contract No. DE-AC05-96OR22464 with Lockheed Martin Energy Research Corp., and the Environmental Management Science Program (EMSP), U.S. Department of Energy, under contract No. DE-FG07-97ER14817, between the University of Tennessee and Oak Ridge National Laboratory.

References

1. Kresge, C.T.; Leonowicz, M.E.; Roth, W.J.; Vartuli, J.C.; Beck, J.S. *Nature* **1992**, *359*, 710.
2. Beck, J.S.; Vartuli, J.C.; Roth, W.J.; Leonowicz, M.E.; Kresge, C.T.; Schmitt, K.D.; Chu, C.T.; Olsen, D.H.; Sheppard, E.W.; McCullen, S.B.; Higgens, J.B.; Schlenker, J.L. *J. Am. Chem. Soc.* **1992**, *114*, 10834.
3. Vartuli, J.C.; Schmitt, K.D.; Kresge, C.T.; Roth, W.J.; Leonowicz, M.E.; McCullen, S.B.; Hellring, S.D.; Beck, J.S.; Schlenker, J.L.; Olsen, D.H.; Sheppard, E.W. *Chem. Mater.* **1994**, *6*, 2317.
4. Vartuli, J.C.; Kresge, C.T.; Leonowicz, M.E.; Chu, A.S.; McCullen, S.B.; Johnson, I.D.; Sheppard, E.W. Chem. Mater. **1994**, *6*, 2070.
5. Beck, J.S.; Vartuli, J.C.; Kennedy, G.J.; Kresge, C.T.; Roth, W.J.; Schramm, S.E. *Chem. Mater.* **1994**, *6*, 1816.
6. Huo, Q.; Margolese, D.I.; Stucky, G.D. *Chem. Mater.* **1996**, *8*, 1147.
7. Bagshaw, S.A.; prouzet, E,; Pinnavaia, T.J. *Science* **1995**, *269*, 1242.
8. Bagshaw, S.A.; Pinnavaia, T.J. *Angew. Chem., Int. Ed. Engl.* **1996**, *35*, 1102.
9. Prouzet, E.; Pinnavaia, T.J. *Angew. Chem., Int. Ed. Engl.* **1997**, *36*, 516.
10. Zhao, D.; Huo,Q.; Feng, J.; Chmelka, B.F.; Stucky, G.D. *J. Am. Chem. Soc.* **1998**, *120*, 6024.
11. Feng, X.; Fryxell, G.E.; Wang, L.Q.; Kim, A.Y.; Liu, J.; Kemner, K.M. *Science* **1997**, *276*, 923.

158

12. Liu, J.; Feng, X.; Fryxell, G.E.; Wang, L.Q.; Kim, A.Y.; Gong, M. *Adv. Mater.* **1998**, *10*, 161.
13. Mercier, L.; Pinnavaia, T.J. *Adv. Mater.* **1997**, *9*, 500.
14. Wulff, G.; Sarhan, A. *Angew. Chem. Int. Ed. Engl.* **1972**, *11*, 341.
15. Robinson, D.K.; Mosbach, K.J. *Chem. Soc., Chem. Commun.* **1989**, 969.
16. Vlatakis, G.; Andersson, L.I.; Muller, R.; Mosbach, K. *Nature* **1993**, *361*, 645.
17. Wulff, G. *Angew. Chem. Int. Ed. Engl.* **1995**, *34*, 1812.
18. Takeuchi, T.; Matsui, J. *Acta Polymer* **1996**, *47*, 471.
19. Kuchen, W.; Schram, J. *Angew. Chem. Int. Ed. Engl.* **1988**, *27*, 1695.
20. Dai, S.; Burleigh, M.C.; Shin, Y.; Morrow, C.C.; Barnes, C.E.; Xue, Z.L. *Angew. Chem. Int. Ed. Engl.* **1999**, *38*, 1235.
21. Dai, S.; Shin, Y.S.; Barnes, C.E.; Toth, L.M. *Chem. Mater.* **1997**, *9*, 2521
22. Dai, S.; Shin, Y.S.; Ju, Y.H.; Burleigh, M.C.; Lin, J.S.; Barnes, C.E.; Xue, Z.L. *Adv. Mater.* **1999**, *11*, 1226.
23. Dai, S.; Burleigh, M.C.; Ju, Y.H.; Gao, H.J.; Lin, J.S.; Pennycook, S.; Barnes, C.E.; Xue, Z.L. *J. Am. Chem. Soc.* **2000**, *122*, 992.
24. Brunauer, S.; Emmett, P.H.; Teller, E. *J. Amer. Chem. Soc.* **1938**, *60*, 309.
25. Barrett, E.P.; Joyner, L.G.; Halenda, P.P. *J. Amer. Chem. Soc.* **1951**, *73*, 373.
26. Smith, R.M.; Martell, A.E. *Critical Stability Constants*; Vol. 2, Plenum: New York. NY, 1975.
27. Banks, D.; Younger, P.L.; Arnesen, R.; Iverson, E.R.; Banks, S.B. *Env. Geol.* **1997**, *32*, 157.

Chapter 11

Extraction of Metals from Soils Using Fluoro-Supported Ligands in CO$_2$

D. L. Apodaca[1], E. R. Birnbaum[2], T. M. McCleskey[1,*], and T. M. Young[3,*]

[1]Los Alamos National Laboratory, MS J514, Los Alamos, NM 87545
[2]Los Alamos National Laboratory, MS E518, Los Alamos, NM 87545
[3]Department of Civil and Environmental Engineering, University of California at Davis, One Shields Avenue, 150 Everson Hall, Davis, CA 95616

The use of supercritical carbon dioxide (SCCO$_2$) for extraction of metals has been demonstrated. The goals of this research were to synthesize fluoro supported ligands that have improved solubility in SC CO$_2$ and to test their efficiency in extracting cadmium and neodymium from surrogate soils. The fluorinated ether chain used in the synthesis of the ligands is commercially known as fomblin or perfluoropolyether vacuum-pump oil. The ligand groups that were synthesized and tested include a carboxylic acid (Fomblin) and a fluorinated amide with two amine groups for metal binding. All extractions were performed at 300 atm, 50°C and a CO$_2$ flow rate of 2 mL/min. Two different soils were spiked with each metal yielding four test soils. One soil is a fine, silty soil that contains 4.3% organic content and the other is a sandy soil with an organic content of 0.13%. The Fomblin extracted an average of 29% of the neodymium from the sandy soil using a ten-fold excess of ligand to metal ratio. When the ligand was increased to a hundred-fold excess, extraction efficiencies grew to over 60% for the cadmium and an average of 28% for the neodymium. When a five hundred-fold excess of fomblin was used, 54% of the cadmium and an average of 18% of the neodymium were extracted. The fluorinated amide had cadmium extraction efficiencies of approximately

15% and neodymium extraction efficiencies of 12% using a ten-fold excess of ligand. No metal was extracted from the fine, silty soil with higher organic matter content.

Introduction

We have investigated metal ion removal from contaminated soils using fluoro-supported ligands in supercritical carbon dioxide (SC CO_2). Removal of radioactive and hazardous metal ions from waste streams, process streams, or contaminated environmental media is one of the highest priorities for the U.S. Department of Energy (DOE). The treatment of heterogeneous wastes or soils contaminated with actinides or fission products is difficult because relatively small amounts of radionuclides are dispersed through a large volume of solid material.

Soil contaminated with radioactive or toxic metals is the single largest "waste" problem within the DOE complex in terms of volume. Alternatives for remediation of toxic or radioactive metal-contaminated soils include 1) excavation of the site and transport of the contaminated soil to a secure repository, 2) immobilization of the metal(s) in place, 3) physical separation (usually by density or size) of the more highly contaminated soil fractions, and 4) soil extraction technologies (*1*). The first three technologies have several problems. Transport and storage of the entire volume of contaminated soil is exceedingly expensive. Immobilization approaches suffer from high ongoing monitoring costs and result in high dilution of wastes. Both of these alternatives have poor public acceptance because they do not solve the problem, but simply transfer it to a new site or form. Physical separation methods are effective only for contaminated soils where most of the contaminant is concentrated in a small volume of soil, which can be separated by size or density. This method does not remove metal compounds that are sorbed to the soil, and in most cases physical separation methods give very limited volume reductions. (*2*).

Two commonly considered soil solution extraction technologies are soil washing and heap leaching. The "soil washing" treatment technologies use high liquid to solid ratios to clean the soil. Typically, short treatment times are achievable, but the high energy (for agitation) and liquid inputs required often make "soil washing" impractical and expensive. Heap leaching is a process adapted from the mining industry. In the mining application, leftover tailings are leached for their precious metals using extraction agents, such as cyanide or bromide. This process is equilibrium driven and has been shown to achieve high removal efficiencies (90-100%) for a single target metal from mining tailings but not from mixed waste materials (*3*). Heap leaching requires time to complete, is used mainly for large sites, and the use of extraction agents such as cyanide or bromide can result in site contamination that is as problematic as the toxic metals themselves (*3*). Soil washing, heap leaching, and in-situ leaching all produce a significant liquid waste stream that must be separately treated or disposed of as waste.

The selective removal and/or separation of radioactive and hazardous metal ions from a variety of waste and process streams is one of the highest priorities for the DOE Environmental Management Program. Currently there is a great deal of plutonium-238 that is either in storage or landfills. For example, Savannah River has retrievably-stored large volumes (4900 m^3) of combustible waste contaminated with ^{238}Pu. These wastes contain 30,000 Ci of activity and represent 64% of the ^{238}Pu in the complex. Similarly, Hanford and INEEL also have significant combustible TRU wastes in storage or landfills that will require treatment prior to ultimate disposal. Rocky Flats currently has over 2000 drums of ^{238}Pu contaminated processing trash (gloves, glassware, towels, etc.) in storage. Thermal treatment that has been proposed for some of these materials is controversial, making novel extraction methods to reduce waste volume and radionuclide content highly desirable.

Supercritical Carbon Dioxide. The use of supercritical fluids in various extraction applications has increased due to restrictions that environmental legislation has placed on conventional solvents (*4*). Compared to conventional solvent techniques, supercritical fluid extraction (SFE) offers several advantages including relatively high sample throughput and the reduction in both solvent usage and solvent waste generation (*5*). Supercritical fluids offer mass transfer rates that are an order of magnitude higher than liquid solvents. They have zero surface tension facilitating micropore penetration and rapid contaminant removal. When using a supercritical solvent, a chemical's solubility can be increased by several orders of magnitude by isothermally increasing the system pressure.

There has been significant interest in utilizing supercritical CO_2 as an environmentally benign medium for a wide range of applications including extractions, materials processing, and chemical synthesis. Supercritical carbon dioxide is the most frequently used solvent in supercritical fluid technology (*6*). Carbon dioxide is one of the few solvents that is not regulated as a volatile organic compound (VOC) by the Environmental Protection Agency (EPA) because it is not harmful to human health or the environment (*7*). Although there are concerns about CO_2 emissions because it is a greenhouse gas, using it as a solvent poses no new atmospheric threat to the environment. Carbon dioxide is captured and purified for use from currently operating industrial processes. CO_2 is nontoxic, nonflammable and abundantly available at low cost. It is now used as a replacement for organic solvents in a variety of applications, ranging from analytical extractions to industrial processes (8).

Extraction of Metals from Soils using Supercritical Carbon Dioxide. In recent years, SC CO_2 solutions containing ligands have been successful at extracting a wide range of metal species from both liquid and solid matrices (5,6,9-13). Sievers and coworkers have demonstrated the solubility of metal ligand complexes in supercritical CO_2 using beta-diketones (14), and Wai and coworkers have pioneered the development of small fluorinated beta-diketones for extracting metals from a variety of heterogeneous surfaces using supercritical CO_2 (4,5,9-11). More recently, Beckman and coworkers have shown that attaching "CO_2-philic" moieties consisting of either highly fluorinated or polysiloxane groups, including oligomer or polymer chains, to ligands can dramatically increase the solubility of the resulting metal complex (6,7,12). These results present the possibility of using a single extraction operation to remove both organic and inorganic species from waste materials. Supercritical carbon dioxide's low surface tension allows for easy separation of metal ligand complexes through pressure and density control. The high diffusivity aids in facilitating transport and extraction of metals from complex heterogeneous waste (e.g. combustibles, sludges, soils, high salt matrices, etc).

Metal extraction with SC CO_2 is often facilitated by converting metal ions into neutral metal complexes that are either soluble in the fluid phase alone or soluble in the fluid phase with the aid of organic solvents (5). Direct extraction of metal ions into CO_2 with supercritical carbon dioxide is not possible because of charge neutralization requirements and weak solvent-solute interactions. Carbon dioxide is non polar, so modifiers such as methanol are often used because they may have better interaction with the solute than carbon dioxide. Addition of modifiers to supercritical CO_2 alters the polarity of the fluid phase, which may also enhance the extraction of metals (11). Complexing agents added to the supercritical fluid phase can neutralize the charge on the metal ion and introduce lipophilic groups to the metal-complex system. Solubilization of the metal complexes is then possible.

In the past, extraction efficiencies have been limited by poor solubility of ligands and ligand/metal complexes in SC CO_2 (6). Tetraalkyl-ammonium dithiocarbamate ligands exhibit solubilities of less than or equal to 60 μM in supercritical CO_2 (15). Although these ligands have been shown to extract cadmium, lead and zinc from dilute solutions (20 ppm metal), a 1000-fold excess of ligand was used in all extractions and the solubility of the metal complex was not measured. It has been shown that fluorine substitution in ligands tends to enhance the solubility of both the ligand and metal/ligand complex (5). Although these workers fluorinated the dithiocarbamate, metal complexes such as lead [bis(trifluoroethyl)-dithiocarbamate)] are still only soluble in the sub-millimolar range.

Tri-n-butylphosphate (TBP) is a ligand that has high solubility being miscible up to 10% v/v in SC CO_2. Once again, a large excess (3500 fold) of ligand was used when a 3 v/v% solution of TBP was used to extract U(VI) from a 10^{-4} M solution (*16*). It has been demonstrated that solubility of the metal ligand complex in supercritical fluids plays a more important role than the solubility of the ligand as long as sufficient amount of ligand is present in the fluid phase (*5*). Beta-diketones are one class of ligand that seem to facilitate high solubilities of complexed metals in SC CO_2, with lanthanide complexes having solubilities close to 0.06 M. This is the highest metal/ligand solubility reported for any metal complex in supercritical CO_2 (*10*). The high solubility of the neutral ML_3 complexes is a reflection of how volatile they are. Because these complexes are extremely volatile, this causes a problem when extracting radioactive and toxic species. Volatile species are extreme inhalation hazards and great caution must be taken when working under these conditions.

While there has been some success by direct fluorination of the above conventional ligands, this method would not be practical as a remediation process. First of all, the large excesses of ligand that are required for extraction and the cost associated with synthesizing these ligands may make this method economically unattractive. Secondly, even though the fluorinated ligands have proven to be soluble, it is essential that the metal-ligand complex be soluble. Finally, addition of fluorine atoms close to ligating heteroatoms can significantly decrease their binding affinity for metal ions by drawing electron density away from the potential metal site.

Our approach is fundamentally different in that a highly soluble fluorosupport was used as the backbone for all ligands. This separates the portion of the molecule that drives CO_2 solubility from that which binds the metal, ideally allowing any type of ligand to be used without decreasing its complexing strength. These extractant-ligand systems should enhance the solubility of both the ligand and metal-ligand complexes as the CO_2-philic backbone carries less soluble metal-ligand complexes into solution.

Expermimental Section

Synthesis. One set of ligands designed and synthesized for this project focused on remediation of actinides, specifically [238]Pu. Because extractions were conducted at a facility not equipped to handle transuranic wastes, neodymium was used as a surrogate for [238]Pu in all extractions. A commercially available CO_2-philic polyfluoroether carboxylic acid, known as Fomblin, was used as the starting material for all ligands that were developed. Fomblin was obtained from Ausimont Chemicals (Italy) and distilled prior to use (see below). The polyfluoroether carboxylic acid Fomblin (Figure 1) was tested on two types of soil. An amine ligand, synthesized from the Fomblin was also tested. Toxic metals such as cadmium, lead and mercury require the introduction of "softer" ligands such as amines. In this research, cadmium was used as the second target metal. The Fomblin was converted to an acid chloride and then reacted with ligand precursors to form a fluorinated amide linkage with two amine ligand end groups. (Figures 2 and 3).

Past experiments show that a polyfluoropolyether carboxylate (PFEC) can be distilled into three different fractions each having a different molecular weight (*17*). The molecular weights of the PFEC fractions were determined using ^{14}N and ^{13}C-NMR relaxation times. Because the PFEC is almost identical to Fomblin, we assumed the same molecular weight distribution in our distillation. Throughout this paper, the molecular fractions of Fomblin are categorized as Fomblin-1, Fomblin-2, and Fomblin-3. Fomblin-1 is a combination of the lowest molecular weight Fomblin molecules ($n = 1$ and 2 or average $n = 1.5$) and the Fomblin-3 being the highest ($n = 4$; Table 2). The fluorinated backbone that was used for the majority of this project was fomblin-2. Distillation of the three Fomblin fractions was done using general distillation techniques. Products were analyzed using NMR and IR to assure that there were three different fractions.

Where n = 1.5, 3 and 4

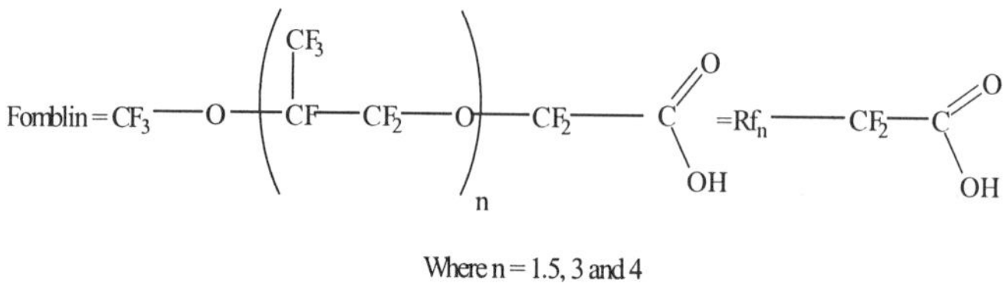

Figure 1. Fomblin (Fluorinated Ether Chain)

The Fomblin-2 was converted to an acid chloride following a published synthetic route (*18*). Addition of thionyl chloride to Fomblin in perfluoro-cyclohexane yields the acid chloride. Spectroscopic analysis of the material in this experiment produced results comparable to those previously published. The literature reported a shift of the carbonyl peak from 1776 cm^{-1} to 1809 cm^{-1}. IR results for the Fomblin show the shift to be from 1780 cm^{-1} to 1800 cm^{-1}. The disappearance of the acid proton at 12 ppm in the proton NMR (^{1}H-NMR) indicates that the acid chloride was quantitatively formed.

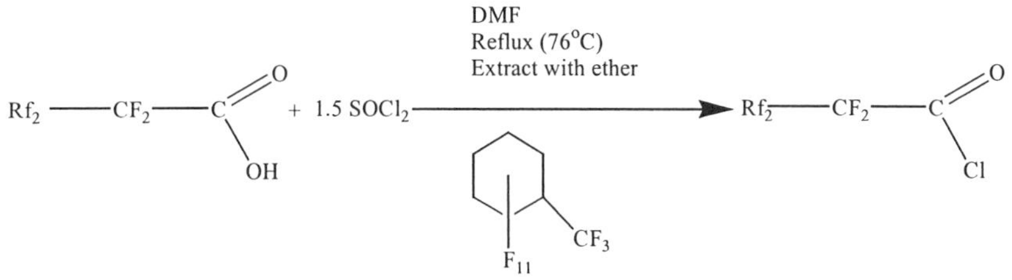

Figure 2. Conversion of Fomblin-2 to Fomblin-2 Acid Chloride

The amide was formed by addition of an excess of 3,3'-iminobis (N,N-dimethylpropylamine) to the Fomblin-2 acid chloride. Four equivalents of amine (11.2 mL) were dissolved in 10 mL of water and the solution was stirred in an ice bath. The Fomblin-2 acid chloride (9.2 g) was slowly added to the mixture and stirred for two hours. The solution was allowed to warm to room temperature while stirring overnight. Water was removed by rotary evaporation at 200 rpm and 60 °C. Water/ether extractions were performed three times to assure that excess reactant (3,3'-iminobis (N,N-dimethylpropylamine)) was removed. Proton NMR of the Fomblin-2 amide-2 showed a shift in the resonances near the amide linkages, indicating that the desired product was formed.

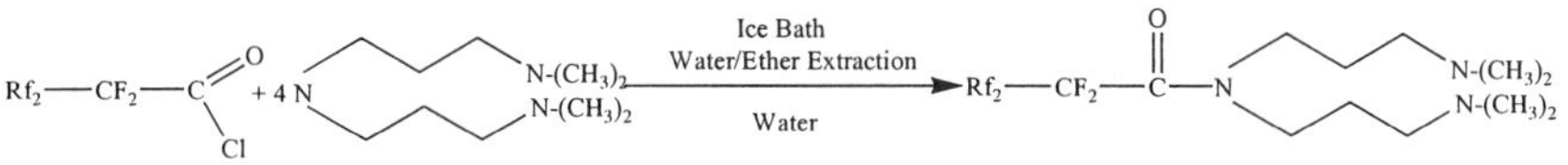

Figure 3. Synthesis of Fomblin-2 Amide-2

Extractions. Lead nitrate and neodymium nitrate salts and nitric acid were purchased from Aldrich (Milwaukee, WI). Two different soils were spiked individually with lead and neodymium salts yielding four test soils. Forbes soil is from the surface horizon and is classified as a loam soil with an organic carbon content of 4.30%. Wurtsmith soil is from the unsaturated zone at 10 to 15 feet depth in an uncontaminated portion of Wurtsmith Air Force Base, Oscoda, Michigan. The Wurtsmith soil is 0.13% by weight organic matter, 0.021 % by weight carbon. (*19*).

The soils were crushed using a mortar and pestle to assure homogeneity before spiking. Two 1000 ppm metal stock solutions were prepared from the neodymium and cadmium salts. For each spike, 100 mL of stock solution was added to 100 g of soil in glass jars. Teflon lined caps were placed on glass jars, covered with Parafilm and tumbled end over end for 72 hours. After 72 hours, the liquid was decanted and soils were air dried under a fume hood and stored in jars for up to 9 months. Microwave digestions were performed and extracts were analyzed for all soils at the DANR Analytical Laboratory at the University of California, Davis. Metal concentrations on each soil are displayed in Table 1.

Table 1. Metal concentrations in soils

Sample	Metal Concentration (ppm)
Wurtsmith Cadmium	377
Wurtsmith Neodymium	323
Forbes Cadmium	720
Forbes Neodymium	725

Extractions were performed using a Suprex Flexprep supercritical fluid extractor with an Accutrap sample collector and MP-1 modifier pump (Isco, Lincoln, NE). SFC-grade carbon dioxide, 99.995% purity cooling carbon dioxide and 99.995% purity nitrogen were used in all extractions. The SFC-grade carbon dioxide was obtained from Air Products & Chemicals (Allentown, PA). The cooling carbon dioxide and nitrogen were purchased from Puritan Bennett (San Ramon, CA).

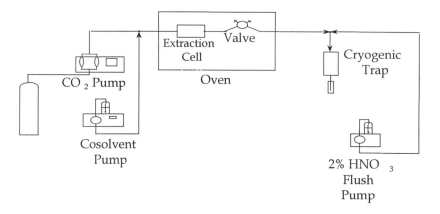

Figure 4. Supercritical Fluid Extraction Unit

The SFE unit consisted of a microprocessor control module, a reciprocating piston pump, a cosolvent addition pump, an oven, an extraction cell, a two-position six-port valve, a heated variable flow restrictor to depressurize the CO_2, a cryogenically cooled solute trap, a pump to recover solute by flushing the trap and a fraction collector (Figure 4). A restrictor, based on an adjustable microbore needle valve, was located before the cryogenic trap.

All extractions were performed using the same parameters. Three milliliter stainless steel cells were used to hold the contaminated soil for each extraction. Each soil sample was between 2 and 3 grams. Extractions were conducted at a pressure of 300 atm, a temperature of 50 °C and a condensed CO_2 flow rate of 2.0 mL/min. The appropriate amount of fluorosupported ligand (10, 100, or 1000 times the nominal moles of mass in the sample) was added to the bottom of the extraction cell immediately before beginning the extraction. The extraction process started with a 15 minute static equilibration step to allow solubilization of the complexing agent, followed by two 5 minute dynamic steps and one 10 minute dynamic step. After the fluid exited the steel cell it expanded through a heated restrictor (100^0 C) into a cryogenic trap (25^0 C) filled with glass beads. The metal was retained on the glass beads and the trap was rinsed after each dynamic step with two percent nitric acid (approximately 10 mL). The trap was then purged with nitrogen gas to remove the metal containing solution from the trap. The aqueous metal-containing samples from each dynamic step were analyzed on a TJA Atomscan 25 Inductively coupled Plasma-Atomic Emission Spectrometer (ICP-AES) following the standard SW846-EPA Method 6010 operation procedure. The values from the individual rinse solutions were summed to obtain the total metal extracted. The individual extracted soil samples were taken to the DANR analytical laboratory for microwave digestion and analyzed for total metal remaining. The extraction percentages shown in Fig. 5 and Fig. 6 are calculated by comparing the metal remaining in the soil (A) relative to the metal measured in the liquid rinses (B), or $100[1-B/(A+B)]$. The only exception to this is the 100X data in figure 5; we were unable to obtain total metal results for these samples, so the percent metal extracted was derived from the average metal concentration shown in Table 1. It was necessary to mass balance the numbers because despite our efforts, the soil was apparently not spiked homogeneously.

Results and Discussion

Synthesis. Three different fractions of Fomblin were observed upon distillation (Table 2). As mentioned previously, the molecular weight separations that were determined by Caboi for perfluoropolyether ammonium carboxylate were assumed to be similar to Fomblin. They reported approximate values for n as 1.5, 3, and 4 (n = 1.5 is a combination of n = 1 and n = 2 fractions) (*16*).

Table 2. Fomblin Distillation

Fomblin Derivative	Temperature of Distillation (°C)	Volume Collected (mL)	Time of Distillation (minutes)	n	Molecular Weight (g/mol)
Fomblin-1	80-100	50	120	1.5	434
Fomblin-2	110-135	100	70	3	722
Fomblin-3	200	50	90	4	923

A ^{19}F-NMR of each fraction demonstrated that different molecular weights were separated (Table 3). A proton NMR was also taken of the Fomblin-2 sample to assure that the carboxylic acid proton was present. The proton showed up at 12 ppm. An IR spectrum showed the carbonyl peak at approximately 1780 cm^{-1}.

Table 3. Fluorine NMR Data for Fomblin Fractions

Sample	CF_2Cl (ppm)	$CFCl$ (ppm)	Ratio	Fraction
1	29.78	11.07	2.7	1
2	32.86	11.86	2.8	1
3	41.45	8.77	4.73	2
4	48.90	7.54	6.49	3

Extractions. Extractions were performed using the Fomblin-2, Fomblin-3 and Amide-2. The extraction efficiency of each ligand was tested on both soils. At least two duplicates were done for each extraction and the ratio of ligand to metal was varied. In no case was a measurable amount of metal extracted from the Forbes soil (see below); the following results are for the Wurtsmith soil. When using a ten-fold (10X) excess of ligand, Fomblin-2 extracts neodymium but not cadmium (Figure 5). Increasing the ratio of ligand to metal from 10:1 to 100:1 increases the extraction efficiency for cadmium. As the excess of ligand is increased to five hundred-fold, the extraction efficiencies slightly decrease for both cadmium and neodymium. Past research has shown that increasing the ratio of ligand to metal increases extraction efficiency at first but the increase in efficiency diminishes at higher ligand to metal ratios (*6*).

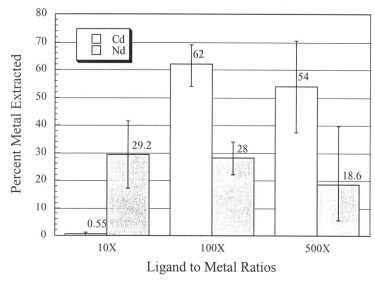

Figure 5. Percent of metal extracted from Wurtsmith using Fomblin-2.

Each extraction was run in duplicate or triplicate; Figures 5 and 6 show the average value (also listed above the column) and the error bars show the range of the values. The deviation between repeat experiments was rather high (RPDs from 20 to 80%) due to two factors. First, despite our attempts to homogenize the soil, it appears that the metal spike was not distributed evenly through the sample. Mass balance calculations performed by digesting the soils after extraction, and adding this result to the amount of metal extracted, did not indicate the same total amount of metal in duplicate experiments.

A second major problem was loss of extracted metal in the system. As mentioned previously, there are three dynamic steps in each extraction. After each step, approximately 10 mL of nitric acid is flushed through the cryogenic trap. The majority of the metal was found to be extracted in the first dynamic step (90-100%). The volume of acid wash collected during these extractions was sometimes less than 10 mL and was more variable with increasing ligand ratios. It was noted that the ligand was partially clogging the system during these extractions. More ligand was flushed from the trap and there was difficulty in rinsing the collection trap when high ligand ratios were used. After each extraction, it was necessary to flush the system with methanol and then trichlorotrifluoroethane to remove residual ligand. The variability in extraction percentages may be due to clogging (either from the ligand or the ligand-metal complex) or it is possible that the ligand did not get mixed throughout the cell.

Adjusting the CO_2 flow may allow better mixing in the cell and overall higher ligand solubility.

The second ligand that showed promise is the amide that was synthesized by adding 3,3'-Iminobis (N,N-dimethylpropylamine) to the fomblin acid chloride (Fomblin-2 Amide-2) (Figure3). Results indicate that the Fomblin-2 Amide-2 is able to extract small amounts of both cadmium and neodymium at a low ligand to metal ratio. As the ratio is increased, the percent of metal extracted slightly decreased for both Cd and Nd. Given the fluctuation between replicate experiments, it is uncertain if the observed decrease is significant. Flush volumes do not necessarily seem to be associated with the variance of results, as in this case, minimal clogging of the lines was observed. It may be that at higher ligand concentrations some of the extracted metal-complex accumulates in the lines rather than being collected in the trap as intended.

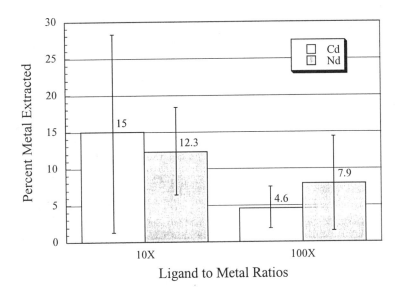

Figure 6. Percent of metal extracted fromWurtsmith soil using the Fomblin-2 Amide-2.

Extractions using a ten-fold excess of Fomblin-1 and the Amide-2 ligands did not remove any metal from the Forbes soil. As the initial metal concentrations indicated (Table 1), both metals sorbed more extensively from an aqueous solution onto the Forbes soil than onto the Wurtsmith soil. The greater affinity of the Forbes soils for the metals is reflected in the poor metal recovery from this soil.

Extraction attempts were also made using the Fomblin-3 (Fomblin with a molecular weight of 922g/mol). The goal was to determine if increasing the length of the CO_2–philic ligand would allow us to extract anything from the Forbes soil. A one hundred-fold excess of Fomblin-3 was used to extract both cadmium and neodymium from the Forbes soil. Two extraction attempts for each metal were made. For each extraction, there was trouble with the ligand clogging the system. Little or no rinse volume was being flushed from the cryogenic trap. The SFE unit was disassembled to locate the clogging. It appeared that the ligand was precipitating out before even reaching the trap. This leads us to believe that the Fomblin-3 is not sufficiently soluble in supercritical carbon dioxide.

Conclusions.

We have demonstrated that fluorinated ligands can be used to extract metals from sandy soils such as Wurtsmith. A good agent for extraction of metal ions using supercritical carbon dioxide requires that it is readily available and that extraction efficiencies are high at low ligand to metal ratios. Results indicate that Fomblin is able to extract neodymium at a ligand to metal ratio of 10 and cadmium at a ratio of 100. All previous research has indicated that a large excess (greater or equal to 1000X) of ligand is required for extraction of metals using SFE. Most of the ligands that have shown promise for metal extraction on a research scale are not readily available; the large excess of ligand needed for complete metal removal makes these ligands impractical to for remediation techniques. It is not surprising that Fomblin, a carboxylic acid, binds cadmium and neodymium efficiently. Carboxylic acids have been known to bind hard metals such as lead and cadmium from soils. A carboxylic-acid derivative of a water-soluble ligand has proven to be successful for removal of lead from soil (*1*). Neodymium, a harder metal than cadmium, could be expected to bind at the lowest ligand to metal ratio in Fomblin extractions.

Fomblin-2 Amide-2, with two amine binding sites, proved to be successful for extracting both cadmium and neodymium when a ten-fold excess of ligand relative to metal was used. In general, softer ligands such as amines are successful at binding toxic metals such as cadmium, lead and mercury; thus, Fomblin-2 Amide-2 was anticipated to bind and remove cadmium from the soils. Neodymium was not expected to bind as well, since it is a harder metal, but similar amounts of metal were extracted. As the ligand to metal ratio was increased to one hundred-fold there was a small decrease in extraction efficiency. It is unclear if this decrease is due to a limitation in the surfactant solubility, loss of ligand-metal complex to the system, or an equilibrium effect.

There was little or no metal extracted from the Forbes soil even when a large excess of ligand was used. Efficient binding of metals by humic substances is well known and the affinity of Forbes soil for the metals likely derives from its higher organic matter content. Metal bound to the soil organic matter will be more difficult to remove, decreasing extraction efficiencies. It is also possible that ligand interactions with the organic matter reduced ligand availability enough to prevent extraction. This type of difficulty is one reason why most of the research done using SC CO_2 for extraction of metals from solids has only performed extractions on non-interacting surfaces such as filter paper, aqueous solutions or sand (5,7,10).

Several authors have demonstrated the use of CO_2-philic ligands in supercritical carbon dioxide to extract metal ions from a variety of matrices (5,6,9-14). Generally this method has been limited by either extremely low solubility of ligand or the resulting metal/ligand complexes in pure carbon dioxide. There are some key differences regarding the ligands that were synthesized in this research compared to ligands that have already been tested. The main difference is that Fomblin was used as a fluorinated backbone and various ligands were attached to it, driving solubility from the perfluoroether rather than the ligand itself. Most other research has tested conventional ligands and has fluorinated them directly. This difference was intended to maintain ligand strength while increasing metal-complex solubility in CO_2.

There are no reports of supercritical carbon dioxide extractions of neodymium, but there has been some research on extraction of U(VI) and lanthandides from a variety of matrices. Again, success has been garnered only with toxic materials (i.e., tri-n-butylphosphate) (16) or volatile ligands (i.e., fluorinated beta-diketones) (10). Fomblin shows promise as it is a chemically inert, non-toxic extraction material.

Although extraction efficiencies were low in some cases, this research is promising in that only a ten-fold excess of ligand to metal ratio has to be used to extract metal from the Wurtsmith soil. As greater than 90% of the total extracted metal was removed in the initial step, it seems that the limiting factor in all of these extractions continues to be the sorption of metal to soil matrix. Higher extraction efficiency would likely be obtained if a non-competitive matrix were attempted. Future work on soils must consider the type of soil (and hence, the possible competition from organic matter) if complete remediation is to be achieved.

The solubility of the ligand is also of concern. Extraction conditions or ligand modifications will have to be further explored to minimize precipitation in the extraction system. Total extraction was limited in these systems by clogging of the tubing with Fomblin, which may be alleviated by modification of the restrictor valve. Conditions such as pressure and temperature could also be altered to increase ligand and metal-ligand complex solubility and therefore metal extraction from the soils.

The use of supercritical fluid extraction technology for remediation of metals from a variety of matrices cannot directly be compared to alternative methods at this date. However, there are several potential benefits of supercritical fluid-based extraction methods over other technologies. It has been demonstrated (*20*) that SFE can be used for the extraction of organics from soils and other matrices. The fact that the method could be used as a single unit operation to extract metals, organics, and radionuclides from contaminated environmental media is the main reason that this research should be continued. Secondly, there would not be a solvent waste stream as compared to other technologies (soil washing or heap leaching) because carbon dioxide could be released to the atmosphere or recycled for additional extractions. The nitric acid rinse generated would be small in volume, and could be reduced and the metal nitrate salts collected for reuse or disposal. The use of supercritical CO_2 as an extraction medium for metals is promising and research in this area should continue.

REFERENCES

1. Ehler, D.; Duran, B.L; Sauer, N.N. *LANL Report* **1995**, 11.
2. EPA, *EPA Report* **1990**, *EPA*/600/S2-891034.
3. Rudd, B.; Hanson, A.; Baca, A., *Final Report for Los Alamos National Laboratory*, **1994**.
4. Smart, N.G.; Carleson, T.; Kast, T.; Clifford A.`A.; Burford, M.D.; Wai, C.M. *Talanta* **1997**, *44*, 137-150.
5. Wai, C.M.; Wang, S.; Liu, Y; Lopez-Avila, V; Beckert, W. F. *Talanta* **1996**, *43*, 2083.
6. Yazdi, A.V.; Beckman, E.J. *Ind. Eng. Chem. Res.* **1997**, *36*, 2368.
7. Yazdi, A.V.; Lefilleur, C.; Singley, E.J.; Liu, W.; Adamsky, F.A.; Enick, R.M.; and Beckman, E.J. *Fluid Phase Equil.* **1996**, *117*, 297.
8. Dooley, K.M.; Brodt, S.R.; Knopf, F.C. *Ind. Eng. Chem Res.* **1987**, *26*, 1267.
9. Laintz, K.E.; Wai, C.M. *J. Supercrit. Fluids* **1991**, *4*, 194.
10. Lin,Y.; Laintz, K.E.; Wai, C.M. *Anal. Chem.* **1993**, *65*, 2549.
11. Wai, C.M.; Lin, Y.; Brauer, R; Wang, S; Beckert, W.F. *Talanta* **1993**, *40* No. 9, 1325.
12. Yazdi, A.V.; Beckman, E.J. *Ind. Eng. Chem. Res.* **1992**, *35*, 3644-3652.
13. Cross, W.; Akgerman, A.; Erkey, C. *Ind. Eng. Chem. Res.* **1996**, *35*, 1765.

14. Lagalante, A.F.; Hansen, B.N.; Bruno, T.J.; Sievers, R.E. *Inorg. Chem.* **1995**, *34*, 5781.
15. Wang, J.; Marshall, W.D., *Anal. Chem.* **1994**, *66*, 1658.
16. Iso, S.; Meguro, Y.; Yoshida, Z. *Chem Letters* **1995,** 365.
17. Caboi, F.; Chittofrati, A.; Monduzzi, M.; Moriconi, C. *Langmuir* **1996**, *12*, 6022-6027.
18. Singley, E.J.; Liu,W.; Beckman, E.J., *Fluid Phase Equil.* **1997**, *128*, 199.
19. Young, T.M. Ph.D. thesis **1996**, University of Michigan, Ann Arbor, MI, 61-66.
20. Laitinen, A.; Michaux, A.; Aaltonen, O. *Envir. Tech.* **1994**, *15*, 715.

Chapter 12

Synthesis, Characterization and Ion Exchange of Novel Sodium Niobate Phases

T. M. Nenoff[1], M. Nyman[1], Y. Su[2], M. L. Balmer[2], A. Navrotsky[3], and H. Xu[3]

[1]Sandia National Laboratory, Org. 1845, M.S. 0710, Albuquerque, NM 87185–0710
[2]Pacific Northwest Laboratories, MSIN K8–93, Battelle Boulevard, Richland, WA 99352
[3]Deparment of Chemistry Engineering and Materials Science, University of California at Davis, Davis, CA 95616

Due to the extreme complexity of DOE mixed wastes (i.e. Hanford tank solutions), a variety of advanced processes are necessary to separate out hazardous radioactive and RCRA metals so that these wastes may be processed and disposed with minimal cost and maximum efficiency. Therefore, there is an ongoing effort to produce and test new phases with novel ion exchange properties for waste cleanup applications. We present here the synthesis of a novel class of thermally and chemically stable, microporous, niobate-based, ion exchanger materials. Ion exchange studies show these new phases are highly selective for Sr^{2+} and other bivalent metals. Additionally, the Sr-loaded ion exchangers undergo direct thermal conversion to perovskite, a highly durable phase with potential waste form applications.

Introduction

The CST ion exchanger, a crystalline silicotitanate material jointly developed by Sandia National Laboratories (SNL) and Texas A & M University, is highly selective for Cs over a broad pH range and in the presence of competitive ions (i.e. Na).(1-5) Additionally, it is stable in extreme radioactive and chemical environments.(4,5) These properties make the CST ion exchanger useful for removal of the ^{137}Cs radionuclide from defense wastes such as those stored at the Hanford site. Additionally, direct thermal conversion of the Cs-loaded IE-911 (a form of CST manufactured by UOP) produces a durable ceramic material, a potential alternative waste form to borosilicate glass.(6,7) Given the great stability of oxide-based ion exchanger materials such as CST in extreme chemical and radioactive environments, related phases are of great interest for selective extraction of other radioisotopes (8-19) (i.e. ^{90}Sr, ^{99}Tc, ^{60}Co) and RCRA metals (20-27) (i.e. Pb) prevalent in DOE mixed wastes. Additionally, the subsequent formation of waste form materials must be considered in the selection of appropriate ion exchanger phases for radioisotope extraction.

We are currently investigating ion exchange and thermal conversion of a novel class of niobate-based microporous materials. These phases have a niobium oxide, octahedral framework with channels containing exchangeable sodium ions. Titanium or zirconium is substituted into the niobium framework. These phases show high selectivity for divalent cations over monovalent cations, which is a useful characteristic for numerous applications including extraction of ^{90}Sr or Pb from mixed wastes. Additionally, direct thermal conversion of the niobate-based ion exchangers produces perovskite, a durable phase and major component of the well-known Synroc ceramic waste form.(28-31) Described here is the synthesis, characterization, ion exchange, ion selectivity, and thermal conversion of these novel materials.

Experimental

Synthesis and Characterization of Sodium Niobate Phases

In a inert atmosphere box, niobium (V) ethoxide {Nb(OC$_2$H$_5$)$_5$} (approximately 2.2 mmol) and the tetravalent transition metal alkoxide are combined to form a homogenous mixture of the alkoxides. In a 23 ml Parr reactor teflon liner, NaOH (33.6 mmol) is dissolved in water by stirring. Concurrently, the alkoxide mixture is added and stirring is continued for 30 minutes, followed by additional water ans stirring. The mixture is heated in the

Parr reactor for 5 - 7 days at 175 °C. The precipitated product is isolated from the parent solution by filtration and is washed with water. Approximately 0.3 g of white, microcrystalline product is collected and analyzed quantitatively for Na, Nb and the tetravalent transition metal. The Na content is determined by atomic adsorption spectroscopy (AAS) and Nb, and the tetravalent transition metal contents are determined by direct coupled plasma spectroscopy (DCP). Water content and high temperature phase changes are determined by thermogravimetric analysis/ differential thermal analysis (TGA-DTA). Product morphology and purity is examined by scanning electron microscopy (SEM). Products are also characterized by powder X-ray diffraction (XRD) and Raman and infrared (IR) spectroscopies.

Ion Selectivity and Exchange Experiments

Ion selectivity experiments are carried out using the following general procedure: 0.05 g ion exchanger is added to a 10 ml solution which contains 50 ppm of the ion of interest (i.e. Sr). The ion exchanger is shaken with the solution at 300 rpm in a capped, 20 ml vial for 20 hrs at room temperature. The solution is filtered using a Whatman's 0.02 µm ANOTOP syringe filter. The solution is diluted and analyzed for the ion of interest by AAS, using standards containing the same matrix elements as the solution. Nitrate salts are used for both the ions for selectivity studies and the competing ions (i.e. Na). Solutions for the pH dependent selectivity experiments are made with constant sodium concentration using combinations of HNO_3, $NaNO_3$ and NaOH to obtain a range of pH's.

Maximum loading of ion exchangers is carried out by shaking the ion exchanger with a 1 molar solution of the metal nitrate for ion exchange at room temperature for 6 - 20 hours at 300 rpm. The ion exchanger is removed from the solution by filtration and washed with 3000 ml hot water. This process is repeated twice. The percent loading is determined by AAS analysis of the loaded ion exchanger.

Results and Discussions

Synthesis and Characterization of Sodium Niobate Phases

Two representative tetravalent transition metal containing phases (Ti-Phase I and Zr-Phase II) were analyzed for metal composition and water content. Thermogravimetry (figure 1) suggests a volatile (OH/H_2O) content for both phases of approximately 7.5 wt %. The weight loss between 50-200 °C is likely the free H_2O (2 wt %) and the weight loss between 200-300 °C is the 5 OH^- groups, volatilized as $2.5H_2O$.

The powder X-ray diffraction spectra of these two phases revealed that they are isomorphous (figure 2). Similar diffraction patterns are recorded for both samples with different relative peak intensities and slight peak shifts as a function of size and concentration of the tetravalent transition metal. Scanning electron microscopy or reflected light microscopy of a typical reaction product reveals that Phase I and Phase II contain no visible impurity phases or amorphous impurities, and the crystals are ~ 1 x 10 - 50 μm fibers (figure 3).

As evidenced by X-ray diffraction, identical structures are obtained from different ratios of Nb:M^{4+}. This data suggests that the tetravalent transition metal substitute isostructurally for niobium. The charge discrepancy created by the tetravalent transition metal substituted for Nb^{5+} is likely balanced by protons in the pores, and will be confirmed as structural data becomes available. The Nb/ M^{4+} coordination site is most likely to be octahedral or distorted octahedral, and is confirmed by Raman spectroscopy, shown in figure 3. There are two types of distorted octahedra observed, as marked in figure 4; a niobium octahedron with equatorial distortion and a niobium octahedron with planar distortion.(32,33) The coordination environment of the tetravalent transition metal is currently under investigation. Mid-IR spectra of Phases I and II were also recorded. From 370 - 1200 cm^{-1}, vibrational modes of the Nb-O and M^{4+}-O bonds are observed. Peaks observed around 1600 and 3000 - 3500 cm^{-1} are vibrational modes of OH^- and internal water. The patterns are very similar for Phases I and II, again indicating isomorphism. In both samples; there are clearly two O-H vibrational modes (3000 - 3500 cm^{-1}) which are associated with the M^{4+} substituents.

Ion Exchange Experiments

In Table I, the distribution coefficients (K_d; ml/g) of a series of monovalent, divalent and trivalent metals for Phase I and II are listed, along with the Shannon 6-coordinate radii.(34) The cations exchanged for sodium are not expected to be 6-coordinate in the ion exchanger; but this common coordination number is used just for direct comparison of size. It is apparent from this survey study that this phase is highly selective for divalent cations over monovalent cations. The high selectivity for Cr (trivalent) requires further study. The distribution coefficients for the heavy metals (i.e. Zn, Cd, Pb) are several orders of magnitude higher than those recorded for a series of silicotitanate phases.(1) However, this is a metal common in industrial waste streams, and therefore Cr recovery is of interest. Two possible explanations for divalent cation selectivity include: 1) The pores of the structure are paired and a M^{2+} cation in one pore charge balances the neighboring empty pore, since one M^{2+} cation replaces two Na^+ cations; 2) The M^{2+} compensates a charge imbalance created by the tetravalent transition metal cation in the Nb^{5+} framework. Structural characteristics which influence this unusual property of

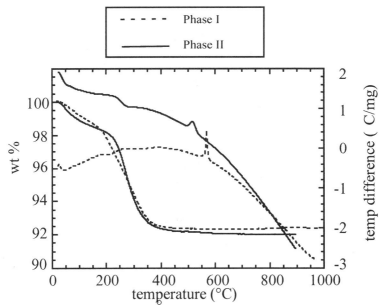

Figure 1. TGA-DTA plots of Phase I and Phase II.

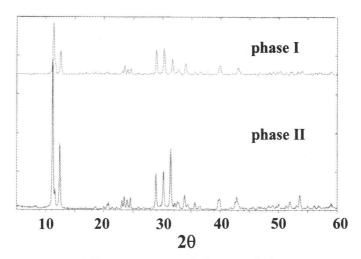

Figure 2. X-ray diffraction patterns of Phase I and Phase II.

Figure 3. Scanning electron micrograph of Phase II.

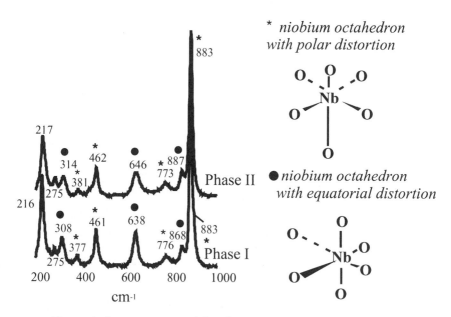

*Figure 4. Raman spectra of the Phase II (top)
and Phase I (bottom).*

divalent metal selectivity will become more apparent once crystallographic data is obtained.

Table I. Distribution Coefficient
of a Variety of Exchange Metals on Niobate Phases

metal	radius in pm (6-coordinate)	Phase I (K_d, ml/g)	Phase II (K_d, ml/g)
Pb^{2+}	133	66,497	22,022
Cr^{3+}	94	> 99,800	> 99,800
Co^{2+}	89	> 99,800	> 99,800
Ni^{2+}	83	> 99,800	> 99,800
Zn^{2+}	88	> 99,800	> 99,800
Cd^{2+}	109	> 99,800	> 99,800
Cs^+	181	150	169
K^+	152	95	153
Li^+	90	8	35
Ba^{2+}	149	> 99,800	> 99,800
Sr^{2+}	132	> 99,800	> 99,800
Ca^{2+}	114	2300	2657
Mg^{2+}	86	226	458

In order to further investigate the ion exchange behavior of Phase I, extensive experiments of Sr selectivity were carried out. Furthermore, strontium selectivity is of interest for cleanup of nuclear wastes such as the Hanford tanks.(10,13-16) Strontium selectivity was studied as: 1) a function of Na concentration as a competing ion, since the Hanford tanks contain up to 5 molar Na concentration (35), and 2) a function of pH. Distribution coefficients for Sr on Phase I as a function of Na concentration as a competing ion is plotted in figure 5. At lower Na concentrations (< 0.1 M Na), the K_d values approach 10^6 ml/g, which is the value obtained for approximately 0.1 ppm Sr remaining in solution; the detection limit of the Sr analytical technique (AAS). As [Na] increases, the K_ds decrease to approximately 500 ml/g. Figure 6 shows a plot of distribution coefficient of Sr on Phase I as a function of pH with 0 M Na, 0.01 M Na and 0.1 M Na as a competing ion. The purpose of using Na as a competing ion for these studies is twofold: 1) to suppress the Sr selectivity so that K_d values may be obtained; and 2) addition of NaOH is necessary for the higher pH experiments, so a consistent concentration of sodium (combination of $NaNO_3$ and NaOH) is added to all the experiments to keep the solution matrices

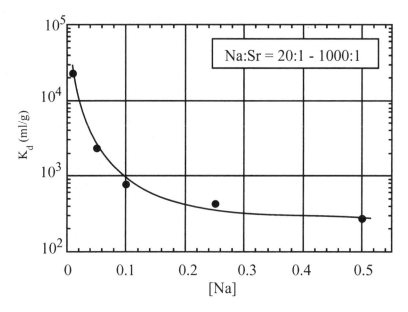

Figure 5. Distribution coefficient of Sr on Phase I as a function of [Na]

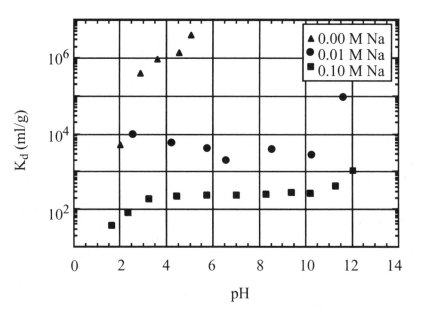

Figure 6. Distribution coefficient of Sr on Phase I as a function of pH

as similar as possible. In general, there is an increase in Sr selectivity with increasing basicity. However, above pH = 12, Sr uptake is decreased. This corresponds with an increase in $Sr(OH)^+$ concentration in solution; above pH = 12.8, $Sr(OH)^+$ becomes the sole Sr specie in solution.(10) This result suggests either 1) supporting evidence is provided for the ion exchanger's selectivity for divalent cations; or 2) the pore size limits uptake of $Sr(OH)^+$. The selectivity experiments with 0 M Na were carried out only in the acidic range, because at pH > 6, all Sr is removed from solution. With lower concentrations of Na (0.01, 0), the selectivity for Sr remains surprisingly high ($>10^4$ml/g), which is several orders of magnitude higher than other Sr selective phases(10). On the other hand, high concentrations of Na inhibits Sr selectivity considerably. However; to compare this effect directly to the performance of other Sr selective ion exchangers such as layered titanates and silicotitanates, experiments using very high hydroxyl concentrations need to be executed.(1) Below pH = 1.5, the niobate ion exchanger undergoes decomposition, and selectivity decreases. Decreased ion adsorption is typical of inorganic ion exchangers in acidic solutions.(9,24,36) The anionic framework preferentially takes up H^+ from the acidic solution over the larger metal cations. Conversely, in highly basic solutions where H^+ concentration is low, the ion exchanger adsorbs metal cations.

Finally, maximum Sr-loading of Phase I was accomplished by twice-repeated contact and agitation of the ion exchangers with 1M $Sr(NO_3)_2$ solutions. The final Na:Sr ratio of the maximum exchanged phases is ~4:1.

High Temperature Phase Transitions of Niobate Ion Exchangers

The TGA-DTA spectra of the Phase I and Phase II are observed in figure 1. Both undergo an exothermic phase transitions between 500 – 600 °C. Analysis of the thermal decomposition product by X-ray diffraction reveals that both phases undergo transition to a perovskite phase. The phase transition of Phase II phase occurs at a lower temperature, and is less exothermic (as evidenced by the shape and size of the exothermic peak on the DTA curve) than the phase transition of Phase I. This suggests; 1) Phase II is slightly less stable than Phase I and 2) the perovskite formed from Phase I is slightly more stable than the perovskite formed from Phase II.

Scanning Electron Microscopy (SEM) / energy dispersive spectroscopy (EDS) analysis of the crystalline perovskite shows that the composition is homogeneous. This phase alteration of the Sr-loaded ion exchangers is an interesting property with regard to radioactive nuclide (i.e. ^{90}Sr) immobilization. Perovskite exhibits great near and long term chemical and radioactive stability, and is a major component of the well-known Synroc ceramic waste form.(28-31) Preliminary PCT leach tests show that Sr does not leach from Phase I.

Micrographs of the Sr-exchanged Phase I, and the perovskite formed by thermal phase transition of the Sr-exchanged Phase I, are shown in figures 7a

Figure 7a. scanning electron
            micrograph of
            Sr-loaded Phase I.

Figure 7b. scanning electron micrograph
            of perovskite from thermal alter-
            ation of Sr-loaded Phase I.

and 7b, respectively. The two phases have an identical needle-like morphology. Perovskite is a cubic mineral which typically exhibits a cubic morphology. The perovskite morphology observed in figure 7b is apparently preserved from the morphology of the Sr-loaded Phase I.

Summary and Conclusions

A novel class of microporous, niobate-based ion exchanger materials which contain tetravalent transition metal substituted for niobium. Two representative phases, Phase I and Phase II have been characterized and their ion exchange behaviors have been studied. Ongoing work involves structural characterization of these isostructural phases, and study of properties which vary as a function of M^{4+} content. Preliminary ion exchange results show these phases are highly selective for Sr^{2+}, and other bivalent cations. Selectivity for Sr increases with increasing pH, exhibiting maximum K_ds > 10^6 ml/g. Selectivity for Sr in acidic solutions containing no competing metal cations is around 10^3 - 10^4 ml/g. Ongoing ion exchange experiments include back-exchange experiments, studies using RCRA solutions and radioactive waste simulant solutions, and more thorough characterization of the exchange properties of Phase II. Other ongoing and future studies of these exciting new materials include synthesis with substitution of other heteroatoms, and study of their catalytic behavior.

Acknowledgements

This work was supported by the U.S. DOE under contract DE-ACO4-94AL8500, and the DOE/ Environmental Management Science Program.
The authors thank Mr. Jeff Kawola and Mr. Jesse Stanchfield for assistance with the ion exchange experiments, and Dr. David Tallant and Ms. Regina Simpson for the Raman data collection.

References

1) Anthony, R. G.; Phillip, C. V.; Dosch, R. G. *Waste Management* **1993**, *13*, 503 - 512.
2) Anthony, R. G.; Dosch, R. G.; Gu, D.; Phillip, C. V. *Industrial Engineering and Chemistry Research* **1994**, *33*, 2702 - 2705.
3) Poojary, D. M.; Cahill, R. A.; Clearfield, A. *Chemistry of Materials* **1994**, *6*, 2364 - 2368.
4) Nenoff, T. M.; Thoma, S. G.; Krumhansl, J. L. "The Stability and Selectivity of TAM5: A Silicotitanate Molecular Sieve for Radwaste Clean-up," Sandia National Laboratories, 1996.
5) Nenoff, T. M.; Krumhansl, J. L. "The Structure of TAM5: A Silicotitanate Molecular Sieve for Radwaste Clean-up," Sandia National Laboratories, 1996.
6) Su, Y.; Balmer, M. L.; Bunker, B. C. *Evaluation of Cesium Silicotitanates as an Alternative Waste Form*: Boston, 1997; Vol. 465, pp 457 - 465.
7) Su, Y.; Balmer, M. L.; Wang, L.; Bunker, B. C.; Nenoff, T. M.; Nyman, M.; Navrotsky, A. *Evaluation of Thermally Converted Silicotitanate Waste Forms*: Boston, 1998.
8) Lazic, S.; Vukovic, Z. *Journal of Radioanalytical and Nuclear Chemistry* **1991**, *149*, 161 - 168.
9) Lehto, J.; Harjula, R.; Leinonen, H.; Paajanen, A.; Laurila, T.; Mononen, K.; Saarinen, L. *Journal of Radioanalytical and Nuclear Chemistry* **1996**, *208*, 435 - 443.
10) Nenoff, T. M.; Miller, J. E.; Thoma, S. G.; Trudell, D. E. *Environmental Science and Technology* **1996**, *30*, 3630 - 3633.
11) Sayed, S. A. *Zeolites* **1996**, *17*.
12) Sylvester, P.; Clearfield, A. *Separation Science and Technology* **1998**, *33*, 1605 - 1615.
13) Behrens, E. A.; Poojary, D. M.; Clearfield, A. *Chemistry of Materials* **1996**, *8*, 1236 - 1244.
14) Behrens, E. A.; Clearfield, A. *Microporous Materials* **1997**, *11*, 65 - 75.
15) Behrens, E. A.; Sylvester, P.; Clearfield, A. *Environmental Science and Technology* **1998**, *32*, 101 - 107.

186

16) Behrens, E. A.; Poojary, D. M.; Clearfield, A. *Chemistry of Materials* **1998**, *10*, 959 - 967.

17) Zheng, Z.; Philip, C. V.; Anthony, R. G. *Industrial Engineering and Chemical Research* **1996**, *35*, 4246 - 4256.

18) Gu, D.; Nguyen, L.; Philip, C. V.; Huckman, M. E.; Anthony, R. G. *Industrial Engineering and Chemical Research* **1997**, *36*, 5377 - 5383.

19) Burham, N.; Abdel-Halim, S. H.; El-Naggar, I. M.; El-Shahat, M. F. *Journal of Radioanalytical and Nuclear Chemistry* **1995**, *189*, 89 - 99.

20) Leinonen, H.; Lehto, J.; Makela, A. *Reactive Polymers* **1994**, *23*, 221 - 228.

21) Mitchenko, T.; Stender, P.; Makarova, N. *Solvent Extraction and ion Exchange* **1998**, *16*, 75 - 149.

22) Janardanan, C.; Smk, N. *Indian Journal of Chemistry A* **1992**, *31*, 136 - 138.

23) Pandit, B.; Chudasama, U. *Indian Journal of Chemistry, A* **1998**, *37*, 931 - 934.

24) Bortun, A. I.; Bortun, L. N.; Clearfield, A. *Solvent Extraction and Ion Exchange* **1997**, *15*, 909 - 929.

25) Bortun, A. I.; Bortun, L.; Clearfield, A.; Jaimez, E.; Villagarcia, M. A.; Garcia, J. R.; Rodriguez, J. *Journal of Materials Research* **1997**, *12*, 1122 - 1130.

26) Singh, D. K.; Mishra, N. K. *Chemia Analityczna* **1994**, *39*, 39 - 46.

27) Varshney, K. G.; Gupta, A.; Singhal, K. C. *Colloids and Surfaces A* **1994**, *82*, 37 - 48.

28) Li, L.; Luo, S.; Tang, B.; Wang, D. *Journal of the American Ceramic Society* **1997**, *80*, 250 - 252.

29) Li, L. *Journal of the American Ceramic Society* **1998**, *81*, 1938 - 1940.

30) Luo, S.; Li, L.; Tang, B.; Wang, D. *Waste Management* **1998**, *18*, 55 - 99.

31) Vance, E. R.; Stewart, M. W. A.; Lumpkin, G. R. *Journal of Materials Science* **1991**, *26*, 2694 - 2700.

32) Xia, H. R.; Yu, H.; Yang, H.; Wang, K. X.; Zhao, B. Y.; Wei, J. Q.; Wang, J. Y.; Liu, Y. G. *Physical Reviews B.* **1997**, *55*, 14892.

33) Jehng, J. M.; Wachs, I. E. *Chemistry of Materials* **1991**, *3*, 100.

34) Shannon, R. D. *Acta Crystallographica* **1976**, *A32*, 751.

35) Agnew, S. F.; Watkin, J. G. "Estimation of Limiting Solubilities for Non-Radioactive Ionic Species in Hanford Waste Tank Supernates," Los Alamos National Laboratory, 1994.

36) Bortun, A. I.; Bortun, L. N.; Clearfield, A. *Solvent Extraction and Ion Exchange* **1996**, *14*, 341 - 354.

Chapter 13

Oxidation of Selected Polycyclic Aromatic Hydrocarbons by the Fenton's Reagent: Effect of Major Factors Including Organic Solvent

Z. Qiang, J. H. Chang, C. P. Huang[*], and D. Cha

Department of Civil and Environmental Engineering,
University of Delaware, Newark, DE 19716

The degradation of PAHs compounds, exemplified by naphthalene, fluorene, phenanthrene, fluoranthene, pyrene and anthracene, by Fenton's reagent was studied using batch, pseudo-continuous and continuous dosing modes. The effect of organic solvent, namely, methanol and ethanol, on the degradation of PAHs was investigated. Results indicate that the Fenton's reagent can effectively degrade all selected PAHs. In the continuous dosing mode, the reaction follows a "time-squared" kinetic expression, i.e., $C = C_0 \exp(-k_{obs}t^2)$. Results also show that methanol and ethanol inhibit the degradation of target PAHs due to competition for hydroxyl radicals.

Introduction

Polycyclic aromatic hydrocarbons (PAHs) belong to a typical group of hazardous organic contaminants commonly found at waste sites. They are major byproducts of incomplete thermal combustion of fossil fuels. Automobile exhausts, power plants, chemical, coke and oil-shale industries, urban sewage (Trapido *et al.*, 1995), cigarette smoke and charcoal grilled foods (Watts, 1998) are the major emission sources of environmental PAHs. PAHs are also found in the heavier fractions of petroleum products, e.g., lubricating oils, asphalt, and tarlike materials (Watts, 1998). Accidental petroleum spills and leaking of underground petroleum storage tanks can seriously contaminate surrounding soils and groundwater with PAHs. It is estimated that the U.S. Air Force alone needs to remediate more than 2000 petroleum contaminated sites (Miller, 1994). Many PAHs are known to be carcinogenic or potentially carcinogenic to animals and probably to humans (Menzie *et al.*, 1992). Sixteen PAHs are listed as the U.S. EPA priority pollutants (Keith and Telliard, 1979). The U.S. EPA has imposed a maximum contaminant level (MCL) for total PAHs in drinking water at 0.2μg/L (Sayre, 1988).

Fenton's reagent is a reaction system consisting of H_2O_2 and Fe^{2+} ion. Under acidic conditions and in the presence of Fe^{2+}, H_2O_2 is decomposed to hydroxyl radical. The hydroxyl radical ($\cdot OH$) is an intermediate species characterized by an unpaired electron. The one-electron deficiency of the hydroxyl radical results in its transient and highly oxidizing characteristics, with a redox potential being only next to elemental fluorine (Huang *et al.*, 1993). The hydroxyl radical can nonselectively oxidize most organic compounds as well as inorganic chemicals through a chain reaction mechanism. The oxidation mechanism of the hydroxyl radical was first proposed by Haber and Weiss (1934), and further clarified by Walling and co-researchers (1970, 1975) using competitive reaction theory. Major reactions taking place in the system are shown as follows (Walling, 1975):

$$Fe^{2+} + H_2O_2 \rightarrow Fe^{3+} + \cdot OH + OH^-, k_1 \qquad (1)$$

$$H_2O_2 + \cdot OH \rightarrow HO_2 \cdot + H_2O, k_2 \qquad (2)$$

$$Fe^{2+} + \cdot OH \rightarrow Fe^{3+} + OH^-, k_3 \qquad (3)$$

$$RH + \cdot OH \rightarrow R \cdot + H_2O, k_4 \qquad (4)$$

$$R \cdot + Fe^{3+} \rightarrow Fe^{2+} + product \qquad (5)$$

$$R \cdot + Fe^{2+} \rightarrow Fe^{3+} + R^- \xrightarrow{H^+} RH \qquad (6)$$

$$R \cdot + R \cdot \rightarrow product \text{ (dimer)} \qquad (7)$$

where, $k_1 = 76$ M^{-1}·s^{-1} (Walling, 1975),
$k_2 = 3 \times 10^8$ M^{-1}·s^{-1} (Walling, 1975),
$k_3 = 2.7 \times 10^7$ M^{-1}·s^{-1} (Buxton *et al.*, 1988),
$k_4 = 10^7$-10^{10} M^{-1}·s^{-1} (Dorfman and Adams, 1973).

Reaction 1 produces hydroxyl radicals, initiating a series of chain reactions. Compared with the reactivity of hydroxyl radical toward most organic compounds (k_4), ferrous ion (k_2) and hydrogen peroxide (k_3), the production rate of hydroxyl radical (k_1) is very small. Therefore, the overall reaction rate is dominated by Reaction 1. Both H_2O_2 and Fe^{2+} compete for hydroxyl radicals at a moderate rate constant. They are known as "scavengers" for hydroxyl radicals. Other common inorganic scavengers in aqueous solution include bicarbonate, carbonate, and chloride. The organic radical produced in Reaction 4 may be oxidized by Fe^{3+} to regenerate Fe^{2+}, or be reduced by Fe^{2+} to reform the parent compound under acid catalysis, or dimerize at a high initial organic concentration.

Generally, most organic compounds react with hydroxyl radicals at a rate constant of $10^9 \sim 10^{10}$ $M^{-1} \cdot s^{-1}$ (Watts, 1998). The high rate constant allows target organic contaminant to be preferably oxidized in the competitive environment. Due to the unselective oxidation nature of the hydroxyl radical, Fenton's reagent has been extensively studied as a technology of advanced oxidation processes (AOP) for the treatment of various hazardous organic compounds. These include chlorinated aromatic hydrocarbons (Sedlack and Andren, 1991), phenolic compounds (Hayek and Dore, 1990), microorganism-refractory organics in wastewater (Spencer *et al.,* 1992), and dye wastewater (Kuo, 1992). Watts and co-researchers (1990, 1991, 1994) have used the Fenton's reagent to oxidize pentachlorophenol, octachlorodibenzo-*p*-dioxin and hexachlorobenzene in silica sands or natural soils.

Extensive research effort has been made to study the oxidation of PAHs with ozone (Sturrock *et al.,* 1963; Bailey *et al.,* 1964; Legube *et al.,* 1984, 1986). Most of their experiments were conducted in organic solvents and at high PAHs concentrations. The extremely low water solubility of PAHs makes the study of PAHs ozonation in water considerably difficult. Therefore, methanol, acetone and octane were frequently used as participating solvents to increase the solubility of PAHs. However, the influence of organic solvent on the degradation of PAHs has not been investigated. The rate constants so obtained may not totally reflect the oxidation behavior of PAHs in the aqueous solution. Many byproducts by the ozonation of naphthalene and phenanthrene have been identified (Legube *et al.,* 1984, 1986; Sturrock *et al.,* 1963). The degree of mineralization of PAHs is considerably low. The Fenton's reagent is a more attractive oxidation agent than ozone. It is stronger and it does not have mass transfer limitation. Therefore, it is expected that a higher degree of mineralization may be obtained.

Some specific DOE sites are significantly contaminated by PAHs. Among the various soil and groundwater remediation processes, in-situ treatments are more attractive than ex-situ applications. Current in-situ remediation technologies include hydrodynamic or physical barrier, solidification/ stabilization, soil vapor extraction (soil venting) and bioventing. Clearly, only the bioventing process can destroy the contaminants. Other processes provide only temporary solutions, i.e., either containment or phase transfer. However, the bioventing process is not suitable for the silt-laden or clay soils that have a low hydraulic conductivity, e.g., $< 10^{-5}$ cm/sec (Destephen *et al.*, 1994). Under such circumstance, the transport of chemical species, i.e., gas, water and nutrients, in the subsurface is significantly limited. Therefore, innovative technologies are urgently needed for in-situ remediation of low permeability soils. This has prompted us to study an electro-chemical technology for in-situ applications. The total project integrates the electro-kinetic process with the electro-Fenton oxidation process. The electro-kinetic process removes hazardous contaminants from subsurface soils, and the followed electro-Fenton process decomposes the released contaminants in aqueous solutions. Additionally, the electro-Fenton process can also be coupled with pump and treat, soil washing and soil flushing processes for the treatment of hazardous organic contaminants on-site or off-site. In this study, we only focus on the degradation of selected PAHs by conventional Fenton process. The optimal conditions and the reaction kinetics of the degradation of selected PAHs, namely, naphthalene, fluorene, phenanthrene, fluoranthene, pyrene and anthracene using Fenton oxidation process were investigated. Several media-assisted dissolution methods for the preparation of reaction solutions were developed. Furthermore, the effect of organic solvent, exemplified by methanol and ethanol, on the degradation of PAHs was also studied.

Materials and Methods

Chemicals

All selected PAHs, namely, naphthalene (99.5%), fluorene (98%), phenanthrene (98%), fluoranthene (98%), pyrene (98%) and anthracene (97%) were purchased from the Aldrich Chemical Company (Milwaukee, WI). They were used without further purification. Hydrogen peroxide and ferrous sulfate were purchased from the Fisher Scientific Company with a content of 31.5% and 98%, respectively. H_2O_2 and $FeSO_4$ solutions were stored in acidic conditions in a refrigerator (4 °C) after preparation. At low pH and low temperature, H_2O_2 and $FeSO_4$ can be sufficiently preserved for several weeks. Table 1 lists important physical-chemical properties of selected PAHs. Naphthalene is bicyclic; fluorene, phenanthrene and anthracene are tricyclic; and fluoranthrene and pyrene are tetracyclic. Furthermore, phenanthrene and anthracene are isomers. Based on water solubility and octanol/water partition coefficient, it is seen that all selected PAHs are hydrophobic. Naphthalene has a relatively large water solubility, around 32 ppm, while all other PAHs have a water solubility of less than 2 ppm. The negative Henry's constants (log)

Table 1. Major physical-chemical properties of selected PAHs (Schwarzenbach *et al.*, 1993).

Compound	Chem. Struct.	Wat. Solub. @ 25 °C (mg/L)	$LogK_H$ @25 °C $(atm \cdot M^{-1})$	$Log\ K_{ow}$ @ 25 °C
Naphthalene		31.9	-0.37	3.36
Fluorene		1.84	-1.14	4.18
Phenanthrene		1.09	-1.59	4.57
Anthracene		0.06	-1.64	4.54
Fluoranthene		0.23	-1.98	5.22
Pyrene		0.13	-2.05	5.13

indicate that the volatilization of PAHs is negligible. Naphthalene is semi-volatile, whereas other PAHs are non-volatile. Experimental conditions such as vigorous stirring may facilitate the volatilization of the semi-volatile compound, i. e., naphthalene. We have discovered that by stirring the naphthalene solution with an initial concentration of 27 ppm in an open beaker, all naphthalene volatizes into air within 20 minutes. Therefore, a closed reactor without headspace is necessary for naphthalene oxidation experiments. For other non-volatile PAHs, volatilization is negligible even under vigorous stirring.

Reaction System

Figure 1 illustrates the laboratory setup of the reaction system. A closed, double-jacketed glass reactor with a total volume of 600 mL was used to conduct oxidation experiments. The reactor contents were completely stirred. A pH controller was used to maintain constant solution pH through intermittent additions of NaOH (1M) or $HClO_4$ (1M) solutions. Since Cl^- is a known hydroxyl radical scavenger, $HClO_4$ was used instead of HCl for pH adjustment. Default experimental conditions were ambient room temperature (23~25°C), 0.05M $NaClO_4$ ionic strength and pH 3, unless otherwise stated. To evaluate the effect of dosing mode on the oxidation of PAHs, 3 different dosing modes, namely, batch, pseudo-continuous and continuous modes, were employed to deliver the Fenton's reagent into the reactor. In the batch mode, both H_2O_2 and Fe^{2+} were singly added at the beginning of reaction. In the pseudo-continuous mode, either H_2O_2 or Fe^{2+} was singly added at the beginning of reaction, whereas the other reagent was continuously delivered using a fine-flowrate dosing pump. In the continuous mode, both H_2O_2 and Fe^{2+} were continuously delivered using two separate dosing pumps. Since the Fenton process consists of a series of competitive reactions, it is expected that the concentrations of H_2O_2 and Fe^{2+} will affect their ability to compete for hydroxyl radicals. As a result, the reaction kinetics of PAHs will vary in the experiments. Another consideration is the volatility of PAHs. Naphthalene is considerably volatile under our experimental conditions. To minimize its evaporation, 600 mL solution was used with zero initial headspace. It is noted that the headspace would vary a little during the reaction period caused by sample withdrawals and chemical additions. However, the volume fluctuation is negligible compared to the relatively large volume of the original solution. All rubber stoppers were wrapped in Teflon tape to avoid loss of organic due to adsorption onto the rubber surface. For other non-volatile PAHs such as fluorene, phenanthrene, fluoranthene, pyrene and anthracene, 500 mL of solution was used with 100 mL of reactor headspace.

Chemical Analysis

The selected PAHs were first extracted by hexane and then analyzed by gas chromatograph / mass spectrometry (GC/MS, Hewlett Packard; GC model 5890, MS model 5972). An HP-1MS capillary column (cross-linked 5% Ph Me silicon, 30m x 0.25mm x 0.25μm) was used to separate organic compounds. The extremely low water solubility of all selected PAHs makes chemical analysis difficult. However, PAHs can produce very stable molecular ions under EI ionization mode in the MS detector due to their stable fused-ring structures. This characteristic makes GC/MS very sensitive for detecting PAHs compounds, with a detection limit of about 100 ppb. The concentrating factor of hexane extraction is compound specific and depends on water solubility. Naphthalene, fluorene, phenanthrene, fluoranthene, pyrene and anthracene were concentrated by 1, 2.5, 2.5, 5, 20 and 20 times, respectively. Preliminary experiments show that the recovery efficiency of hexane extraction is greater than 95% for all PAHs studied. Chemical oxygen demand (COD) was determined using a Hach spectrometer (Hach DR/2000, Loveland, CO) at an absorbance wavelength of 420 nm. The concentrations of H_2O_2 and Fe^{2+} were determined by measuring the light absorbance after chelating with certain reagents using a HP diode array spectrophotometer (8452A) at 410nm and 510nm, respectively. H_2O_2 chelates with Ti^{4+}, forming an orange complex. Fe^{2+} ion chelates with 1,10-phenanthroline, forming an orange-red complex (Standard Methods for the Examination of Water and Wastewater, 3500-Fe D, 1995).

Media-assisted Dissolution Methods

To accelerate the dissolution process of selected PAHs, three media-assisted dissolution methods were investigated and compared. Phenanthrene was selected as a model compound for dissolution study. Methanol, glass-beads, and hexane were separately used to enhance phenanthrene dissolution. Methanol, miscible with water, can easily dissolve hydrophobic organic compounds. Hexane is an organic solvent, but immiscible with water. Glass beads (No. 50 sieve size) provide large surface area for the partitioning of phenanthrene, thus accelerating phenanthrene dissolution.

In the methanol method, 1mL phenanthrene/methanol solution (1.5mg phenanthrene /mL methanol) was spiked in 1L distilled water. After vigorous stirring for 1 hour, followed by vacuum filtration to remove solid phenanthrene residue, the reaction solution was obtained. All methanol remained in the solution with a volume fraction of 0.1% (about 790 ppm). In the glass beads method, 3 mL phenanthrene/methanol solution was spiked onto 10g glass beads. After methanol completely evaporated, the glass beads with fine phenanthrene crystals adsorbing on the surface were transferred into 1L distilled water. The reaction solution was obtained by vigorous stirring followed by vacuum filtration. No methanol was detected by GC/FID in the final solution. In the hexane method, 1mL phenanthrene/hexane solution (4.5mg phenanthrene /mL hexane) was spiked in 1L distilled water which was preheated to 75~80 °C. Preheating not only increases the evaporation rate of hexane but also increases the initial dissolution rate of phenanthrene. After vigorous stirring for 20 minutes in an open flask, all hexane evaporated. By determining the aqueous concentration of phenanthrene at different times, the dissolution kinetics could be obtained. For the purpose of comparison, a blank experiment without any media assistance was also conducted.

Figure 2 shows the kinetics of phenanthrene dissolution with or without media assistance. Results indicate that the dissolution of phenanthrene in the blank experiment is very slow. It may take weeks to obtain a nearly saturated solution. However, both glass beads and hexane significantly enhance the dissolution rate of phenanthrene. After 2 hours, a relative concentration of 85% can be obtained with both methods. In the hexane method the vacuum filtration step is very fast, while in the glass-beads method the filtration is very difficult because the broken glass fragments clog the filter paper (0.45 μm).

Based on above results, hexane is a better medium for preparing PAHs solutions than glass beads. In subsequent experiments, all PAHs solutions without organic solvent were prepared by the hexane method. Meanwhile, the methanol method is suitable for preparing PAHs/methanol (analogously PAHs/ethanol) solutions when investigating the influence of organic solvent.

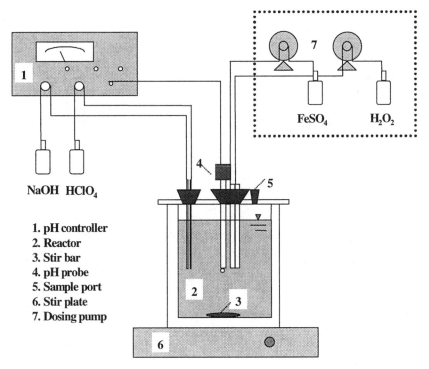

Figure 1. Schematic diagram of Fenton oxidation system.

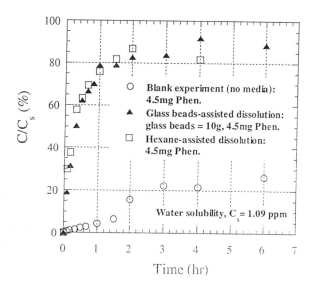

Figure 2. Dissolution of phenanthrene by different methods.

Results and Discussion

Effect of pH

The effect of pH on the oxidation of naphthalene was evaluated at pH 2, 3, 5, 9 and 12 with continuous dosing mode. Results in Figure 3 show that the optimal pH value is pH 3. In alkaline conditions, the oxidation efficiency decreases dramatically. It is well known that the solubility of both Fe^{2+} and Fe^{3+} ions decreases markedly as pH increases. Moreover, at high pH values Fe^{2+} may be readily oxidized to Fe^{3+} by dissolved oxygen and loses its catalytic ability. Another possibility is that H_2O_2 may be unstable at alkaline pH values. However, our preliminary results show that H_2O_2 only self-decays by about 10% at an initial concentration of 150 mg/L, pH 12.75, and temperature 23°C in 3 days. Apparently, H_2O_2 self-decomposition can be ruled out as a significant factor. Therefore, the low efficiency of Fenton's reagent at alkaline pH values can be attributed to insufficient iron(II) in the solution. It is also noted that the oxidation efficiency decreases again below pH 3. In highly acidic conditions, the aromatic radicals generated by the hydroxyl radical (Reaction 4) may be reduced by Fe^{2+}, and then under acid catalysis reform the parent compound (Reaction 6). Therefore, less substrate is decomposed. Schumb (1949) reported that the decomposition rate of H_2O_2 reaches the maximum at pH 3.5. With the progressive hydrolysis of ferric ions, the hydrolysis product provides a relatively large catalytically active surface for contact with H_2O_2. Therefore, the decomposition of H_2O_2 is accelerated, and more hydroxyl radicals are produced. Watts *et al.* (1990) also pointed out that the Fenton's reaction proceeds most effectively at pH 2-4. The acidic condition maintains iron stability and lowers the redox potential of the system, which promotes the most efficient generation of hydroxyl radicals (Watts *et al.*, 1990). Latt *et al.* (1999) investigated the effect of pH on the pseudo-first-order rate constant (k_{obs}) for the initial rate of decomposition of H_2O_2 by Fe (III). Their results indicate that k_{obs} reaches the maximum at pH 3.2, approximately 100 times the rate at pH 1, and 5 times the rate at pH 2. Therefore, the optimal pH value for the Fenton's reagent oxidation is generally around pH 3.

Degree of Mineralization

The chemical oxygen demand (COD) is used as an indicator for the degree of mineralization. The COD removal efficiency versus naphthalene degradation is shown in Figure 4. It should be mentioned that both H_2O_2 and Fe^{2+} can be oxidized by potassium dichromate, thus exerting a certain amount

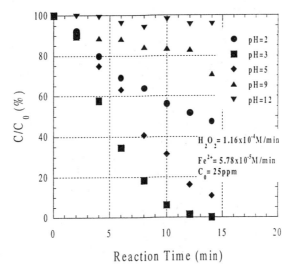

Figure 3. Effect of pH on the oxidation of naphthalene.

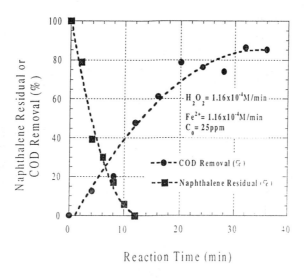

Figure 4. COD removal versus naphthalene degradation.

of COD. The corresponding COD equivalence is 0.471 mg O_2 / mg H_2O_2 and 0.143 mg O_2 / mg Fe^{2+}. The COD removal efficiencies in Figure 4 have been corrected by subtracting the contributions from H_2O_2 and Fe^{2+}. Results indicate that most of naphthalene can be oxidized to carbon dioxide and water. The COD removal rate lags behind the naphthalene degradation rate due to the formation of reaction intermediates. It also means that the intermediates are less reactive and possibly refractory toward hydroxyl radicals. When naphthalene is completely oxidized, only 50% COD is removed. The COD removal efficiency increases further when additional Fenton's reagent is added. The final degree of mineralization is around 85%. There is still 15% COD remaining in the solution that can not be further decomposed under the current experimental conditions.

Effect of Dosing Mode

The Fenton oxidation process is a competitive reaction system. Target organic contaminant, H_2O_2 and Fe^{2+} all compete for hydroxyl radicals. The dosing mode determines the time-dependent concentrations of H_2O_2 and Fe^{2+}, thus affecting the oxidation kinetics of the target organic compound. If H_2O_2 is singly added, there is an excess amount of H_2O_2 initially which will significantly compete for hydroxyl radicals. Likewise, if Fe^{2+} is singly added, the excessive Fe^{2+} will significantly compete for hydroxyl radicals during the initial phase of reaction. By comparing the second-order rate constants of H_2O_2 (2.7×10^7 $M^{-1} \cdot s^{-1}$) and Fe^{2+} (3×10^8 $M^{-1} \cdot s^{-1}$) toward hydroxyl radicals, it is seen that Fe^{2+} scavenges hydroxyl radicals approximately 10 times faster than H_2O_2. Figure 5 shows the oxidation kinetics of fluorene using various dosing modes, i.e., batch, pseudo-continuous and continuous modes. The total dosages of hydrogen peroxide and ferrous ion were maintained constant in all modes, i.e., 7.92×10^{-4} and 9.9×10^{-5} M ($[H_2O_2]/[Fe^{2+}]=8:1$), respectively. Results indicate that the oxidation kinetics varies significantly with different dosing modes.

In the batch mode, the largest initial oxidation rate of fluorene is observed. Since both H_2O_2 and Fe^{2+} were singly added at the beginning of reaction, most hydroxyl radicals were generated during the initial reaction period. This results in about 90% fluorene removal within the first minute. The solution became cloudy brown immediately, suggesting the rapid oxidation of ferrous to ferric ion by both hydrogen peroxide and hydroxyl radicals. As ferrous ion disappears rather quickly, the production of hydroxyl radicals dramatically decreases. Only additional 10% naphthalene is oxidized in the next 5 minutes. So, the batch mode is characterized by a rapid initial organic removal followed by a much slower successive removal.

In the pseudo-continuous mode, either H_2O_2 or Fe^{2+} was batch dosed while the other was continuously dosed. Results show that the oxidation kinetics differs markedly. When H_2O_2 is batch dosed, the oxidation rate lags behind when Fe^{2+} is batch dosed. The rate then reverses as reaction proceeds. The excessive Fe^{2+} initially present in the batch Fe^{2+} dosing mode efficiently produces hydroxyl radicals; then as Fe^{2+} becomes rapidly depleted, the reaction rate slows down. In contrast, in the batch H_2O_2 dosing mode, hydroxyl radicals are gradually produced and consumed. The slower reaction rate of H_2O_2 toward hydroxyl radicals makes a significant amount of residual H_2O_2 available throughout the reaction. Therefore, the latter system exhibits a higher organic removal efficiency.

In the continuous mode, the production rate of hydroxyl radicals was slower since both H_2O_2 and Fe^{2+} were continuously added. The competition from the Fenton's reagent is minimized. H_2O_2 and Fe^{2+} are available throughout the course of reaction with hydroxyl radicals being continuously produced. The relatively slow reaction rate is compensated by high removal efficiency.

Similar experiments were conducted to oxidize tetrachloroethylene (PCE) by the Fenton's reagent using different dosing modes (data not shown). Results indicate that while all PCE is removed in the continuous mode, only 75% removal efficiency can be obtained by the batch mode. Furthermore, the reaction stops immediately upon the dosing of the Fenton's reagent (within 0.5 minutes). However, in the fluorene oxidation experiments, reaction still proceeds after 0.5 minutes. This suggests different reaction mechanisms for PCE and PAHs. In PAHs oxidation, a small portion of ferric ion may be reductively converted to ferrous ion by some intermediate organic radicals, as shown by Reaction 5. In PCE oxidation, no intermediate organic radicals possess this ability. Additionally, it should be mentioned that most previous researchers used the pseudo-continuous mode, specifically, batch Fe^{2+} dosing and continuous H_2O_2 dosing. Results of our experiments, either in PCE or PAHs oxidation, have proved that this dosing mode yields the lowest organic removal rate. Therefore, in subsequent experiments, we employed the continuous mode for all selected PAHs.

Kinetics of PAHs Oxidation

The intrinsic reaction kinetics of organic substrate toward hydroxyl radicals can be described by a second-order expression:

$$-\frac{d[RH]}{dt} = k_{\text{int}}[RH][\cdot OH] \tag{8}$$

where RH represents organic substrate and k_{int} is the second-order intrinsic rate constant. Most k_{int} data have been supplied by radiation chemists (Walling, 1975). In contrast to most other oxidants, reactions of hydroxyl radicals with organic compounds containing unsaturated structures, such as double bonds and aromatic rings, generally proceed with rate constants approaching the diffusion-controlled limit ($\sim 10^{10} M^{-1} \cdot s^{-1}$). Many researchers (Sedlak and Andren, 1991; Dong, 1993; Huang et al., 1993) observed pseudo-first-order reaction kinetics for the oxidation of organic substrates by the Fenton's reagent, with batch Fe^{2+} dosing and continuous H_2O_2 dosing. In those cases, a steady-state assumption for hydroxyl radicals can be applied to the reaction system. Since the concentration of hydroxyl radicals remains constant during the course of reaction, one can obtain a pseudo-first-order kinetic expression (Equation 9) by transforming Equation 8:

$$-\frac{d[RH]}{dt} = k_{obs}[RH] \qquad (9)$$

$$k_{obs} = k_{int}[\cdot OH] \qquad (10)$$

From Equation 10, it is clear that the observed rate constant, k_{obs} depends on both k_{int} and the concentration of hydroxyl radicals. Though k_{int} is very large, k_{obs} is relatively small due to the extremely low concentration of hydroxyl radicals in the reaction system. In natural water systems, hydroxyl radical is produced by photolysis of nitrate ions (Schwarzenbach et al., 1993). Its concentration usually ranges from 10^{-18} to 10^{-16} M. From the viewpoint of engineering application, k_{obs} is more meaningful than k_{int} since hydroxyl radical concentration varies markedly with different reaction systems.

However, the concentration of hydroxyl radicals is no longer constant in the continuous dosing mode since both H_2O_2 and Fe^{2+} are continuously supplied to the reactor. Either H_2O_2 or Fe^{2+} must accumulate in the solution, thus increasing the generation rate of hydroxyl radicals. This in turn increases the concentration of hydroxyl radicals in the reaction solution. According to Reaction 1, the generation of hydroxyl radicals is first-order with respect to H_2O_2 and Fe^{2+}, individually. By simultaneously measuring the concentrations of H_2O_2 and Fe^{2+} at different reaction times in the fluorene oxidation experiments, we can achieve useful kinetics information. Figure 6 indicates that at a molar ratio ($[H_2O_2]/[Fe^{2+}]$) of 4:1, the concentration of Fe^{2+} remains constantly low ($3.0 \sim 3.5 \times 10^{-5} M$) during the course of reaction, while the concentration of H_2O_2 increases linearly with the reaction time (Figure 6 insert: $C_{H_2O_2} = 1.19 \times 10^{-4} \times t$, R = 0.989). The linear increase in H_2O_2 concentration will have the effect of increasing the generation rate of hydroxyl radicals.

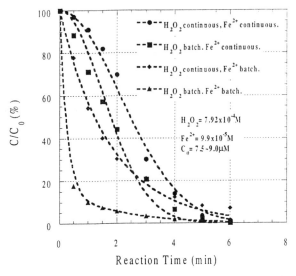

Figure 5. Effect of dosing mode on the oxidation of fluorene.

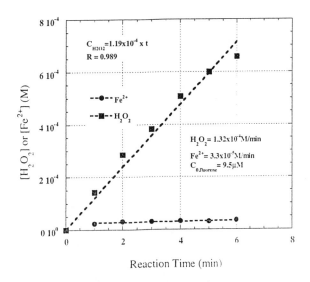

Figure 6. Concentration change of H_2O_2 and Fe^{2+} with reaction time in the oxidation of fluorene.

Therefore, we approximate that the concentration of hydroxyl radicals increases as a linear function of reaction time, i.e., $[\cdot OH] = at$. After substituting this equation into Equation 8, one gets:

$$-\frac{d[RH]}{dt} = ak_{int}t[RH] \qquad (11)$$

By merging a and k_{int}, and integrating Equation 11, one obtains:

$$[RH] = [RH]_0 \exp(-k_{obs}t^2) \qquad (12)$$
$$a = 2k_{obs} / k_{int} \qquad (13)$$

where a is the slope of the concentration increase of hydroxyl radicals (M·min⁻¹), k_{obs} is the observed rate constant (min⁻²), and $[RH]_0$ is the initial concentration of organic substrate (M). k_{obs} can be computed by model fitting of experimental data, and k_{int} is available in the literature. Thus, the slope a can be calculated from Equation 13. Consequently, the concentration of hydroxyl radicals can be calculated at any reaction time. Equation 12 can be rewritten in a more common form, i.e., $C = C_0 \exp(-k_{obs}t^2)$. This "time-squared" kinetic expression is used to fit our experimental data.

The results of naphthalene, fluorene, phenanthrene and fluoranthene oxidation at various $[H_2O_2]/[Fe^{2+}]$ molar ratios are shown in Figure 7. For the oxidation of naphthalene (Figure 7a), 600mL solution was used without initial headspace. The dosage rate of H_2O_2 was constant at 1.16×10^{-4} M/min. The dosage rate of Fe^{2+} was adjusted to obtain the desired $[H_2O_2]/[Fe^{2+}]$ molar ratios ranging from 1:1 to 1:1/16. For the oxidation of fluorene (Figure 7b), phenanthrene(Figure 7c) and fluoranthene (Figure 7d), 500 mL solution was used with 100mL headspace. H_2O_2 dosage was maintained at 1.32×10^{-4}M/min. Results indicate that as H_2O_2 dosage remains constant, the reaction rate increases with increasing dosage of Fe^{2+}. In Figure 7, the solid symbols represent experimental data points, and the dashed lines show the model fitting results using the "time-squared" kinetic equation. Results indicate that all experimental data in the continuous mode can be well-fitted by the "time-squared" model, especially at high Fe^{2+} dosages. The k_{obs} values at various Fe^{2+} dosages are summarized in Figure 8. Due to the high hydrophobicity of selected PAHs, the initial concentration of each reaction solution varied a little which would somewhat affect the k_{obs} values. Therefore, a plot of k_{obs} against $[Fe^{2+}]/[RH]_0$, is preferred to $[Fe^{2+}]$. Results in Figure 8 indicate that k_{obs} increases with increasing ratios of $[Fe^{2+}]/[RH]_0$. As the dosage of Fe^{2+} increases, more hydroxyl radicals are competitively consumed by Fe^{2+}. Thus,

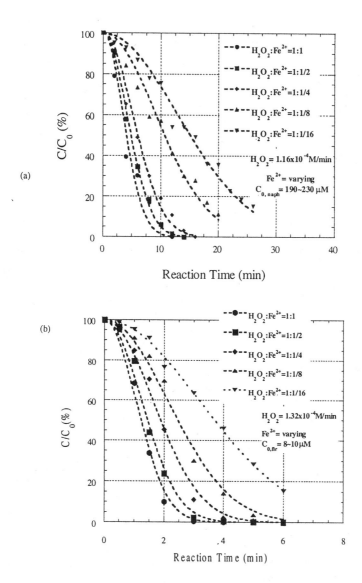

Figure 7. Oxidation of naphthalene (a), fluorene (b), phenanthrene (c) and fluoranthene (d).

Solid symbols: experimental data; dashed lines: "time-squared" model fitting.

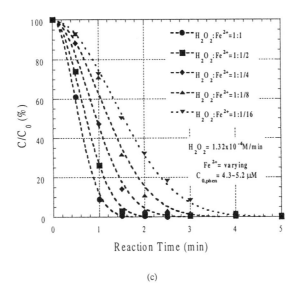

(c)

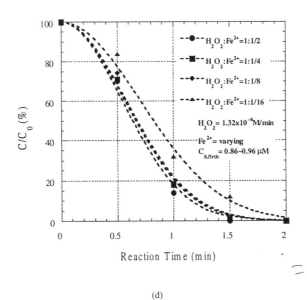

(d)

Figure 7. *Continued.*

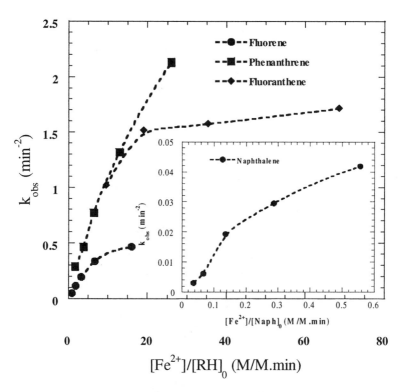

Figure 8. Plot of k_{obs} versus $[Fe^{2+}]/[RH]_0$.

the increment of k_{obs} slows down at high ratios of $[Fe^{2+}]/[RH]_0$. The intrinsic rate constant of PAHs toward hydroxyl radicals is about 10^{10} $M^{-1} \cdot s^{-1}$ (Watts, 1998). From the k_{obs}, one can calculate the a values via Equation 13, and then the time-dependent concentration of hydroxyl radicals using the linear equation, $[\cdot OH] = at$. For example, in the naphthalene oxidation experiments the concentration of hydroxyl radicals ranges from 2.1 x 10^{-13} to 2.8 x 10^{-12} M at the end of the reaction. It is clear that the Fenton oxidation system generates hydroxyl radicals more effectively than the natural water systems (10^{-18} to 10^{-16} M) by a factor of 4 to 5 orders of magnitude.

To compare the reaction rate constants of selected PAHs, a solution containing fluorene, phenanthrene, fluoranthene, pyrene and anthracene was prepared with a nearly identical concentration for each compound. Figure 9 shows the k_{obs} values of five PAHs compounds. Anthracene exhibits the largest k_{obs} value, about 4.5 times that of fluorene. The observed rate constants differ among different PAHs and follow the order: fluorene < phenanthrene < fluoranthene ≈ pyrene < anthracene. It is noted that the order of rate constant is in reverse order of water solubility. It has been reported that various PAHs compounds vary considerably in their reactivity toward ozone (Bailey, 1988). For example, the order of reaction rate of some PAHs compounds with ozone follows: anthracene > pyrene > phenanthrene (Razumovskii and Zaikov, 1971). This is in agreement with our results. Since both ozonation and Fenton oxidation are AOP processes that in principle use hydroxyl radicals as the major oxidant, these results are comparable.

Effect of Organic Solvent

As described above, the majority of PAHs oxidation experiments were carried out in organic solvent or in water/organic solvent mixture. However, the effect of organic solvent on the oxidation of PAHs has not been reported. Moreover, organic solvents are frequently employed to enhance the desorption of hydrophobic organic compounds, e.g., PAHs and PCBs, from soil particles in pump and treat, soil flushing and electro-kinetic processes (Khodadoust et al., 1999; Raghavan et al., 1991; Walker et al., 1998; Atalay et al., 1996). Since the Fenton oxidation process can be integrated with any of the above processes as a means to destroy hazardous organic contaminants on-site or off-site, the effect of organic solvent should be investigated. In this study, methanol and ethanol were selected as model organic solvents. Results in Figure 10 compare the degradation efficiencies of fluorene in three different media: water, water/methanol and water/ethanol. In the water medium, fluorene can be effectively oxidized. All fluorene is removed within 6 minutes. However, in the presence of methanol or ethanol, the removal efficiencies decrease to 30%

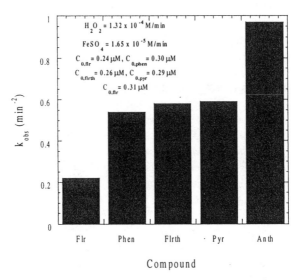

Figure 9. The observed rate constant, k_{obs} of PAHs.

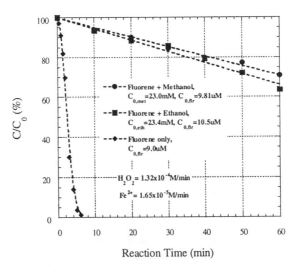

Figure 10. Effect of alcohol on the oxidation of fluorene.

and 35%, respectively, even after 1 hour of reaction. It is obvious that methanol or ethanol significantly inhibits fluorene oxidation due to strong competition for hydroxyl radicals. Walling *et al.* (1970) investigated the oxidation of alcohols (methyl, ethyl, isopropyl alcohols) by the Fenton's reagent, and found that all of them can be oxidized by hydroxyl radicals. They reported that the rate constants of methanol and ethanol are 1.2×10^9 and 2.1×10^9 $M^{-1} \cdot s^{-1}$, respectively (Walling, 1970). In our experiments, the initial concentration ratios of methanol/fluorene and ethanol/ fluorene range from 2,000 to 2,500. Though methanol and ethanol are less reactive than fluorene, they are able to capture most of hydroxyl radials due to their high concentrations. Figure 10 also indicates that methanol is a stronger hydroxyl radical scavenger than ethanol. Therefore, the oxidation of fluorene is inhibited more severely in the water/methanol media than in the water/ethanol media. Since organic solvent exerts significant influence on the degradation of PAHs, the rate constants obtained in the presence of organic solvent must be much smaller than those obtained in the water medium only.

Conclusion

The Fenton's reagent can effectively degrade all selected PAHs, i.e., naphthalene, fluorene, phenanthrene, fluoranthene, pyrene and anthracene. The optimal pH value is pH 3. The extent of mineralization exemplified by naphthalene is around 85%. A "time-squared" kinetic model, $C = C_0 \exp(-k_{obs} t^2)$, can be used to describe the degradation kinetics of selected PAHs in the continuous dosing mode. This model assumes that the concentration of hydroxyl radicals increases linearly with reaction time. The observed rate constant, k_{obs}, follows the order: fluorene < phenanthrene < fluoranthene $\approx$ pyrene < anthracene, which reverses the order of water solubility. Methanol or ethanol significantly inhibits the degradation of PAHs by competing for hydroxyl radicals.

Acknowledgement

This study was supported by the U.S. Department of Energy (Grant No. DE- FG07-96ER14716). Contents of this paper do not necessarily reflect the views of the funding agency.

References

1. Atalay, A.; Hwang, K. J. *Water Air and Soil Pollution*; **1996**, 90 (3-4), 451-468.
2. Bailey, P. S.; Kolsaker, P.; Sinha, B.; Ashton, J. B.; Dobinson, F.; Batterbee, J. E. *J. Org. Chem.*, **1964**, 29, 1400-1409.
3. Bailey, P. S. *Ozonation in Organic Chemistry*; Academic Press, Inc.: New York, NY, 1982; Vol. II, pp 46.
4. Buxton, G. V.; Greenstock, C. L.; Helman, W. P.; Ross, A. B. *J. Phys. Chem. Ref. Data*, **1988**, 17 (2), 513-886.
5. Destephen, R. A.; Benson, C. P. In *Hazardous and Industrial Wastes: Proceedings of the Twenty-sixth Mid-Atlantic Industrial Waste Conference*; Huang, C. P. Ed.; Technomic Publishing Company, Inc.: Lancaster, PA, 1994; pp 20-27.
6. Dong, C. Ph. D. Thesis, University of Delaware, Newark, DE, 1993.
7. Dorfman, L. M.; Adams, G. E. *National Bureau of Standards Report No. NSRDS-NBS-46*; U. S. Government Printing Office, Washington D.C., 1973; pp 59.
8. Haber, F.; Weiss, J. *J. Proc. R. Soc. London*, **1934**, A147, 332-336.
9. Hayek, N. A.; Dore, M. *Water Res.*, **1990**, 24 (8), 973-982.
10. Huang, C. P.; Dong, C.; Tang, Z. *Waste Management*, **1993**, 13, 361-377.
11. Keith, L. H.; Telliard W. A. *Environ. Sci. Technol.*, **1979,** 13 (4), 416-423.
12. Khodadoust A. P.; Sorial, G. A.; Wilson, G. J.; Suidan M. T.; Griffiths, R. A.; Brenner, R. C. *J. of Environ. Eng.-ASCE*, **1999**, 125 (11), 1033-1041.
13. Khodadoust A. P.; Suidan M. T.; Acheson, C. M.; Brenner, R. C. *J. of Hazardous Materials*, **1999**, 64 (2), 167-179.
14. Kuo, W. G. *Water Res.*, **1992**, 26 (7), 881-886.
15. Latt, J. D.; Gallard, H. *Environ. Sci. Technol.*, **1999**, 33 (16), 2726-2732.
16. Legube, B.; Guyon S.; Sugimitsu, H.; Dore, M. *Ozone: Sci. & Eng.*, **1983**, 5 (3), 151-170.
17. Legube, B.; Guyon S.; Sugimitsu, H.; Dore, M. *Water Res.*, **1986**, 20 (2), 197-208.
18. Menzie, C. A.; Potocki, B. B.; Santodonato, J. *Environ. Sci. Technol.*, **1992**, 26 (7), 1278-1284.
19. Miller, R. *Bioventing, Performance and Cost Summary*; U. S. Air Force Center for Environmental Excellence, Brooks AFB, TX, 1994.
20. Raghavan, R.; Coles E.; Dietz, D. *J. of Hazardous Materials*, **1991**, 26 (1), 81-87.
21. Razumovskii, S. D.; Zaikov, G.E. *Bull. Acad. Sci. USSR, Div. Chem. Sci. (Engl. Transl.)*, **1971**, 2524.
22. Sayre, I. M. *J. Am. Water Works Assoc.*, **1988**, 80 (1), 53-60.

23. Schumb, W. C. *Ind. Eng. Chem.*, **1949,** 41, 992.
24. Schwarzenbach, R. P.; Gschwend, P. M.; Imboden, D. M. *Environmental Organic Chemistry*; John Wiley & Sons, Inc.: New York, NY, 1993; pp 480-481, 621.
25. Sedlak, D. L.; Andren, A. W. *Environ. Sci. & Technol.*, **1991**, 25 (4), 777-782.
26. Spencer, C. J.; Stanton, P. C.; Watts, R. J. *Water Res.*, **1992**, 26 (7), 976-978.
27. *Standard Methods for the Examination of Water and Wastewater*; Eaton, A. D.; Clesceri, L. S.; Greenberg, A. E., Eds; 19th edition, American Public Health Association: Washington D.C., 1995; pp 3-68~3-70.
28. Sturrock, M. G.; Cline, E. L.; Robinson, K. L. *J. Org. Chem.*, **1963**, 28 (9), 2340-2343.
29. Trapido, M.; Veressinina, Y.; Munter, R. *Environ. Technol.*, **1995**, 16 (8), 729-740.
30. Walker, R. C.; Hofstee, C.; Dane, J. H.; Hill, W. E. *J. of Contaminant Hydrology*; **1998**, 34(1-2), 17-30.
31. Walling, C.; Kato S. *J. Am. Chem. Soc.*, **1970**, 93 (17), 4275-4281.
32. Walling, C. *Acc. Chem. Res.*, **1975**, 8 (4), 125-131.
33. Watts, R. J.; Udell, M. D.; Rauch, P. A.; Leung, S. W. *Hazardous Waste & Hazardous Materials*, **1990**, 7 (4), 335-345.
34. Watts, R. J. *Chemosphere*, **1991**, 23 (7), 949-955.
35. Watts, R. J.; Kong, S; Dippre, M; Barnes, W. T. *Journal of Hazardous Materials*, **1994**, 39 (1), 33-47.
36. Watts, R. J. *Hazardous Wastes: Sources, Pathways, Receptors;* John Wiley & Sons, Inc.: New York, NY, 1998; pp 69, 360.

Geochemistry

Chapter 14

Interaction of Water and Aqueous Chromium Ions with Iron Oxide Surfaces

G. E. Brown, Jr.[1,2], S. A. Chambers[3], J. E. Amonette[3], J. R. Rustad[3],
T. Kendelewicz[1], P. Liu[1], C. S. Doyle[1], D. Grolimund[1],
N. S. Foster-Mills[3], S. A. Joyce[3], and S. Thevuthasan[3]

[1]Surface and Aqueous Geochemistry Group, Department of Geological
and Environmental Sciences, Stanford University, Stanford, CA 94305–2115
[2]Stanford Synchrotron Radiation Laboratory, SLAC, Stanford, CA 94309
[3]Environmental Molecular Sciences Laboratory, Pacific Northwest National
Laboratory, 908 Battelle Boulevard, Richland, WA 99352

To gain a more fundamental understanding of abiotic
processes controlling reduction reactions of aqueous chromate
and dichromate ions ($Cr(VI)_{aq}$) in subsurface environments,
we carried out molecular-level experimental and modeling
studies of the interaction of water and $Cr(VI)_{aq}$ with well-
characterized single crystal samples of synthetic and natural
hematite and magnetite. A reductionist approach was adopted
in which simplified model systems of increasing complexity
were studied. Photoemission spectroscopy (PES), photoelectron diffraction, and vacuum STM were used to
characterize the composition, atomic structure, and morphology of clean surfaces of $\alpha\text{-}Fe_2O_3(0001)$ and $Fe_3O_4(100)$
grown by molecular beam epitaxy on single crystal substrates
of $\alpha\text{-}Al_2O_3(0001)$ and $MgO(100)$, respectively. A simple
surface autocompensation model is adequate to predict the
stable termination(s) of these surfaces. In addition, interlayer
relaxations were experimentally determined and found to
compare favorably with predictions from molecular
simulations for hematite (0001), but not for magnetite (100).
Similar synthetic model surfaces, as well as clean (0001) and
(1-102) surfaces of natural hematites and clean (100) and
(111) surfaces of natural magnetites were reacted with
controlled, sequential doses of water vapor (ranging from 10^{-9}

to 10 torr p(H$_2$O) at 3 min. exposure times) under UHV conditions. These surfaces were characterized following each dose using O 1s PES. In each case, a threshold p(H$_2$O) of 10^{-4} to 10^{-3} torr was observed, below which water vapor reacted primarily with defect sites and above which water vapor reacted with terrace sites to produce extensive surface hydroxylation. The kinetics of reduction of Cr(VI) to Cr(III) and the nature of the reacted surface layer on surfaces initially containing Fe(II) were determined by Cr and Fe L-edge and O K-edge x-ray absorption and Cr and Fe 2p and O 1s PES on clean natural magnetite (111) samples reacted with 50 µM Na$_2$CrO$_4$ solutions for 5 to 120 minutes. Reduction rates were found to be pH dependent, with the fastest rates occurring at pH values below the point of zero charge of magnetite (6.6). At the highest rate, the surface redox reaction is ~95% complete within 10 minutes. The reacted magnetite surfaces were found to consist of a thin (15±5 Å) layer of poorly crystalline Cr(III)-(oxy)hydroxide that passivates the surface and prevents further reduction of Cr(VI)$_{aq}$. Iron in the surface region of the magnetite substrates was simultaneously oxidized to Fe(III). The reaction rates and products are similar to those previously observed for the reaction of Cr(VI)$_{aq}$ with "zero-valent" iron. Grazing-incidence XAFS spectroscopy on Cr(III)$_{aq}$- and Cr(VI)$_{aq}$-reacted hematite (0001) and reduced hematite (0001) surfaces under ambient conditions showed that the reduction products, Cr(III)-(oxy)hydroxide surface complexes, are bound dominantly as inner-sphere edge-sharing, tridentate complexes on the oxidized surface. Competitive effects of the common aqueous oxoanions HPO$_4^{2-}$ and SO$_4^{2-}$ on Cr(VI)$_{aq}$ sorption on iron oxides were examined by measuring the extent of uptake and uptake kinetics of Cr(VI)$_{aq}$ reacting with hematite powder in the presence of these other oxoanions using laser photoacoustic spectroscopy and batch uptake experiments. Sorption of Cr(VI)$_{aq}$ and HPO$_4^{2-}$ occurs even at low pH where monovalent anions predominate in solution. As a consequence, the fraction of total species in solution available in the form of divalent anions is a more important determinant of the relative amount of Cr(VI)$_{aq}$ or phosphate sorbed than the intrinsic selectivity of the anions for the surface. In mixed-anion systems typical of groundwater, significantly different sorption affinities may be observed than predicted solely on the basis of relative single-anion sorption constants.

Redox-sensitive inorganic and organic contaminants present in sediments at DOE sites can be altered or destroyed by electron transfer reactions occurring abiotically on mineral surfaces. One such contaminant is chromium, which can exist in 6+, 3+, 2+, and 0 oxidation states (*1*), depending upon conditions. Over the pH range of most natural groundwaters, Cr(III) is thermodynamically more stable as a sorbed surface complex or a solid (oxy)hydroxide precipitate than as an aqueous complex (*2*). Therefore, the small environmental hazard posed by Cr(III) tends to be localized to the region where it has sorbed or precipitated. In contrast, $Cr(VI)_{aq}$ is thermodynamically stable as anionic solution species, in the absence of a reductant such as Fe(II), over the same pH range where Cr(III) species tend to precipitate. These differences make $Cr(VI)_{aq}$ the more mobile and, potentially, the more bioavailable of the two most common oxidation states of chromium. $Cr(VI)_{aq}$, is also the more toxic form of Cr (*3*).

A fundamental understanding of abiotic surface redox processes provided by molecular-level studies on structurally and compositionally well-defined mineral surfaces is needed to improve predictive models of the fate and transport of chromium in geochemical systems and to optimize manipulation of these processes for remediation purposes. Toward this end, we have undertaken a program of research designed to answer the following questions:

(1) What are the structures of clean iron oxide surfaces that might react with $Cr(VI)_{aq}$?
(2) How does water interact with these surfaces?
(3) What are the reaction products when $Cr(VI)_{aq}$ interacts with reduced hematite and magnetite surfaces, and what effect do they have on the Cr(VI) to Cr(III) reduction process?
(4) What are the kinetics of this reduction reaction and how are they affected by solution pH and the build-up of reaction products on the surface?
(5) Do other common aqueous oxoanions such as HPO_4^{2-} and SO_4^{2-} interfere with $Cr(VI)_{aq}$ sorption on iron oxides?

We have utilized both synthetic and natural hematites and magnetites in this work and have characterized their surfaces using photoemission spectroscopy, LEED, RHEED, photoelectron diffraction, and vacuum STM. The clean model surfaces were reacted with water vapor under UHV conditions and the reaction products were characterized by O 1s photoemission. Molecular simulations were used to predict the structures of hematite and magnetite surfaces before and after interaction with water. Reaction products resulting from the interaction of $Cr(VI)_{aq}$ with iron oxide surfaces were studied under ambient conditions using grazing-incidence Cr K-edge XAFS spectroscopy and under UHV conditions using Cr and Fe L-edge and O K-edge x-ray absorption and O 1s, Cr 2p, and Fe 2p photoemission. Reaction kinetics were also studied by monitoring changes in the Cr 2p, Fe 2p, and O 1s photoemission spectra. Finally, the competitive effects of aqueous HPO_4^{2-} and SO_4^{2-} on CrO_4^{2-} sorption on powdered hematite were studied by laser photoacoustic spectroscopy and batch uptake experiments.

Epitaxial Growth and Characterization of Model Iron Oxide Surfaces by Oxygen-Plasma-Assisted Molecular Beam Epitaxy

In order to produce the full range of iron oxide phases and crystal surface orientations required for studies of $Cr(VI)_{aq}$ reduction reactions, we used oxygen-plasma-assisted molecular beam epitaxy (OPA-MBE) to prepare well-defined epitaxial films ($\approx$ 300-500 Å thick) on two kinds of oxide substrates – α-Al_2O_3 for α-Fe_2O_3 (*4-10*) and MgO for Fe_3O_4, α-Fe_2O_3 and $Fe_{1-x}O$ (*4,7,8,10-13*). The lattice mismatches for α-Fe_2O_3 on α-Al_2O_3, Fe_3O_4 on MgO, α-Fe_2O_3 on MgO and $Fe_{1-x}O$ on MgO are +5.8%, −0.38%, −0.89%, and +2.30%, respectively, after taking into account the factor-of-two difference in lattice parameter between Fe_3O_4/α-Fe_2O_3 and MgO. Fe_3O_4 can be used as an interlayer between $Fe_{1-x}O$ and MgO to distribute the strain (*12*). $Fe_3O_4(111)$ has been grown on α-$Al_2O_3(0001)$, although the sizeable lattice mismatch (+8.1%) results in the nucleation of rotational domains with equal populations (*4*). It is, in principle, possible to grow Fe_3O_4, α-Fe_2O_3 and $Fe_{1-x}O$ on MgO(110), although facetting on this surface results in poor film quality and rough surfaces. The Fe_3O_4/MgO and α-Fe_2O_3/MgO interfaces are thermodynamically unstable, leading to cation interdiffusion. The full range of iron oxide stoichiometries can be reached in OPA-MBE by adjusting the relative Fe and activated oxygen fluxes at the substrate. Figure 1 shows an empirically derived phase diagram that illustrates these values (*12*). Points on the border between Fe_2O_3 and Fe_3O_4 represent growths that resulted in a mixed phase in the film.

The relatively large compressive lattice mismatch for α-Fe_2O_3 on α-Al_2O_3 leads to the formation of a heavily strained wetting layer of fully stoichiometric α-Fe_2O_3 that is buckled along the [1-102] direction, provided the first several tens of Å's are grown at an exceedingly slow rate (~1 Å/min) at a substrate temperature of 450°C (*9*). This strained layer then transforms to 3-D nanocrystals above a coverage of a few ML. A higher initial growth rate (such as 6 Å/min) kinetically impedes the critically important transition from strained layer-by-layer to relaxed 3-D island formation that leads to the nucleation of well-ordered epitaxial α-$Fe_2O_3(0001)$. Instead, a γ-Fe_2O_3–like epilayer is nucleated and exhibits domains with 180° rotational twinning and surface orientations that deviate substantially from (111) (*14*). The formation of single-phase α-Fe_2O_3 does not occur once this γ-Fe_2O_3–like layer has nucleated. These 3-D islands, or nanocrystals, gradually coalesce with film thickness into what looks by all probes except high-resolution x-ray diffraction to be a single crystal with a reasonably flat surface. STM images of a partially conductive α-$Fe_2O_3(0001)$ film that was doped with Fe(II) (not shown) reveal typical terrace widths of a few tens of Å's, and terraces that are separated by $\approx$ 2 Å high steps. This step height is the difference between structurally equivalent planes in the α-$Al_2O_3(0001)$ structure. Thus, the surface consists of a single termination.

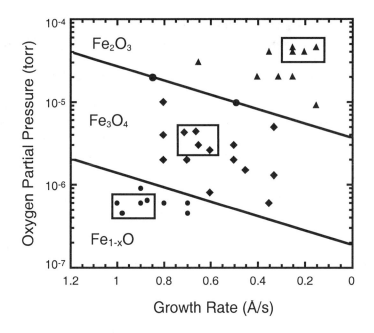

Figure 1. Phase diagram for the growth of iron oxides by OPA-MBE

X-ray photoelectron diffraction results established that the terminating plane is that of a single 1/3 ML of Fe, which is the only autocompensated and, therefore, stable termination of α-Fe_2O_3(0001) (*15*). This result is at odds with that of Wang et al., whose spin-DFT calculations and STM images suggest that both the 1/3 ML Fe and O terminations can be stabilized on this surface (*16,17*). MBE growth of Fe_3O_4(100) on MgO(100) has produced surfaces that are readily imaged by STM with atomic resolution due to the high electrical conductivity of Fe_3O_4 at room temperature (*18-24*). However, there has been some controversy and ambiguity concerning the interpretation of these images. For instance, Gaines et al. observed elongated images at the corners of a primitive square that these authors interpreted as being due to tetrahedral Fe(III) dimers in the terminal layer (*21*). Similar elongated images were seen earlier on the surface of natural Fe_3O_4(100) crystal by Terrach et al. (*25*). However, this surface structure requires a full ML of tetrahedral Fe(III). Such a surface is not autocompensated and, therefore, has a non-zero surface dipole and the associated instability. In contrast, we did not see evidence for elongation in the Fe atom images and thus interpreted them as being due to single tetrahedral Fe(III) cations in the top layer

(*23*). This interpretation requires that the top layer be composed of 1/2 ML of tetrahedral Fe(III) in which every other Fe is missing. This surface structure constitutes a ($\sqrt{2}$x$\sqrt{2}$)R45° surface net, which matches both RHEED and LEED patterns. Such a surface is autocompensated and has no dipole moment. These results suggest that the most stable surface is that terminated with tetrahedral Fe(III). However, Voogt first suggested that the octahedral Fe/tetrahedral O terminal layer would also be autocompensated and, therefore, stable if there were a modification in the distribution of Fe oxidation states in the octahedral plane, and an ordered array of O vacancies (*26*). Subsequently, Stanka et al. observed that such a surface can be routinely stabilized with MBE-grown Fe$_3$O$_4$(100) specimens that are transferred through air and cleaned under somewhat oxidizing conditions (*24*). Figure 2 shows STM images of Fe$_3$O$_4$(100) illustrating the two terminations – (a) tetrahedral Fe, representative of the as-grown surface (*23*) and, (b) octahedral Fe/tetrahedral O termination on specimens transferred under vacuum from the MBE chamber to an STM chamber at Tulane University in the laboratory of Professor Ulrike Diebold (*24*). It is currently not understood why the different terminations seem to arise rather exclusively under different preparation conditions. What can be said at this point is that both terminations are autocompensated and, therefore, expected to be stable in vacuum.

Interaction of Water Vapor and Bulk Water with Iron Oxide Surfaces

Prior to exposure of the hematite and magnetite surfaces to Cr(VI)$_{aq}$, we undertook a study of the interaction of deionized water vapor with the clean surfaces of hematite (0001) and magnetite (100) and (111) using O 1s photoemission on wiggler beam line 10-1 at the Stanford Synchrotron Radiation Laboratory (SSRL). The purpose of these experiments was to gain insights about the electronic structure and nature of the water-exposed vs. clean magnetite and hematite surfaces without the complication of Cr(VI)$_{aq}$ ions. The hematite samples were (0001) surfaces of thin-film α-Fe$_2$O$_3$ ($\approx$ 300 Å thick) grown on an α-Al$_2$O$_3$ (0001) substrate by the MBE methods described earlier, whereas the magnetite (100) samples were cut and polished (using 300 µm grit α-alumina) from natural single crystal magnetites. The magnetite (100) surfaces were cleaned using several cycles of Ar$^+$ sputtering and annealing in vacuum at 700°C for 1 to 10 hours. These surfaces showed normal LEED patterns following cleaning and no XPS evidence of adventitious carbon. The hematite (0001) surfaces were cleaned by several cycles of annealing in 10^{-5} torr oxygen at 500-800°C for 30 min., without Ar$^+$ sputtering, and showed good 1 x 1 LEED patterns following cleaning and no XPS evidence of adventitious carbon. The water dosing experiments were conducted by exposing each surface to sequential doses of water vapor at pressures, p(H₂O), ranging from 10^{-9} torr to 10 torr, at a constant reaction time (3 min.) and, in separate experiments, at

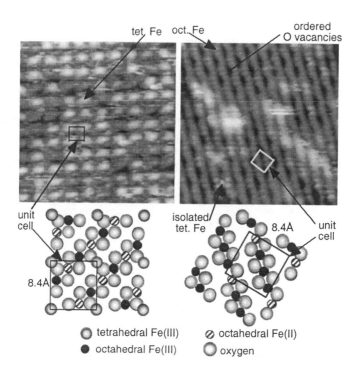

Figure 2. STM images of (√2x√2)R45° Fe₃O₄/MgO(100) grown by OPA-MBE showing: (a) the tetrahedral Fe termination, and (b) the octahedral Fe/tetrahedral O termination, with an ordered array of oxygen vacancies.

longer reaction times (≤ 30 min., with total exposures ranging from 1.8 Langmuir (1L = 10^{-6} torr sec) to 1.8 x 10^{10} L) at constant p(H₂O). The doubly deionized water used in these experiments was degassed by several freeze-pump-thaw cycles and was introduced into the UHV chamber using a precision leak valve. These water-dosing experiments were conducted in a preparation chamber attached to our main UHV analysis chamber, which operated at a base pressure of 1 x 10^{-10} torr. The O 1s photoemission spectra were obtained using a 1000 line/mm grating, entrance and exit slits of 50 μm, and incident photon energies of 630 or 650 eV, which are near the minimum of the universal curve, thus are very surface sensitive. Additional details about these experiments can be found in (27) for the magnetite surfaces and in (28) for the hematite surfaces.

Changes in the electronic structure of magnetite (100) and (111) surfaces were examined after reaction with water vapor at $p(H_2O)$ ranging from 10^{-9} to 9 torr and with liquid water at 298 K using chemical shifts in the O 1s core-level photoelectron spectra measured on BL 10-1 at SSRL. Oxygen 1s photoemission data are shown for the magnetite (100) surface in Figure 3 and consist of a lattice oxygen feature at higher kinetic energy and a second feature shifted by 1.2 to 2.5 eV to lower kinetic energy. Similar results were obtained for the magnetite (111) surface. We attribute this lower kinetic energy feature to OH groups resulting from dissociative chemisorption of water on the magnetite surface. The range of chemical shifts observed is attributed to a range of surface hydroxyl sites with different binding energies (27). We found that the $p(H_2O)$ at which water first reacts with magnetite is similar for the two surfaces ($\leq 10^{-5}$ torr for 3 min. exposures, corresponding to doses of $\leq 1.8 \times 10^3$ L) and is consistent with a small sticking coefficient and hydroxylation of defect sites. The $p(H_2O)$ for the onset of an extensive hydroxylation reaction is $\approx 10^{-3}$ Torr (3 min. dose or $\approx 1.8 \times 10^5$ L). Magnetite (100) and (111) surfaces exposed to higher $p(H_2O)$ react more extensively, with hydroxylation extending several layers (≈ 8 Å) deep into the bulk. Comparison of O KVV Auger K-edge absorption spectra of water vapor-exposed magnetite (100) and (111) surfaces with the corresponding total yield spectra of goethite (α-FeOOH), limonite (FeOOH • nH$_2$O), and hematite (α-Fe$_2$O$_3$) (not shown) shows that the reaction product on the magnetite surfaces is not goethite, limonite, or hematite (27). In addition, similarity of the Fe $L_3M_{23}M_{23}$ Auger yield L-edge absorption spectra before and after exposure of the magnetite (111) surface to liquid water (not shown) indicates that the oxidation state of iron is unchanged (27). We also measured O 1s chemical shifts on magnetite (111) surfaces that had been immersed in liquid water. These spectra also showed evidence for extensive surface hydroxylation.

Annealing experiments to temperatures of 700°C did not cause significant loss in intensity of the OH photoemission feature from the water-dosed magnetite surfaces. These results indicate that the hydroxyls are strongly bound to the surface. We also observed a shift in the centroid of the OH component to higher kinetic energies with annealing, and interpret this result as indicating a loss of the most weakly bound surface hydroxyls (27).

Our O 1s photoemission results for hematite(0001) are similar to those for the two magnetite surfaces. A two-stage reaction was observed for the hematite (0001) surface, with dissociative chemisorption of water occurring mainly at defect sites below a threshold pressure of $\approx 10^{-4}$. The percentage of defects on this surface estimated from the low water dosage experiments is 5 - 10%. Extensive hydroxylation of this surface was found to occur above the threshold pressure. Longer water vapor exposures below the threshold pressure did not result in significant increases in hydroxylation; however, longer exposures above the threshold pressure resulted in increased hydroxylation. The measured

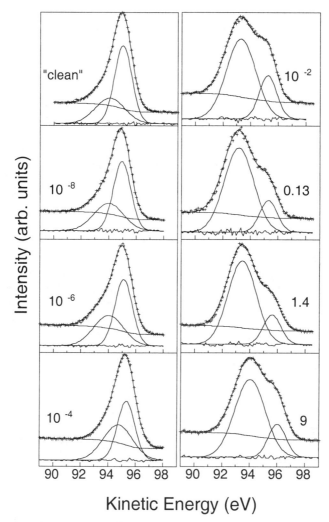

Figure 3. Oxygen 1s photoemission spectra of the (100) magnetite surface, using 630 eV incident photons, before and after sequential doses of water vapor, at the doses indicated in torr (3 min. exposures at each dose). Reproduced from reference 27, Elsevier Science, 2000.

threshold pressure for hematite is more than five orders of magnitude lower than the threshold pressure for the conversion of hematite to α-FeOOH or Fe(OH)$_3$ predicted from equilibrium thermodynamics (**28**). This difference between observation and prediction for hematite (0001) is not completely understood, but may be due to the presence of a passivating layer of Fe-(oxy)hydroxide that reduces the surface energy of the hydroxylated hematite (0001) surface (**28**).

These results, coupled with those from our studies of the interaction of water with MgO(100) (**29,30**), and α-Al$_2$O$_3$(0001) (**28**) and with molecular mechanics studies of the interaction of water molecules with MgO (**31**) and α-Al$_2$O$_3$ (**32**), lead to the suggestion that extensive dissociative chemisorption of water on terrace sites on these surfaces is, in part, a function of the water coverage. The "threshold" p(H$_2$O) values observed for these metal oxide surfaces (10^{-4} to 10^{-3} torr) correspond to coverages of about 0.5-0.6 monolayers, which is about the same coverage found in the theoretical studies that resulted in significant interaction between adjacent water molecules. Such intermolecular interactions appear to be necessary for the dissociation of water on these surfaces.

Molecular Simulations of Iron Oxide Surfaces Before and After Interaction with Water

In an effort to simulate the surface structures of clean and hydroxylated magnetite and hematite, molecular modeling calculations were carried out, and the results are compared with those from our photoemission and x-ray photoelectron diffraction studies discussed in previous sections.

Potential Model

The model for pair potentials employed in our molecular simulations is an ionic one that was originally designed to calculate structures and energies for hydroxylated/hydrated ferric oxide surfaces. The O-H potential functions were taken from the polarizable, dissociating water model of Halley and co-workers (**33**), which is a variant of the Stillinger-David model (**34**). The Fe-O parameters, including a short-range repulsion and charge-dipole cutoff functions, were fit to the Fe^{3+}-H$_2$O potential surface of Curtiss and co-workers (**35**). This model has been used in a variety of applications including simulations of ion hydrolysis in solution (**36**), ferric oxide and (oxy)hydroxide crystal structures (**37**), the vacuum termination of hematite (0001) (**38**), monolayers of water on hematite (1-102) (**39,40**), and the surface charging behavior of goethite (α-FeOOH) and hematite (**41,42**).

Simulation of the Hydroxylation of the Hematite Surfaces

In our simulation of the hematite (0001) surface (*42*), excellent predictions were obtained for surface relaxation in good agreement with experiment (*6*) and later high-level LAPW *ab initio* calculations (*43*). The surface energy from our simulations also compared favorably with the LAPW *ab initio* calculations. The extent of hydroxylation of the hematite (1-102) surface was also in good agreement with experimental studies. The adsorption energy for water as a function of extent of hydroxylation, taken from a search of over 1200 configurations is shown in Figure 4.

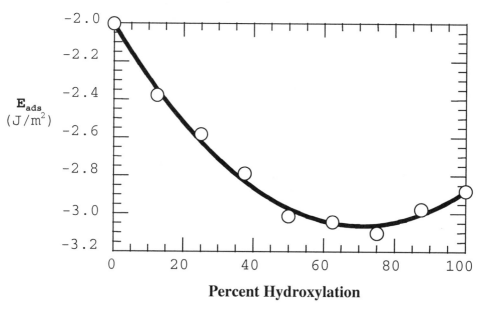

Percent Hydroxylation

Figure 4. Calculated adsorption energy of water molecules at 100 % coverage on the (1-102) surface of hematite as a function of the amount of hydroxylation. Adapted from reference 40, Elsevier Science, 1999.

Simulation of Magnetite Surfaces

More recently, the same potentials were applied to magnetite (*44*). In this work, it was shown that a reasonable structural model for magnetite and wustite could be obtained simply by changing the Fe charge to 2.5+ or 2+, keeping the same short-range repulsive and charge-dipole cutoff function parameters. As is the case for the ferric oxide structures, the Fe-O bond was found to be approximately 5 percent too long. For example, the Fe^{2+}-O distance in $Fe(H_2O)_6^{2+}$ was predicted to be 2.21 Å, as opposed to the experimental value of 2.15 Å. High-level quantum mechanical calculations give 2.19 Å (*45*). It appears that going beyond our rather crude description of the Fe-O bond may be difficult for *ab initio* methods. Given our good agreement with experiment as outlined above, and in view of the electronic structural complexity of magnetite (resulting in formidable challenges in applying *ab initio* methods), it is justifiable to continue to explore the predictions of this simple model.

Simulations on the "A" Termination of Magnetite (100)

The relaxed structure for the "A" termination of magnetite (100) was calculated in (*44*). The surface is predicted to undergo significant surface relaxation involving the rotation of the two-fold coordinated irons into the adjacent octahedral vacancies exposed at the surface. At the same time the bulk tetrahedral ferric ions, which share a face of their coordination polyhedron with the surface half-octahedron newly occupied by the relaxing surface ferric irons, are themselves pushed to the surface, yielding a sequence of ferric "dimers" as shown in Figure 5. The driving force for the relaxation is the reestablishment a Pauling bond order of 2.0 v.u. for the surface oxygens, some of which have become significantly over and undercoordinated at the unrelaxed surface. This is essentially the same reasoning as that discussed earlier involving the autocompensation concept. The relaxation energy associated with the reconstruction is approximately 0.72 J/m^2, which is quite significant given that the unrelaxed surface energy is about 2.3 J/m^2. The dimers formed on the surface provide a compelling interpretation of STM images showing "dimeric" structures.

Simulations on the "B" Termination of Magnetite (100)

The "B" or octahedral termination of magnetite (100) has been discussed in (*26*), and predictions of its structure stability were made in (*46*). The simplest way to envision this termination is (within each unit cell):

1) remove 1/2 e^- from two of the $Fe^{2.5+}$ in the octahedral sites to give Fe^{3+} sites.
2) place the electron on the tetrahedral Fe^{3+} site to make an Fe^{2+} site.
3) remove FeO from the system by taking the tetrahedral Fe^{2+} site and creating an oxygen vacancy.

The surface is neutral; stoichiometric FeO was removed. The "B" termination is therefore oxidized relative to Fe_3O_4 and the relative stabilities of the "A" and "B" terminations will depend on the oxygen fugacity in the system. Note that both surfaces maintain the $(\sqrt{2}\times\sqrt{2})R45°$ unit cell characteristic of the neutral stacking units. This cell is in fact observed in LEED patterns of magnetite (100) for both "A" and "B" terminations (**11,47**).

If the "B" surface is generated from the "A" surface using the three-step recipe listed above, the octahedral sites are treated as a mixture of equal numbers of $Fe^{2.5+}$ and Fe^{3+} sites, as shown in Figure 5. Maintaining the policy, for the present, of keeping the electrons delocalized, this mixture of $Fe^{2.5-}$ and

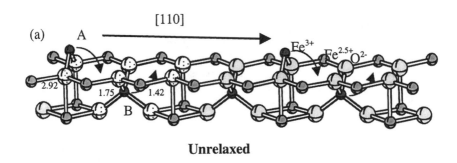

Unrelaxed

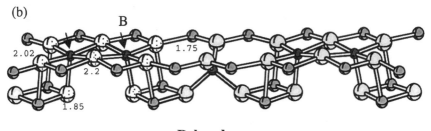

Relaxed

*Figure 5. Atomic rearrangements during relaxation of the vacuum-terminated magnetite (100) surface. Pauling bond strengths in valence units (v.u.) are given on the figure for the "active" oxygen atoms. In the bulk, the net contribution to each oxygen atom is 2 v.u. Reproduced from reference **44**, Elsevier Science, 1999.*

Fe^{3+} sites is treated as if all surface irons have a charge of +2.75. Each of the three different possibilities for creation of the oxygen vacancy was tested. In the lowest energy structure, the O atom adjacent to the vacancy moves into a bridge position with respect to the two Fe^{3+} irons at the surface. This small relaxation contributes less than 0.1 J/m^2 to the slab energy. The $(\sqrt{2} \times \sqrt{2})R45°$ symmetry is maintained during the surface relaxation.

Relative Energetics of the "A" and "B" Terminations of Magnetite (100)

Vacuum Terminated Surfaces

In (*46*) the relative stabilities of the "A" and "B" terminations of magnetite (100) were assessed. It was necessary to model the surface with a charge-ordered slab, rather than using a non-integral charge on the octahedral sites, to maintain consistency with energy calculations on Fe^{3+} and Fe^{2+} solids. For example, the energy of charge-ordered magnetite (with Fe^{3+} and Fe^{2+} B sites) will always be lower than the energy of charge-disordered magnetite (with $Fe^{2.5+}$ B sites). This arises from the lack of quantum mechanical effects favoring charge delocalization (such as electron kinetic energy). The model system described here therefore differs from that described in (*44*) in that the system is charge-ordered. Because no attempt was made to self-consistently calculate the charge distribution, the simple charge-ordering scheme in Hamilton (*48*) was used, with Fe^{3+} and Fe^{2+} alternating in layers perpendicular to [100]. In either the "A" or "B" terminations, the terminating surface could consist of either Fe^{2+} or Fe^{3+} ions. Because, within the context of our model, the Fe^{2+} ions prefer to be at the surface, we choose the former arrangement. For this arrangement, the relaxation of the surface tetrahedral iron atoms is the same as in the charge-delocalized arrangement as described in (*44*). The structure of the "B" termination is also essentially unchanged from the charge-disordered case. We note that if the higher energy Fe^{3+} termination is used to model the surface, the relaxation mechanism identified in (*44*) does not occur due to crowding of ferric iron in the first layer.

For non-stoichiometric systems, calculation of the surface energies cannot be carried out in the standard way, according to the formula $\gamma = 1/2A(E_{slab}-E_{bulk})$, where A is the area of the slab, E_{slab} is the energy of the slab and E_{bulk} is the bulk energy of an equivalent number of formula units in the bulk. The issue, of course, is that for the nonstoichiometric "B" surface, there is no "bulk" value against which to reference the slab energy. A similar problem was addressed by Wang and coworkers (*43*) in their study of nonstoichiometric terminations of hematite (001). In (*46*) a similar approach was taken, but was modified to overcome the constraints of using the ionic model. Energies for O_2 or metallic iron cannot be calculated with an ionic model, and the oxygen fugacity was fixed through calculation of the energies of hematite and wustite and knowledge of the experimental oxygen fugacities at the hematite-magnetite, and metastable magnetite-wustite, and hematite-wustite buffers. This allows calculation of the surface energy of the model slab as a function of oxygen pressure (*44*) and, hence, of the relative energies of the "A" and "B" terminated slabs over the entire range of accessible conditions. As shown in Figure 6, the "A" termination is stable relative to the "B" termination over the entire range of P_{O2} examined. As a caveat, we remark that this calculation will depend to some extent on the possibility of rearranging charge in going from the "A" to the "B" termination. We have ignored this possibility.

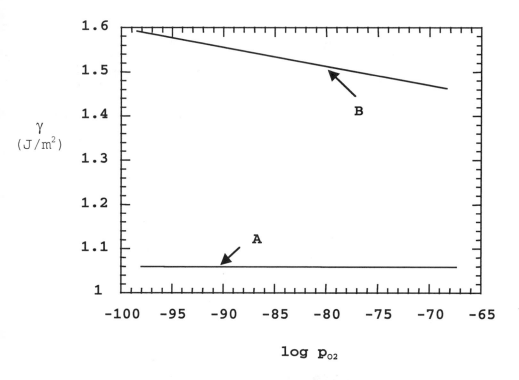

Figure 6. Relative energies of the "A" and "B" terminated magnetite (100) surfaces using the charge ordering scheme from (48) with Fe^{2+} ions in the surface layer.

Hydroxylated Surfaces of Magnetite (100)

In (*46*) calculations were carried out on three sets of hydroxylated slabs, including both the relaxed and unrelaxed "A" terminations, and the "B" termination. For the "A" termination, four waters were added per unit cell to the octahedral sites and two waters per unit cell were added to the tetrahedral sites. An exhaustive search through the possible surface tautomers yielded the structure shown in Figure 7a. An analogous investigation for the "B" terminated surface yielded the structure shown in Figure 7b. Total water binding energies for both surfaces are about 2.32 J/m^2, indicating that the water will have little effect, at least in a thermodynamic sense, on which surface is observed. It was also shown in (*46*) that hydroxylation stabilizes the unrelaxed magnetite (100) structure as opposed to the relaxed structure as identified in (*44*).

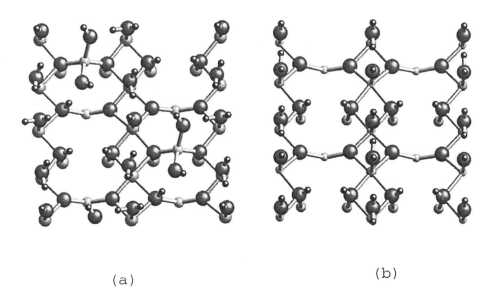

(a) (b)

Figure 7. Minimum energy tautomeric forms for the hydroxylated "A" (a) and "B" (b) terminations of the magnetite (100) surface. Note that in the minimum energy structure for "A", tetrahedral sites revert back to the unrelaxed configuration (see Fig. 5).

Chromium Sorption Products on Reduced Hematite and Magnetite Surfaces

One of the main objectives of our DOE-EMSP-funded research was to gain an improved molecular-level understanding of the reduction of $Cr(VI)_{aq}$ on the surfaces of Fe(II)-containing oxides under near-neutral pH conditions. In acidic solutions, $Cr_2O_7^{2-}$ is the dominant $Cr(VI)_{aq}$ species, whereas in basic solutions, CrO_4^{2-} is dominant (*49*). A number of past studies have shown that Cr(VI) can be reduced to Cr(III) on such oxides via abiotic pathways (e.g., (*50-68*)). Eary and Rai (*53*) reviewed some of the most commonly used abiotic reductants and their efficiencies in reducing $Cr(VI)_{aq}$ to Cr(III) under a range of conditions in aqueous wastes. The rates of these reactions are similar to or faster than those involving biotic pathways, thus abiotic reactions could provide dominant pathways for chromate reduction (*69*). For example, Fendorf et al. (*69*) recently summarized rate data for the reduction of $Cr(VI)_{aq}$ to Cr(III) by various bacteria, organic reductants, $Fe(II)_{aq}$, and Fe(II)-containing minerals, and pointed out that magnetite has a rate similar to that of *Desulfovibrio vulgaris* (the most rapid biological reductant of $Cr(VI)_{aq}$ (*70*)). Thus, the surfaces of magnetite grains in sediments and soils can act as reductants for $Cr(VI)_{aq}$, with rates that are competitive with those of *D. vulgaris* at near-neutral pH, and maybe faster, thus more dominant, at lower pH (*69*).

Cr K-edge grazing-incidence (GI)-XAFS and Cr L-edge NEXAFS spectroscopy were used to investigate the types of surfaces complexes formed when aqueous Cr(VI) and Cr(III) at μM and mM concentrations react, respectively, with partially reduced (containing Fe^{2+}) and unreduced (containing only Fe^{3+}) single crystal (0001) hematite surfaces. Partially reduced hematite surfaces were produced by annealing hematite (0001) at ~500°C in vacuum. Partially reduced hematite was exposed to a 5 mM Na_2CrO_4 solution (pH 6.0, 0.1 M $NaNO_3$) for 10 min. in a N_2 atmosphere, then immediately transfer-red to the UHV chamber for Cr L-edge NEXAFS data collection. Subsequently, the same sample was investigated by GI-XAFS under ambient conditions. In addition, unreduced thin-film, MBE-grown hematite(0001) was exposed to a 10^{-5} M $Cr(NO_3)_3$ solution (pH 4.8) in a N_2 atmosphere and analyzed by GI-XAFS under ambient conditions. Experimental conditions (pH, total Cr concentrations, and background electrolyte) were chosen to avoid the formation of multinuclear Cr complexes or supersaturation of Cr species in solution with respect to known hydroxides, carbonates, or basic salts (*49,71*). In addition, attempts were made to minimize the possibility of photochemical reduction of $Cr(VI)_{aq}$ during the experimental procedure. X-ray photoelectron spectroscopy was used to ensure surface cleanliness prior to the experiment and to estimate surface coverage after reaction, which is < 0.1 ML. Based on this estimated coverage and the fact that the reduced hematite (0001) surfaces have thin rafts of magnetite (111) structure (*72*), the samples examined in our GI-XAFS experiments have far less extensive reaction of $Cr(VI)_{aq}$ on the surface than the $Cr(VI)_{aq}$-reacted natural magnetite (111) samples characterized using core-level PES and L-edge absorption spectroscopy (see next section). This provides an opportunity to examine the initial stages of $Cr(VI)_{aq}$ interactions with magnetite (111) surfaces.

GI-XAFS experiments were performed at SSRL on wiggler magnet beam line 6-2, and NEXAFS experiments were performed on wiggler magnet beamline 10-1, with the SPEAR ring operating at 3 GeV and 60-100 mA. The SSRL grazing-incidence apparatus was used to collect GI-EXAFS data in the specular geometry with the incident x-ray angle set slightly below the critical angle of the hematite substrate ($\approx 0.2°$). Analysis of the GI-XAFS data was performed using EXAFSPAK (**73**); k^3-weighted EXAFS were fit over a k-range of $\approx$ 3-11 Å^{-1}. Phase and amplitude functions were calculated with FEFF7 (**74**) and verified by comparison with model mineral compounds. The accuracy of the optimized parameters was estimated based on fits of crystalline model compounds (first shell: N $\pm15\%$ and R ±0.03 Å; more distant shells: N $\pm30\%$ and R ±0.07 Å). The Debye-Waller term (σ^2) for each shell was estimated based on fits of both model compounds and single-shell backtransforms and was kept fixed during the final fitting procedure.

The L-edge spectrum of the Cr(VI) sorption sample (Fig. 8a) reveals features characteristic of both Cr(III) (cf. chromite) and Cr(VI) species (cf. crocoite), indicating that both Cr oxidation states are simultaneously present at the partially reduced hematite-solution interface in a ratio of ~25% Cr(III)-75% Cr(VI). The observed Cr(III) is thought to be the result of electron transfer from three Fe(II) in the partially reduced hematite to produce one Cr(III) at the hematite-aqueous solution interface.

We also carried out a Cr K-edge NEXAFS analysis of the same sample after it had been analyzed by XPS, and the result is shown in Figure 8b. The intensity of the pre-edge feature (Fig. 8b) suggests ~80% Cr(III) and ~20% Cr(VI) based

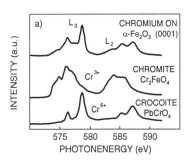
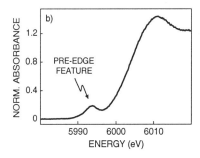

Figure 8. a) Normalized Cr L-edge NEXAS spectra of a Cr(VI)-exposed, partially reduced hematite sample compared to Cr(VI) and Cr(III) model compounds, b) Corresponding Cr K-edge NEXAFS spectrum. Reproduced from reference **75***. International Union of Crystallography, 1999.*

on the calibration spectra of Peterson et al. (*63*). The apparent difference between these two experiments is most likely due to further reduction of Cr(VI) on the partially reduced hematite (0001) surface during XPS analysis, which involved the use of a low energy electron flood gun to charge neutralize the surface. Thus the Cr(VI):Cr(III) ratio from the L-edge spectrum is more reliable for this Cr(VI)/hematite sorption sample. The simultaneous presence of both oxidation states at the interface, even after extended exposure times, points to a limited reducing capacity of the partially reduced hematite which is depleted during the course of the experiment.

Figure 9 shows the GI-EXAFS spectra and resulting Fourier transform (FT) of an unreduced hematite (0001) surface exposed to a Cr(III)-containing solution (Fig. 9, top) in comparison with a partially reduced sample exposed to Cr(VI)$_{aq}$ (Fig. 9, bottom). Despite the different reaction pathways, the least-squares fitting results in a similar local structure of the observed Cr surface species for both samples. The first feature in the FT is due to Cr-O correlations while the second FT feature is most likely due to Cr-Cr correlations (see discussion below). In addition, there is evidence for a more distant (~3.5 Å) metal-metal shell (Cr-Fe or Cr-Cr), which, however, is difficult to resolve unambiguously. This third shell is best fit by a Cr-Fe correlation, and could be evidence for direct bonding of Cr(III) to the hematite surface in a tridentate fashion. It is important to note that the GI-XAFS spectra taken with the E-vector parallel to the hematite surface should be particularly sensitive to Cr-metal correlations parallel to the hematite surface, and much less sensitive to Cr-metal correlations perpendicular to the surface. GI-XAFS spectra taken with the E-vector perpendicular to the hematite (0001) surface are needed to confirm the presence of Cr-Fe$_{surface}$ correlations. None of the hypothetical surface complexes representing monomer adsorption is consistent with the local structure obtained in the GI-XAFS analysis (*75*). Consequently, we propose a multinuclear surface complex rather than a mononuclear complex, which is often assumed in surface complexation modeling (e.g., (*76*)). The observed metal-metal distance of ~3.0 Å is consistent with edge-sharing metal-oxygen (MeO$_6$) octahedra with either Fe or Cr as the central metal atom (*77*). The fitted number of second-shell metal neighbors is approximately two, suggesting that the Cr-metal clusters are fairly limited in size or are arranged in a disordered fashion. The lack of more pronounced next-nearest neighbor metal shell features is consistent with this finding and further emphasizes the limited size of the proposed "two-dimensional" multinuclear species on partially reduced hematite (0001).

The different metal backscatterer correlations can not be identified unambiguously due to the similarity of the Cr and Fe phase shift and amplitude functions, respectively. Nevertheless, the GI-XAFS information obtained regarding the local structure of the Cr species provides relatively restrictive geometrical constraints due to the polarized nature of the spectra. Multinuclear Cr complexes, in particular dimers and trimers sorbed in a two-dimensional fashion to the atomically flat hematite (0001) surface, are consistent with the observed spectra. The more distant metal-metal shell at 3.5 Å is consistent with

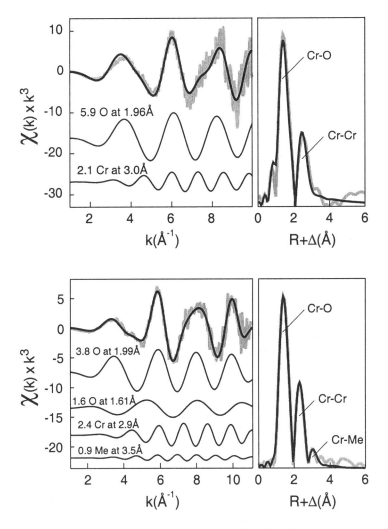

*Figure 9. Grazing-incidence Cr K-edge XAFS data (gray) of chromium sorbed on α-Fe₂O₃ single crystals (0001). Top: Unreduced hematite (0001) reacted with Cr(III)aq. Bottom: Partially reduced hematite (0001) exposed to Cr(VI)aq. Least squares fits of the EXAFS including shell-by-shell deconvolution and Fourier transforms (uncorrected for phase shift) are shown. Reproduced from reference **75**. International Union of Crystallography, 1999.*

Fe backscatterers from the hematite (0001) surface arranged in a somewhat disordered fashion. In this case, the Cr-Fe bonds would have a significant component in the E-vector parallel spectra. In addition, evidence for a true Cr(III) sorption complexes on the surface of reduced hematite comes from a consideration of solution conditions, which do not favor multinuclear species or precipitates [see (75)].

The existence of Cr at the hematite-water interface in multinuclear complexes even at a surface coverage of < 0.1 monolayer was not anticipated based on the findings of earlier EXAFS studies of metal ion surface complexation as a function of metal loading. For example, the EXAFS studies of Chisholm-Brause et al. (78) and Towle et al. (79,80) found evidence for mononuclear complexes of Co(II) on alumina at the lowest surface coverages examined (~ 0.05 μM/m^2), with multinuclear complexes or precipitates forming at higher surface coverages. Nevertheless, our suggestion of multinuclear Cr(III) surface complexes is consistent with observations elaborated in previous studies involving Cr(III) sorption on iron oxides (77,81).

The Nature of the Reacted Layer and Kinetics of Cr(VI)$_{aq}$ Reduction Reaction on Magnetite Surfaces

The grazing-incidence XAFS study of Cr(VI)$_{aq}$ sorption on reduced hematite (0001) surfaces discussed above provided information about the structure, composition, and mode of attachment of Cr(III) sorption complexes on this surface at submonolayer Cr surface coverages. To obtain quantitative information about the composition, structure, and thickness of the overlayer produced by the reaction of Cr(VI)$_{aq}$ with magnetite, we used a combination of soft x-ray core-level photoemission and adsorption spectroscopies (82). Samples were prepared by reacting 50 μM aqueous Na$_2$CrO$_4$ solutions at pH 6 and 8.5 with clean surfaces of magnetite(111) prepared under UHV conditions. The spectra were measured on beam line 10-1 at SSRL. Chromium 2p photoemission and Cr L-edge absorption (Fig. 10) show that tetrahedrally coordinated Cr(VI)$_{aq}$ is reduced by a heterogeneous redox process to octahedral Cr(III) on the magnetite (111) surface. The thickness of the reacted overlayer at pH 6 was found to increase with immersion time in the chromate solution for up to $\approx$ 10 min. and to remain unchanged for longer doses. The reaction rate at pH 6 was found to be initially fast and to follow a logarithmic law (Fig. 11). At saturation, the passivating overlayer is 15±5 Å thick and consists of chromium (oxy)hydroxide with only trace amounts of iron.

Evidence for extensive hydroxylation in the overlayer was provided by a chemically shifted component in the O 1s photoemission (Fig. 12) and O K-edge absorption spectra taken with surface-sensitive O KVV Auger yield (feature A in Fig. 13). The overlayer appears to lack long-range order based on loss of the first EXAFS-like feature in the O K-edge spectrum with increasing dosing time (feature B in Fig. 13). Clear evidence for oxidation of Fe(II) to Fe(III) in the surface region of the magnetite substrate during reduction of Cr(VI)$_{aq}$ to Cr(III) was provided by Fe 2p photoemission and Fe L-edge absorption spectra (not shown). Strong attenuation of the Fe 2p signal during the first 10 minutes of the redox reaction indicated that iron does not outdiffuse to the overlayer. At pH 8.5 the reaction follows a similar path, but its rate is slower and Cr(VI) reduction continues for immersion times of up to one hour (Fig. 11). This difference is due in part to the lower affinity of the CrO_4^{2-} and $Cr_2O_7^{2-}$ oxoanions at pH 8.5 vs. pH 6 due to the fact that the magnetite surface is negatively charged at

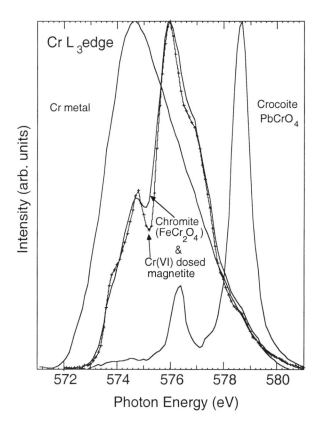

*Figure 10. Cr L₃-edges of Cr(VI)$_{aq}$-reacted magnetite (111) compared with the Cr L₃-edges of Cr metal, chromite, and crocoite. Reproduced from reference **82**, Elsevier Science, 2000.*

pH values above its point of zero charge, 6.6 (*83*), and positively charged below the pH$_{pzc}$.

The results of this study compare well with those from an earlier XPS study of the passivation of "zero-valent" iron by Cr(VI)-containing solutions (*84*). The resulting overlayer compositions, thicknesses, and reaction rates are very similar for the two systems. Once the magnetite or "zero-valent" iron surface is passivated by a relatively thin layer of disordered Cr-(oxy)hydroxide, the electron transfer reaction responsible for the reduction of Cr(VI)$_{aq}$ should stop.

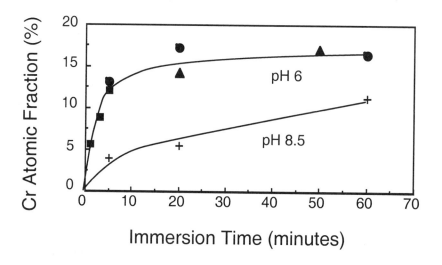

Figure 11. Atomic fraction of Cr for the overlayers studied plotted as a function of immersion time in 50 μM sodium chromate solutions at pH 6 and pH 8.5. These results were obtained from the intensities of the Cr 2p, Fe 2p, and O 1s core levels. Reproduced from reference 82, Elsevier Science, 2000.

Cr(VI)$_{aq}$ Sorption on Hematite Surfaces in the Presence of Phosphate and Sulfate

Powdered hematites present several crystalline faces for surface reactions and thus represent a system that is intermediate in complexity between the single-crystal surfaces described above and the heterogeneous mixture of surfaces found in soils and sediments. We conducted several experiments with hematite powder to understand the thermodynamics and kinetics of Cr(VI)$_{aq}$

sorption, with particular emphasis placed on competitive sorption by $Cr(VI)_{aq}$, phosphate, and sulfate oxoanions.

Batch sorption experiments with 20 mg of hematite powder (specific surface of 9.7 m^2g^{-1} by BET-N_2) in a 15 mL solution were conducted at room temperature with initial chromate, phosphate, or sulfate concentrations ranging from 2.7 to 54 μM (i.e., 140 ppb to 2.8 ppm). Preliminary experiments

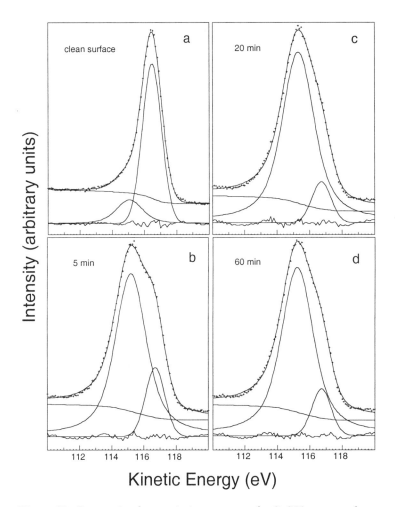

*Figure 12. Oxygen 1s photoemission spectra of a $Cr(VI)_{aq}$-reacted magnetite (111) sample as a function of dosing time. Reproduced from reference **82**, Elsevier Science, 2000.*

established that equilibration times of one hour were sufficient, and the concentrations of the anions remaining in solution after this time were

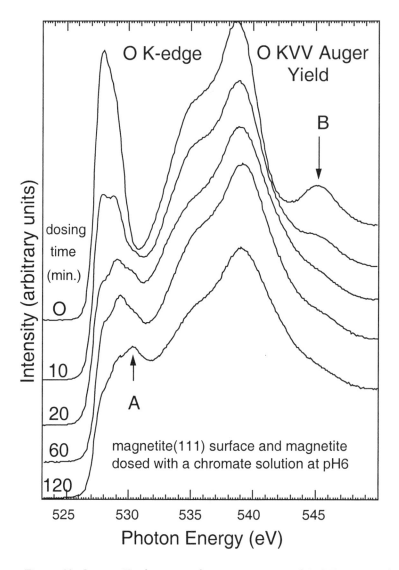

*Figure 13. Oxygen K-edge x-ray absorption spectra of Cr(VI)$_{aq}$-reacted magnetite(111) as a function of dosing time (in minutes). Reproduced from reference **82**, Elsevier Science, 2000.*

determined by ion chromatography. The solutions were not buffered for pH to eliminate interferences from the anions associated with buffering agents, and as a consequence, final sorption pH's ranged from 6.0 to 6.8. Despite the range in final pH, the results of these sorption experiments (Fig. 14) could be described well by the linearized Langmuir isotherm equation:

$$C_{final}/C_{sorbed} = (1/C_{max})C_{final} + 1/(K_{ads} C_{max})$$

where C_{final} and C_{sorbed} are the equilibrium concentrations in solution and on the hematite surface, respectively, C_{max} is the maximum sorption capacity of the mineral under the conditions of the experiment, and K_{ads} is the Langmuir adsorption constant.

Values for K_{ads}, which is a surface-complexation constant and thus a measure of the specific binding affinity of a surface site for the anion, were 0.27,

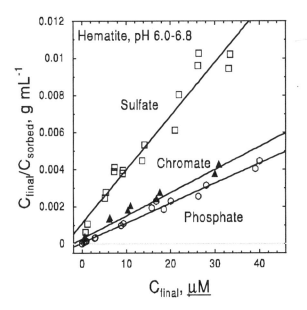

Figure 14. Linearized Langmuir isotherm plots of sorption of mono-anionic solutions of sulfate, Cr(VI)$_{aq}$, and phosphate on hematite powder.

0.25, and 2.4 for sulfate, Cr(VI)$_{aq}$, and phosphate, respectively. These results can be interpreted as suggesting an outer-sphere (i.e., hydrated anion) binding mechanism for sulfate and Cr(VI)$_{aq}$, and an inner-sphere mechanism for phosphate. Maximum sorption capacities of the hematite for sulfate, $Cr(VI)_{aq}$, and phosphate were 3.5, 8.3, and 9.3 μmol g^{-1}, respectively, indicating that more sorption sites for $Cr(VI)_{aq}$ and phosphate were available than for sulfate. The low C_{max} values for sulfate could be related to the absence of a protonated species, and suggest that hydrogen bonding of $Cr(VI)_{aq}$ and phosphate to neutral oxygen surface sites could increase the total number of sites available for sorption of these anions. In fact, C_{max} is strongly correlated with pK_2, the negative log of the second acid dissociation constant for each anion (Fig. 15).

For competitive sorption between these anions, however, other factors such as aqueous speciation must also be considered. Simple electrostatic considerations suggest that divalent anionic species will be more strongly attracted to positively charged surface sites than monovalent species. When both types of species are present in solution and in excess of the available surface sites, divalent species (i.e., SO_4^{2-}, CrO_4^{2-}, $Cr_2O_7^{2-}$, and HPO_4^{2-}) should dominate the observed sorption behavior. This premise can be tested by competitive sorption experiments because the amounts of divalent anion available for sorption will vary differently for each anion when pH's close to the

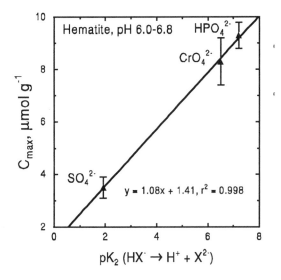

Figure 15. Correlation between observed C_{max} values for sorption by mono-anionic solutions of sulfate, $Cr(VI)_{aq}$, and phosphate on hematite powder and the acidities of the anions.

pK_2 values are selected. Although experiments at pHs near the pK_2 of sulfate are impractical, such experiments with $Cr(VI)_{aq}$ and phosphate are both practical and directly relevant to groundwater chemistry because the pK_2 values for these anions are near 7.

The expectation is that the relative sorption of $Cr(VI)_{aq}$ and phosphate would be directly proportional to the aqueous concentrations of $Cr(VI)_{aq}$ and HPO_4^{2-}. Speciation calculations show that the ratio of these two species in solution (i.e., $HPO_4^{2-}/Cr(VI)_{aq}$) varies from about 0.2 to 1 in going from pH 4 to 9 (Fig. 16, solid curve) and suggest that sorption of $Cr(VI)_{aq}$ would be favored over phosphate over much of this range. The linear relationship between C_{max} and pK_2 suggests that the relative numbers of available sites for $Cr(VI)_{aq}$ and HPO_4^{2-} are constant regardless of pH, provided that the same sorption mechanisms are involved. Combining the speciation and site-availability data, we can estimate the relative amount of total phosphate needed to achieve 50% sorption [i.e., equivalent sorption with $Cr(VI)_{aq}$] by

$$[P_{total}/Cr(VI)_{total}]50\% = [C_{max}(Cr(VI))/C_{max}(P)][CrO_4^{2-}/HPO_4^{2-}].$$

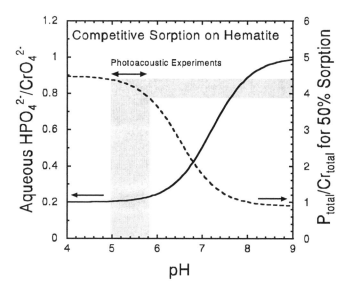

Figure 16. Influence of pH on aqueous speciation of phosphate and Cr(VI)$_{aq}$ (solid line-left axis) and the predicted ratio of total phosphate to Cr(VI)$_{aq}$ needed to achieve equal sorption on hematite powder (dashed line-right axis).

This result is also plotted in Figure 16 (dotted curve, right axis). On this basis, we would predict that about 4.4 times as much phosphate as $Cr(VI)_{aq}$ would be needed at pH 5 for equal sorption by the two anions on powdered hematite, whereas at pH 8 the same concentration of phosphate as $Cr(VI)_{aq}$ would yield equal sorption. The prediction at pH 5 is in reasonable agreement with the results of competitive sorption experiments carried out at low ionic strength using laser photoacoustic spectroscopy (LPAS, shown by the shaded portion of the graph). Experiments carried out at pH 5 (± 0.1) in 10 mM KCl to minimize pH drift, however, routinely require ratios of phosphate to $Cr(VI)_{aq}$ near 7 for equivalent sorption to occur. A 40-fold excess of phosphate completely inhibits $Cr(VI)_{aq}$ sorption under these conditions.

The competitive sorption results are different when the phosphate contacts the hematite after the $Cr(VI)_{aq}$ has reacted with it. A delay in phosphate addition results in incomplete desorption of Cr(VI). Between 25 and 50% of the Cr(VI) seems irreversibly bound to the hematite if 15-30 minutes elapse between initial sorption and the addition of phosphate. This result suggests substantial alteration or reorganization of the Cr(VI) once sorbed. Examples of the types of mechanisms that could be responsible for this behavior include conversion from monodentate to bidentate sorption, coalescence/polymerization of isolated chromate ions on the surface, or conversion from outer- to inner-sphere bonding.

Using the same approach as just outlined for phosphate to predict competitive sorption between sulfate and $Cr(VI)_{aq}$ suggests that at pH 5, a $S_{total}/Cr(VI)_{total}$ ratio of about 0.07 is sufficient to achieve 50% sorption, largely because at this pH only about 3% of the Cr(VI) is in the form of CrO_4^{2-}, whereas all the sulfate is divalent. Experimental verification of this prediction, however, shows that a $S_{total}/Cr(VI)_{total}$ ratio of about 500 is required for equivalent sorption of $Cr(VI)_{aq}$ and sulfate. Clearly, different mechanisms are involved in sorption of $Cr(VI)_{aq}$ and sulfate, the most likely difference being a different suite of available sorption sites and the possible stabilizing behavior from hydrogen bonding of Cr(VI) and surface oxygens. Delayed addition of sulfate after $Cr(VI)_{aq}$/hematite reaction yielded results similar to those for phosphate.

Based on these results, which will be reported in detail elsewhere (*85*), we can draw the following conclusions. Sorption of $Cr(VI)_{aq}$ and phosphate species on hematite is dominated by divalent anions even at low pHs where monovalent anions predominate in solution. As a consequence, aqueous speciation (i.e., the fraction of the total species in solution available in the form of divalent anions) is a more important determinant of the relative amount of $Cr(VI)_{aq}$ or phosphate sorbed than the intrinsic selectivity of the anions for the surface. Although completely divalent, sulfate is not very competitive with $Cr(VI)_{aq}$ for sorption sites on hematite, probably as the result of their preference for different types of sorption sites and the potential for hydrogen bonding with $Cr(VI)_{aq}$ species. Delayed addition of phosphate or sulfate after initial $Cr(VI)_{aq}$ sorption, however, shows that some type of rearrangement of the Cr(VI)-surface bond occurs in the first 15-30 minutes after sorption, which renders a large fraction of the Cr(VI) tightly bound to the surface. In mixed-anion systems typical of groundwater, significantly different sorption affinities may be observed than predicted solely on the basis of relative single-anion sorption constants.

An important outcome of this work is the development of a powerful new tool with which competitive adsorption of anions on minerals in suspension can be investigated in real time. We have demonstrated that nontoxic anions present in the subsurface environment, such as phosphate, may compete with toxic anions, such as $Cr(VI)_{aq}$, for sorption sites on redox-active minerals to an extent that is highly pH dependent. Significantly, LPAS can be used for studies of pure mineral phases, such as hematite in the present experiments, as well as for multi-component soil mixtures. Thus, controlled and "real-world" experiments can be carried out to determine the extent to which competing nontoxic anions can prevent adsorption and reduction of toxic anions by minerals or redox-active remediation agents, such as "zero-valent" iron.

Conclusions

The following conclusions can be drawn from our study of the interaction of $Cr(VI)_{aq}$ with magnetite and reduced hematite surfaces:

1. Metal oxide-aqueous solution interfaces are extremely complex and poorly understood at the molecular level. Spectroscopic and molecular modeling studies of the reaction of aqueous sorbates with well-characterized, single crystal, metal oxide surfaces can provide some of this understanding.
2. Oxygen-plasma-assisted MBE permits the growth of iron oxides with well-defined stoichiometries and crystal orientations for use in studies of the interaction of water and $Cr(VI)_{aq}$ with iron oxides.
3. The geometric and electronic structures of clean iron oxide surfaces are better understood than the water-exposed surfaces. In both cases, significant but different reconstructions occur. The structure of the vacuum-iron oxide interface is different from that of the solution-iron oxide interface.
4. Water dissociates and chemisorbs on iron oxide surfaces at ambient temperature, initially on defect sites, then at a threshold $p(H_2O) > 10^{-4}$ torr on terrace sites, resulting in ≈ 0.5 ML of hydroxyls. At $p(H_2O) \geq 10^{-2}$ torr, ≈ 1 ML of hydroxyls is found on hematite (0001) and magnetite (111).
5. XAFS and photoemission spectroscopies clearly show that $Cr(VI)_{aq}$ is reduced to Cr(III) on hydrated iron oxide surfaces that contain Fe^{2+}, but not on hydrated surfaces that contain only Fe^{3+}.

6. Cr and Fe L-edge XAFS and Cr 2p and Fe 2p photoemission spectroscopy show that the magnetite surface is quickly passivated, and the Cr(VI)→Cr(III) reduction reaction ceases after the formation of an electrically insulating layer of 15±5 Å in thickness. This finding has major implications for the use of magnetite, other Fe(II)-containing solids, or "zero-valent" iron for remediating Cr(VI)$_{aq}$-contaminated groundwater.

7. The passivating layer on magnetite consists initially of a poorly crystalline Cr-(oxy)hydroxide phase on a thin Fe(III)-oxide substrate. Comparison of Cr 2p, Fe 2p, and O 1s photoemission data for Cr(VI)$_{aq}$-reacted magnetite (this study) and "zero-valent" iron (*84*) suggests that similar phases form on "zero-valent" iron after reaction with Cr(VI)$_{aq}$.

8. The Cr(VI)→Cr(III) reduction reaction goes to completion on hydrated magnetite surfaces in ≈ 10 minutes at pH 6.0 and within ≈ 1 hour at pH 8.5, the difference being related to the higher affinity of CrO_4^{2-} and $Cr_2O_7^{2-}$ oxoanions for the magnetite surface below its point of zero charge (pH 6.6), where it is positively charged.

9. Common soil anions have some potential to interfere with Cr(VI)$_{aq}$ uptake on iron oxides in soils and sediments. Photoacoustic spectroscopy and batch adsorption isotherm measurements indicate the following:
   - Sulfate does not interfere significantly with Cr(VI)$_{aq}$ sorption on iron oxides until present in 500-fold excess.
   - Phosphate species requires a 4- to 7-fold excess to interfere significantly with Cr(VI)$_{aq}$ sorption below pH 6, but will interfere at equal concentrations above pH 7.

10. The fraction of total species in solution available in the form of divalent anions is a more important determinant of the relative amount of Cr(VI)$_{aq}$ or phosphate sorbed on iron oxides than the intrinsic selectivity of the anions for the surface.

11. Sorbed Cr(VI) species are much more difficult to remove by competing anions when these anions are added after Cr(VI)$_{aq}$ has reacted with hematite than if they are added concurrently with Cr(VI)$_{aq}$.

Acknowledgments

We wish to acknowledgment the Department of Energy Environmental Management Science Program for financial support through Grants DE-FG07-97ER14842 (Stanford University) and DE-AC06-76RLO 1830 (PNNL). Thomas Trainor is thanked for his help with the grazing-incidence XAFS experiments on Cr(III) sorbed on hematite. We also wish to thank the staff of the Stanford Synchrotron Radiation Laboratory for their technical support of the experimental work involving synchrotron-based spectroscopy. SSRL is supported by the Department of Energy and the National Institutes of Health.

References

1. Greenwood, N.N.; Earnshaw, A. *Chemistry of the Elements*. Pergammon Press, Oxford, U.K., 1984.
2. Richard, F.C.; Bourg, A.C.M. *Water Res.* **1991**, *25*, 807.
3. Gauglhofer, J.; Bianchi, V. In: *Metals and Their Compounds in the Environment*, E. Merian, ed., VCH, Weinheim, Germany, 1991.
4. Gao, Y.; Kim, Y.J.; Chambers, S.A.; Bai, G. *J. Vac. Sci. Technol.* **1997**, *A15*, 332.
5. Yi, S.I.; Liang, Y.; Chambers, S.A. *J. Vac. Sci. Technol.* **1999**, *A17*, 1737.
6. Thevuthasan, S.; Kim, Y.J.; Yi, S.I.; Chambers, S.A.; Morais, J.; Denecke, R.; Fadley, C.S.; Liu, P.; Kendelewicz, T.; Brown, G.E., Jr., *Surf. Sci.* **1999**, *425*, 276.
7. Chambers, S.A.; Kim, Y.J.; Gao, Y. *Surf. Sci. Spectra* **1998**, *5*, 219.
8. Kim, Y.J.; Gao, Y.; Chambers, S.A. *Surf. Sci.* **1997**, *371*, 358.
9. Yi, S.I.; Liang, Y.; Chambers, S.A. *Surf. Sci.* **1999**, *443*, 212.
10. Gao, Y.; Chambers, S.A. *J. Cryst. Growth* **1999**, *174*, 446.
11. Chambers, S.A.; Joyce, S.A. *Surf. Sci.* **1999**, *420*, 111.
12. Gao, Y.; Kim, Y.J.; Chambers, S.A. *J. Mater. Res.* **1998**, *13*, 2003.
13. Gao, Y.; Kim, Y.J.; Thevuthasan, S.; Chambers, S.A.; Lubitz, P. *J. Appl. Phys.* **1997**, *81*, 3253.
14. Yi, S.I.; Liang, Y.; Thevuthasan, S.; Chambers, S.A. unpublished.
15. Chambers, S.A.; Yi, S.I. *Surf. Sci.* **1999**, *439*, L785.
16. Wang, X.-G.; Weiss, W.; Shaikhutdinov, S.K.; Ritter, M.; Peterson, M.; Wagner, F.; Schlogl, R.; Scheffler, M. *Phys. Rev. Lett.* **1998**, *81*, 1038.
17. Shaikhutdinov S.K.; Weiss, W. *Surf. Sci.* **1999**, *432*, L627.
18. Bloemen, P.J.H.; van der Heijden, P.A.A.; Wolf, R.M.; aan de Stegge, J.; Kohlhepp, J.T.; Reinders, A.; Jungblut, R.M.; van der Zaag, P.J.; de Jonge, W.J.M. *Mater. Res. Soc. Symp. Proc.* **1996**, *401*, 485.
19. Anderson, J.F.; Kuhn, M.; Diebold, U.; Shaw, K.; Stroyanov, P.; Lind, D. *Mater. Res. Soc. Symp. Proc.* **1997**, *474*, 265.
20. Gaines, J.M.; Kohlhepp, J.T.; Van Eemeren, J.T.W.M.; Elfrink, R.J.G.; Roozeboom, F.; De Jonge, W.J.M. *Mater. Res. Soc. Symp. Proc.* **1997**, *474*, 191.
21. Gaines, J.M.; Bloemen, P.J.H.; Kohlhepp, J.T.; Bulle-Lieuwma, C.W.T.; Wolf, R.M.; Reinders, A.; Jungblut, R.M., van der Heijden, P.A.A.; van Eemeren, J.T.W.M.; aan de Stegge, J.; de Jonge, W.J.M. *Surf. Sci.* **1997**, *373*, 85.
22. Gaines, J.M.; Kohlhepp, J.T.; Bloemen, P.J.H.; Wolf, R.M.; Reinders, A.; Jungblut, R.M. *J. Magn. Magn. Mater.* **1997**, *165*, 439.
23. The specimen was transferred from the MBE chamber at PNNL to an STM chamber down the hall through air in ~10 minutes. For a complete

description of the experiment, see Chambers, S.A.; Thevuthasan, S.; Joyce, S.A. *Surf. Sci.* **2000** (in press).

24. The specimen was transferred from the MBE chamber at PNNL to an evacuated shipping container through air, and then from the shipping container to the STM chamber at Tulane again through air. The total time in air was ~90 minutes. For a complete description of the experiment, see Stanka, B.; Hebenstreit, W.; Diebold, U.; Chambers, S.A. *Surf. Sci.* **2000**, *448*, 49.

25. Terrach, G.; Burgler, D.; Schaub, T.; Wiesendanger, R.; Guntherodt, H.-J. *Surf. Sci.* **1993**, *285*, 1.

26. Voogt, F.C. Ph.D. thesis, Departments of Chemical Physics and Nuclear Solid State Physics, University of Groningen, The Netherlands, 1998.

27. Kendelewicz, T.; Liu, P.; Doyle, C.S.; Brown, G.E., Jr.; Nelson, E.J.; Chambers, S.A. *Surf. Sci.* **2000**, *453*, 32.

28. Liu, P.; Kendelewicz, T.; Brown, G.E. Jr.; Nelson, E.J.; Chambers, S.A. *Surf. Sci.* **1998**, *417*, 53-65.

29. Liu, P.; Kendelewicz, T.; Brown, G.E., Jr.; Parks, G.A. *Surf. Sci.* **1998**, *412/413*, 287.

30. Liu, P.; Kendelewicz, T.; Brown, G.E., Jr. *Surf. Sci.* **1998**, *412/413*, 315.

31. Giordano, L.; Goniakowski, J.; Suzanne, J. *Phys. Rev. Lett.* **1998**, *81*, 1271.

32. Hass, K.C.; Schneider, W.F.; Curioni, A.; Andreoni, W. *Science* **1998**, *282*, 265.

33. Halley, J.W.; Rustad, J.R.; Rahman, A. *J. Chem. Phys.* **1993**, *98*, 4110.

34. Stillinger F.H.; David, C.W. *J. Chem. Phys.* **1978**, *69*, 1473.

35. Curtiss, L.A.; Halley, J.W.; Hautman, J.; Rahman, A. *J. Chem. Phys.* **1987**, *86*, 2319.

36. Rustad, J.R.; Hay, B.P.; Halley, J.W. *J. Chem. Phys.* **1995**, *102*, 427.

37. Rustad, J.R.; Felmy, A.R.; Hay, B.P. *Geochim. Cosmochim. Acta* **1996**, *60*, 1553.

38. Wasserman, E.; Rustad, J.R.; Hay, B.P.; Halley, J.W. *Surf. Sci.* **1997**, *385*, 217.

39. Henderson, M.A.; Joyce, S.A.; Rustad, J.R. *Surf. Sci.* **1998**, *417*, 66.

40. Rustad, J.R.; Wasserman, E.; Felmy, A.R. *Surf. Sci.* **1999**, *424*, 28.

41. Rustad, J.R.; Felmy, A.R.; Hay, B.P. *Geochim. Cosmochim. Act*a **1996**, *60*, 1563.

42. Felmy, A.R.; Rustad, J.R. *Geochim. Cosmochim. Acta* **1998**, *62*, 25.

43. Wang, X.-G.; Weiss, W.; Shaikhutdinov, S.K.; Ritter, M.; Petersen, M.; Wagner, F.; Schlogl, R.; Scheffler, M. *Phys. Rev. Lett.* **1998**, *81*, 1038.

44. Rustad, J.R.; Wasserman, E.; Felmy, A.R. *Surf. Sci.* **1999**, *432*, L583.

45. Akesson, R.; Pettersson, L.G.M.; Sandstrom, M.; Wahlgren, U. *J. Am. Chem. Soc.* **1994**, *116*, 8691.

46. Rustad, J.R.; Wasserman, E.; Felmy, A.R. *Proc. 2nd Internat. Alloy Conf.*, Davos, Eds. Gonis, A.; Meike, A.; Turchi, P.E.A. **2000** (in press).

47. Stanka, B.; Hebenstreit, W.; Diebold, U.; Chambers, S.A. *Surf. Sci.* **2000**, *448*, 49.

48. Hamilton, W.C. *Phys. Rev.* **1958**, *110*, 1050.

49. Baes F.B.; Mesmer, R.E. *The Hydrolysis of Cations*. Krieger Publishing Co., Malabar, Florida, 1996.

50. Mayer, L.M.; Schick, L.L. *Environ. Sci. Technol.* **1981**, *15*, 1482.

51. James, B.R.; Bartlett, R.J. *J. Environ. Qual.* **1983**, *12*, 177.

52. Music', S.; Ristic, M.; Tonkovic, M. *Z. Wasser-Abwasser-Forsch.* **1986**, *19*, 186.

53. Eary, L.E.; Rai, D. *Environ. Sci. Technol.* **1988**, *22*, 972.

54. Eary, L.E.; Rai, D. *Am. J. Sci.* **1989**, *289*, 180.

55. Eary, L.E.; Rai, D. *Soil Sci. Soc. Am. J.* **1991**, *55*, 676.

56. Eckert, J.M.; Stewart, J.J.; Waite, T.D.; Szymczak, R.; Williams, K.L. *Anal. Chim. Acta* **1990**, *236*, 357.

57. Bidoglio, G.; Gibson, P.N.; O'Gorman, M.; Roberts, K.J. *Geochim. Cosmochim. Acta* **1993**, *57*, 2389.

58. Anderson, L.D.; Kent, D.B.; Davis, J.A. *Environ. Sci. Technol.* **1994**, *28*, 178.

59. Weng, C.H.; Huang, C.P.; Allen, H.E.; Cheng, A.H-D.; Sanders, P.F. *Sci. Total. Environ.* **1994**, *154*, 71.

60. Weckhuysen, B.M.; Wachs, I.E.; Schoonheydt, R.A. *Chem. Rev.* **1996**, *96*, 3327.

61. Fendorf, S.E.; Li, G. *Environ. Sci. Technol.* **1996**, *30*, 1614.

62. White, A.F.; Peterson, M.L. *Geochim. Cosmochim. Acta* **1996**, *60*, 3799.

63. Peterson, M.L.; Brown, G.E., Jr.; Parks, G.A. *Colloids Surf. A* **1996**, *107*, 77.

64. Peterson, M.L.; White, A.F.; Brown, G.E., Jr.; Parks, G.A. *Environ. Sci. Technol.* **1997**, *31*, 1573.

65. Peterson, M.L.; Brown, G.E., Jr.; Parks, G.A.; Stein, C.L. *Geochim. Cosmochim. Acta* **1997**, *61*, 3399.

66. Patterson, R.R.; Fendorf, S.; Fendorf, M. *Environ. Sci. Technol.* **1997**, *31*, 2039.

67. Szulczewski, M.D.; Helmke, P.A.; Bleam, W.F. *Environ. Sci. Technol.* **1997**, *31*, 2954.

68. Jardine, P.M.; Fendorf, S.E.; Mayers, M.A.; Larsen, I.L.; Brooks, S.C.; Bailey, W.B. *Environ. Sci. Technol.* **1999**, *33*, 2939.

69. Fendorf, S.E.; Wielinga, B.W.; Hansel, C.M. *Geol. Internat.* 2000 (in press).

70. Lovley, D.R.; Phillips, E.J.P. *Appl. Environ. Microbiol.* **1994**, *60*, 726.

71. Smith, R.M.; Martell, A.E. *Critical Stability Constants*. Plenum Press, New York, 1976.

72. Condon, N.G.; Murray, P.W.; Leibsle, F.M.; Thornton, G.; Lennie, A.R.; Vaughan, D.J. *Surf. Sci.* **1994**, *310*, L609.

73. George, G.N.; Pickering, I.J. *EXAFS-PAK*. Technical Report, Stanford Synchrotron Radiation Laboratory, Stanford, CA, 1995.

74. Ankudinov, A.L.; Ravel, B.; Rehr, J.J.; Conradson, S.D. *Phys. Rev. B* **1998**, *58*, 7565.

75. Grolimund, D.; Kendelewicz, T.; Trainor, T.P.; Liu, P.; Fitts, J.P.; Chambers, S.A.; Brown, G.E., Jr. *J. Synchrotron Rad.* **1999**, *6*, 612.

76. Dzombak, D.A.; Morel, F.M.M. *Surface Complexation Modeling: Hydrous Ferric Oxide*, Wiley-Interscience, New York, 1990.

77. Charlet, L.; Manceau, A.A. *J. Coll. Interf. Sci.* **1992**, *148*, 443.

78. Chisholm-Brause, C.J.; O'Day, P.A.; Brown, G.E., Jr.; Parks, G.A. *Nature* **1990**, *348*, 528.

79. Towle, S.N.; Bargar, J.R.; Brown, G.E., Jr.; Parks, G.A.; Barbee, T.W., Jr. *Mater. Res. Soc. Symp. Proc.* **1995**, *357*, 23.

80. Towle, S.N.; Bargar, J.R.; Brown, G.E., Jr.; Parks, G.A. *J. Colloid Interface Sci.* **1999**, *217*, 312-321.

81. Eggleston, C.M.; Stumm, W. *Geochim. Cosmochim. Acta* **1993**, *57*, 4843.

82. Kendelewicz, T.; Liu, P.; Doyle, C.S.; Brown, G.E., Jr. *Surf. Sci.* (submitted).

83. Parks, G.A. *Chem. Rev.* **1965**, *65*, 177; Sverjensky, D.A. *Geochim. Cosmochim. Acta* **1994**, *58*, 3123.

84. McCafferty, E.; Bernett, M.K.; Murday, J.S. *Corrosion Sci.* **1988**, *28*, 559.

85. Amonette, J.E.; Foster-Mills, N.S. (in preparation).

Chapter 15

Electroosmotic Flow Rate: A Semiempirical Approach

J. H. Chang, Z. Qiang, C. P. Huang[*], and D. Cha

**Department of Civil and Environmental Engineering,
University of Delaware, Newark, DE 19716**

Experiments were conducted to study the electro-osmotic (EO) flow rate as affected by various factors, including solution pH, electric field strength, soil water content, and electrolyte concentration. Results indicate that the EO flow rate is proportional to the electrokinetic charge density and the electric field strength. The presence of inert-electrolytes has no significant effect on the EO flow rate. A parabolic relationship between the EO flow rate and the water content was observed. Based on the finite plate model, the EO flow rate (Q), can be expressed as a function of electrokinetic charge density (σ_e), field strength (E_o/L), and soil water content (ω), i.e. $Q = K\sigma_e(E_o/L)\omega^2$. The characteristic coefficient, K, is a collection of several physical properties of the soil-water system including the fluid density (ρ), the specific surface area (Σ), width of the water layer (w), and the fluid viscosity (μ), i.e., $K = kfw/\mu\rho^2\Sigma^2$. The K value was 57 $cm^3 \cdot \mu C^{-1} \cdot V^{-1}$.

Introduction

Electrokinetics has found its applications in sludge dewatering, grout injection, in-situ generation of chemical reagents in soil remediation, groundwater flow barrier, and soil and groundwater decontamination (Mitchell,

1986). Since its development at the beginning of the 19th century, many theories have been advanced to explain the mechanism of electro-osmotic (EO) flow in the capillary and the porous media. Reuss (1809) was the first to discover that a water flow could be induced through the capillary by an external electric field. Later, Helmholtz (1879) incorporated electrical components into the electro-osmotic flow equation. Although the electric double layer theory can not totally explain electro-osmotic behaviors, it is as a useful means to describe the electro-osmotic phenomenon in most capillary materials (Burgreen and Nakache, 1964).

Figure 1 illustrates the distribution of electro-osmotic flow based on the electric double layer theory. Since the surface charge of most soils is negative in typical soil-water system, it induces an excess counter-ions (cations) distribution in a thin water layer at the vicinity of the soil surface. When an electric field is imposed to the soil-water system, the cations will migrate to the cathode and the anions to the anode. As the cations and anions are always hydrated, a water momentum (or frictional drag) is produced by the movement of these ions (Yeung and Datla, 1995). In the presence of excess positive charges on the soil surface, a net electric driving force transports the water layer from the anode to the cathode (flow 1). In other words, the electro-osmotic flow that results from the fluid surrounding the soil particles is induced by ionic fluxes (Lyklema and Minor, 1998).

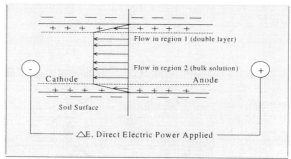

Figure 1. The electro-osmotic flow as predicted by the electric double layer theory.

In addition, the water molecules in bulk phase (region 2) can be carried along with "region 1" in the same flow direction. Interaction between "region 1" and "region 2" enables the movement of water in the bulk phase (i.e., by hydrodynamic drag). Therefore, the total observed electro-osmotic flow is attributed to the movement of these two water layers. In the absence of electrolyte or when both the cationic and anionic components are equivalent in concentrations, there will be no flow in region 1. As a result, flow in region 2 does not occur. Consequently, the overall electro-osmotic flow would be zero.

Based on above arguments, the zeta potential and the charge distribution in the fluid adjacent to the capillary wall play key roles in determining the electro-osmotic flow. Many researchers have reported that the zeta potential of the soil particle correlates well with the electro-osmotic flow (Quincke, 1861, Dorn, 1878, Helmholtz, 1879, Smoluchowski, 1921, Neal and Peters, 1946). The Helmholtz-Smoluchowski equation is one of the earliest and still widely used models of electrokinetic processes (Helmholtz, 1879, Smoluchowski, 1921). According to the Helmholtz-Smoluchowski equation, the pore radii were relatively large compared to the thickness of the diffuse double layer and all of the mobile ions were concentrated near the soil-water interface. These assumptions are valid as long as soils with large pores are saturated with water or dilute electrolyte solutions. The Helmholtz-Smoluchowski equation has the following expression:

$$v_e = \frac{D\zeta}{4\pi\mu} \frac{E}{L}$$

(1)

where v_e = velocity of electro-osmotic flow in soils (m·s^{-1}),
   E = electric potential (V),
   D = dielectric constant of the soil water (C·V^{-1}·m^{-1}),
   ζ = zeta potential of soil particle (V),
   μ = viscosity of soil water (N·s·m^{-2}),
   L = electrode spacing (m).

For small capillaries or unsaturated soils, where the thickness of the electric double layer and the water layer radius are of the same order of magnitude, the Helmholtz-Smoluchowski equation is no longer applicable.

Furthermore, the zeta potential alone can not totally predict the EO flow rate. At the onset of EO flow, the electrostatic force is the only driving mechanism for water movement, i.e. "flow in region 1" (Vane and Zang, 1997). According to the Coulomb's law, the electrostatic force is a function of the electric field strength and the electrical charge. Therefore, the applied field (potential gradient) and excess charge at the soil-water interface must be included in modeling the osmotic water flow. The zeta potential, measured at the shear plane, does not directly represent the charge condition at the surface of the soil particle. Consequently, zeta potential may not be a reliable parameter for predicting the electro-osmotic flow.

The surface charge density of the soil surface has also been considered a contributing parameter to the electro-osmotic flow rate. Esrig and Majtenyi (1966) applied the Buchingham π theorem to describe the electro-osmotic flow in porous media. They suggested that the average velocity is proportional to the surface charge density of the porous media. Schmid (1950) developed a quantitative model based on the concept of volume charge density. He assumed that the excess cations countering the negative charge of clay particles are uniformly distributed through the entire pore cross-section area. The Schmid equation has the following expression:

$$v_e = \frac{r^2 \rho_o F}{8\mu} \frac{E}{L}$$

(2)

where r = radius of pore (m),
ρ_o = volume charge density in the pore ($C \cdot m^{-3}$),
F = Faraday constant (96,500 $C \cdot eq^{-1}$).

In dealing with vadose zone (i.e., unsaturated soils), the above models may not be applicable because these equations were derived for saturated conditions. Practically, the most widely used electro-osmotic flow equation for unsaturated soils is proposed by Casagrande (1949):

$$Q = k_e i_e A$$

(3)

where Q = electro-osmotic flow rate ($m^3 \cdot s^{-1}$),
k_e = coefficient of electro-osmotic conductivity ($m^2 \cdot V \cdot s^{-1}$),
i_e = applied electrical gradient ($V \cdot m^{-1}$),
A = gross cross-sectional area perpendicular to water flow (m^2).

The hydraulic conductivity of different soils can vary by several orders of magnitude; however, the coefficient of electro-osmotic conductivity is generally between 1 x 10^{-9} and 10 x 10^{-9} m^2/(V· s) and is independent of the soil type (Mitchell 1993). Therefore, an electrical gradient is a much more effective driving force than a hydraulic gradient. Although Equation (3) is considered a practical one, it can not account for the effect of various physical-chemical parameters of a soil-water system. Moreover, it is not capable of predicting the EO flow rate in unsaturated soils.

There is need to develop a simple and reliable electro-osmotic water flow equation that can be applied to unsaturated soils in terms of fundamental soil-water parameters, such as, the water content of the soil, the soil surface charge density, and the electrolyte concentration. The objective of this study was to develop a semi-empirical equation for the prediction of electro-osmotic flow rate in unsaturated soils. Several operational parameters were studied. These included the solution pH, the electric field strength, the soil water content, and the electrolyte concentration. It was hoped that the semi-empirical equation can provide a better utility for in-situ electrokinetic remediation of contaminated soils in the vadose zone.

Theoretical Approach

Water flow behavior of an unsaturated soil is totally different from that of a saturated system. In the presence of an electrical field, a friction force is created when water molecules begin to move in the soil pores. The frictional stress decreases as the thickness of the water layer increases. For an unsaturated soil-water system, the water layer is extremely thin, usually ranging from 10^{-10}cm to 10^{-8}cm. Under such circumstance, all water molecules exhibit strong frictional interaction with the soil surface. In the case of a saturated water-capillary system, the radii of capillaries are relatively large, ranging from 10^{-1}cm to 10^{-3}cm. As a result, most capillary water molecules do not interact physically or chemically with the capillary wall (Iwata et al., 1995). The circular capillary model would be useful for the prediction of EO flow rate, if the soils were saturated. However, for unsaturated soil system, the finite plate model is better able to predict the EO flow.

Figure 2. Shows the basic concept of the finite plate model. According to the finite plate model, the soil matrix is conceptualized as a finite plate and all water molecules are evenly spread on the soil surface (surface/water). The water layer can be divided into two portions. One portion of the water layer between shear plane and soil surface is stagnant and its thickness is χ. The other water layer under the influence of a shear stress, τ, is moving at a velocity gradient, dv/dz, as a Newtonian fluid. The thickness of this layer is θ. The Newtonian fluid concept has been used to derive the EO flow in a saturated capillary (Esrig and Majtenyi, 1966). To apply the Newtonian fluid concept, the thickness of water layer must be smaller than the radius of capillary. The proportionality constant is the coefficient of viscosity of the fluid. That is:

$$\tau = \mu \, (dv/dz) = \mu \, (v_o/\theta) \qquad (4)$$

where
τ = shear stress (erg·m^{-2}),
μ = viscosity of fluid (erg·s·cm^{-2}),
dv/dz = velocity gradient of the fluid (s^{-1}),
v_o = the surface velocity of the water layer (cm·s^{-1}),
θ = average thickness of water layer between water surface and shear plane (cm).

When an aqueous solution moves tangentially across a charged surface, the thin water layer becomes a two-dimensional gel which macroscopically behaves as a rigid body (Lyklema et al., 1998). Figure 2 shows the conceptualized dimensions of the water layer of finite length, L, and width, w. The EO flow rate can be expressed by the following equation:

$$Q = \tilde{v}A = \tilde{v}\,\theta w \qquad (5)$$

where Q = Electro-osmotic flow rate $(cm^3 \cdot s^{-1})$,
$\tilde{v}$ = average velocity of the fluid $(cm \cdot s^{-1})$,
A = the cross section area of the water layer (cm^2),
w = width of the water layer (cm).

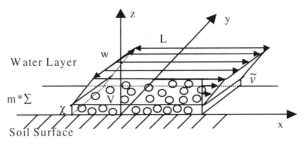

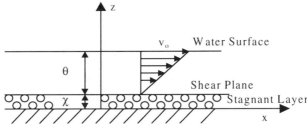

Figure 2. Illustration of EO flow behavior by the finite plate model.

The thickness of water layer can be expressed by the following equation:

$$\theta = V/m\Sigma \qquad (6)$$

where V = total volume of the fluid (cm^3),
m = total mass of soils (g)
Σ = specific surface area of soils $(cm^2 \cdot g^{-1})$.
By substituting Equation (6) into Equation (5), one obtains the following expression:

$$Q = \tilde{v}Vw / m\,\Sigma \qquad (7)$$

The average velocity can be related to the maximum velocity at the water surface, v_0. That is,

$$\tilde{v} = fv_o \tag{8}$$

where   f = a dimensionless ratio.
By substituting Equation (8) into Equation (7), one has:

$$Q = fv_o V w/m\Sigma \tag{9}$$

By substituting Equation (6) into Equation (4), yields:

$$\tau = \mu \, (v_o / \theta) = \mu v_o m\Sigma/V \tag{10}$$

Rearranging the above equation gives the following expression:

$$v_o = \tau V/\mu m\Sigma \tag{11}$$

Substituting Equation (11) into (9), the EO flow rate is:

$$Q = fv_o V w/m\Sigma = fV^2 \tau w/\mu m^2\Sigma^2 \tag{12}$$

The "rigid" water layer represents water molecules in the region covering beyond the shear plane into the bulk phase. Therefore, the shear force is proportional to the electric force induced by the electrokinetic charge density (the accumulated space charges from the bulk phase to the zeta potential plane) and the applied electric field, that is:

$$k\sigma_e(E_o/L) = \tau \tag{13}$$

where   k = frictional coefficient,
        σ_e = electrokinetic charge density (C·cm^{-1}),
        E_o = the electric field strength (V),
        L = total length of water layer (cm).
By substituting Equation (13) into Equation (12), the EO flow rate is written as follows:

$$Q = k\sigma_e(E_o/L)fV^2 w/\mu m^2\Sigma^2 = k'\sigma_e(E_o/L)V^2 \tag{14}$$

where k' = kfw/$\mu m^2\Sigma^2$.
Since $\omega = \rho V/m$, where ω, ρ, and m are water content of the soil, fluid density and soil mass, respectively, the EO flow rate equation finally has the following form:

$$Q = k'\sigma_e(E_o/L)\omega^2 m^2/\rho^2 = K\sigma_e(E_o/L)\omega^2 \qquad (15)$$

where K is a characterized coefficient, i.e., $K = kfw/\mu\rho^2\Sigma^2$.
This characterized coefficient, K, collects several physical properties of the soil-water system such as the fluid density (ρ), the specific surface area (Σ), width of the water layer (w) and fluid viscosity (μ).

Methods and Materials

According to Equation (15), major variables controlling the EO flow rate are the electrokinetic charge density of the soil surface, electric field strength, and the water content. In order to verify Equation (15), experiments were conducted to measure the EO flow rate at various values of these parameters. The electrokinetic charge density of the soil surface is related to the zeta potential which can be measured at different pH values and electrolyte concentrations. Therefore the effect of surface charge density on EO flow rate was determined under various pH values and electrolyte concentrations. Based on the EO flow rate data obtained under different pH, electrolyte concentration, and zeta potential, Equation (15) can be established. Table 1 lists the experimental conditions.

Soil Samples and Chemicals:

Soil samples were collected from a DOE waste site. Prior to experiments, all soil samples were air-dried, and then sieved through a #10 standard sieve (2 mm openings). This soil has composition of 14% sand, 38% silt, and 48% clay. According to the U.S. Department of Agriculture soil classification standard, the soil can be classified as clay. This soil texture has a low hydraulic conductivity, 2.5×10^{-8} cm/sec, which is suitable for EO application. The soil pH value 7.6 measured in 0.01 M $CaCl_2$ solution. The soil organic matter was around 1.7% (w/w) determined by the loss of mass on ignition (L. O. I.) method. The pH_{zpc} was determined in diluted electrolyte solution (10^{-3} M NaCl) by a zetameter (Pen-Kem Inc., Hudson, NY), pH_{zpc} of this soil is 2.2. The other soil characteristics such as hydraulic conductivity and specific surface area were also analyzed. Table 2 shows results of physical-chemical characteristics of the soil sample with corresponding analytical methods used.

Table 1. Experimental conditions of electro-osmotic flow tests

Test No.	pH of Solution	Field Strength (V/cm)	Electrolyte	Water Content (%)
1	4	12/10	1×10^{-3}M NaCl	20
2	6	12/10	1×10^{-3}M NaCl	20
3	7	12/10	1×10^{-3}M NaCl	20
4	9	12/10	1×10^{-3}M NaCl	20
5	*	06/10	1×10^{-2}M CH$_3$COONa	20
6	*	12/10	1×10^{-2}M CH$_3$COONa	20
7	*	24/10	1×10^{-2}M CH$_3$COONa	20
8	*	12/10	1×10^{-3}M NaCl	20
9	*	12/10	1×10^{-3}M CH$_3$COONa	20
10	*	12/10	1×10^{-2}M CH$_3$COONa	20
11	*	12/10	5×10^{-2}M CH$_3$COONa	20
12	*	12/10	1×10^{-1}M CH$_3$COONa	20
13	*	12/10	1×10^{-2}M CH$_3$COONa	5
14	*	12/10	1×10^{-2}M CH$_3$COONa	10
15	*	12/10	1×10^{-2}M CH$_3$COONa	15
16	*	12/10	1×10^{-2}M CH$_3$COONa	20
17	*	12/10	1×10^{-2}M CH$_3$COONa	25

*: uncontrolled pH

Table 2. Physical-chemical characteristics of the soil sample

Physical-chemical characteristics	Result	Method
Sand (%)	14.0	Hydrometer
Silt (%)	38.0	Hydrometer
Clay (%)	48.0	Hydrometer
pH	7.6	In 0.01M CaCl$_2$
ECEC (meq/100g)	20.5	Σ(exchangeable K,Ca,and Mg)
Organic Matter(%)	1.7	Heating at 105 $^\circ$C for 2 hours, then at 360 $^\circ$C for 2 hours
Moisture (%)	13.6	Heating at 105 $^\circ$C for 24 hours
Hydraulic Conductivity (10^{-8}cm/s)	2.5	Constant-head
Specific Surface Area (m^2/g)	10.7	BET
pH$_{zpc}$	2.2	pH meter and Zetameter

ECEC: effective cation exchange capacity

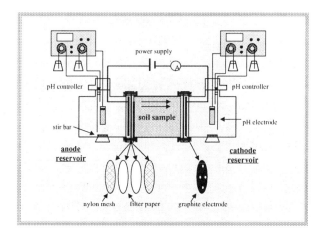

Figure 3. Schematic diagram of the electro-osmosis reactor.

All chemicals, sodium chloride, sodium acetate, and sodium hydroxide were purchased from the Aldrich Co (Milwaukee, WI). The purity of chemicals are 99.0% for sodium chloride and sodium acetate, and 97% for sodium hydroxide.

Electrokinetic Experiments:

Figure 3 shows a diagram of the laboratory electro-osmosis reactor. The electro-osmosis cell consists of an acrylic unit with a central cylinder of 11.5 cm in length and 8.9 cm in internal diameter. The volume of both the cathode and the anode compartments are 700 mL. To separate the soil from the water solution, a pair of nylon meshes (Spectrum model PP, mesh opening 149 μm) and a filter paper (Whatman No. 1) were placed between the soil sample and electrodes. Graphite disks (Carbon of America; grade 2020; 3.25 inches in diameter) were used as the electrodes and placed at each electrolytic compartment right behind the membranes.

As indicated above, the electro-osmotic flow experiments were conducted by changing four operational factors including the pH of the working solution, the electrolyte concentration, the water content of soils, and the electric field strength (Table 1). Strong acid (1M HCl) was added to the cathode reservoir and strong base (1M NaOH) was added to the anode reservoir (Model pH-22, New Brunswick Scientific) to control the pH of solution in Test 1 through 4. Soil cores with 10 % - 25% water content were prepared by mixing 100 - 300 ml solution with 900 g dried soil. The porosity of the soil cores in the EO system is around 58 %. At this range of water content, the thickness of water layer is 10^{-6} ~ 2.5×10^{-6} cm. This value is much smaller than the radius of most capillaries, therefore the Newtonian fluid concept can be readily applied.

The electrodes were connected to a DC power supply (Model WP-705B, Vector-VID Instrument Division). During tests, parameters such as the water flow rate, the intensity of electric current and field strength, pH of catholyte and anolyte were monitored continuously. At the end of a pre-selected time period, the soil sample in the cell were sliced into 10 equal sections. The pH value and the water content of the soil matrix were measured.

Results and Discussions

To gain insight into the mechanism of the electro-osmotic flow the first experiments investigated the effect of surface charges on the flow rate. Several operational parameters such as pH, electrolyte type, and electrolyte concentration can affect the surface charge. The change in soil surface charges can dramatically affect EO flow rate. Other parameters studied were the electric field strength and the soil water content.

Effect of pH

Figure 4 presents the EO flow rate as a function of time at various pH values. Results show that the EO water flow rate decreases with decreasing solution pH and ceases at pH around 2.2. The decrease in EO flow rate can be attributed to the decreasing surface charge density of soils. The soil surface charge is rendered more positive as the pH decreases due to the preferential adsorption of hydronium ions (H^+) on the soil surface (Stana-Kleinschek, 1998). The electrostatic force driving the water molecules reduces and the EO flow decreases as the surface density approaches neutrality. According to Equation (13), the transport of water is attributed to the charge of the water layer between shear layer and water surface, i.e., electrokinetic charge density. Since the electrokinetic charge density is balanced by the charge between the shear plane and soil surface, the electrokinetic charge density would change as the surface charge changes. The magnitude of the electrokinetic charge density at different pH can be readily obtained by the zeta potential measurements. For symmetrical electrolytes, the electrokinetic charge density can be written as follows (Hunter, 1981):

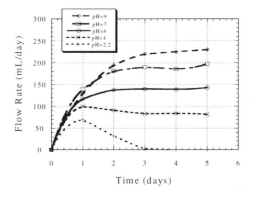

Figure 4. Effect of pH on the electro-osmotic flow rate. Experimental conditions: electrolyte concentration = 10^{-3} M NaCl; voltage = 12 V; water content = 23% (w/w).

$$\sigma_e = 11.74c^{1/2} \sinh(19.46z\zeta) \tag{16}$$

where σ_e = electrokinetic charge density ($\mu C \cdot cm^{-2}$)
    c = electrolyte concentration (M)
    z = valance of electrolyte ion
    ζ = zeta potential (V)

Figure 5 presents the zeta potential of soil particles as a function of pH at various electrolyte concentrations. Results show the absolute value of zeta potential decreases with decreasing the pH value and the pH_{ZPC} is around 2.2 at an electrolyte concentration of 10^{-3}M. Figure 6 presents the pH value of the soil as a function of normalized distance. The average pH value of soil can be obtained from this graph, then, the zeta potential at different pH value can be indicated from Figure 5. According to Equation (16), then, the electrokinetic charge density of soil can be determined at the various pH conditions. The relationship between the EO flow rate and the electrokinetic charge density (at different pH) is shown in Figure 7. Results indicate that the EO flow is proportional to the electrokinetic charge density as expected in the EO flow model.

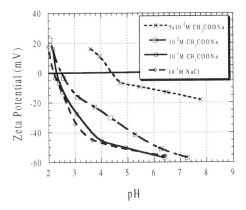

Figure 5. The zeta potential of soil particles as a function of pH at various electrolyte concentrations.

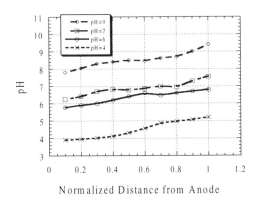

Normalized Distance from Anode

Figure 6. The pH value of the soil as a function of normalized distance. Experimental conditions: electrolyte concentration = 10^{-3} M NaCl; voltage = 12 V; water content=23% (w/w).

Effect of Electric Field Strength

Figure 8 presents the daily flow as a function of time under various applied electric voltages. Results show that the EO flow rate increases with

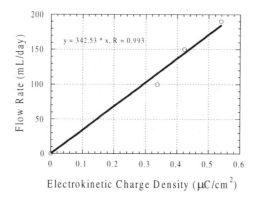

Figure 7. The electro-osmotic flow rate as a function of the electrokinetic charge density.

increasing the electric field strength. As predicted by Equation (13), EO flow rate is proportional to the electric potential. Figure 9 shows the linear relationship between the flow rate and the electric field strength. However, it is seen that the EO flow rate decreases with increasing operation time (Figure 8). Since the solution pH was not controlled during tests, the protons produced at the anode by elecdtrolysis neutralize the surface charge density of the soil. Meanwhile, the electrokinetic charge density decreases which results in the EO flow in the soil decreasing gradually.

Effect of Electrolyte Concentration

Figure 10 presents the influence of the electrolyte concentration on the electro-osmotic flow rate. Results show that the electrolyte concentration has no effect on the EO flow rate at least during the one-day treatment. Moreover, the EO flow rate decreases with increase of operation time. This can be attributed in part to the acidification of soils. According to Figure 4, it is seen that the EO flow rate approaches a steady state as the electrolyte concentration increase due to the addition of the strong acid/base for pH control. This observation also implies that the electrolyte concentration has no effect on the EO flow rate as long as the electrolyte ions do not interact specifically with the soil (i.e., indifferent ions). However, the electrolytes that specifically adsorb on the soil can change the zeta potential and which in turn modify its surface charge density.

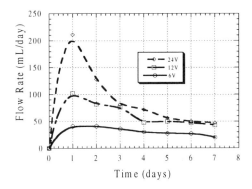

Figure 8. Effect of the electric field strength on the electro-osmotic flow rate. Experimental conditions: electrolyte concentration = 10^{-2} M CH_3COONa; water content = 20% (w/w).

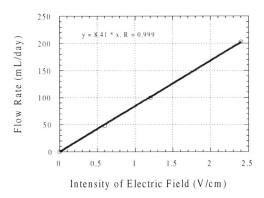

Figure 9. The electro-osmotic flow rate as a function of the electric field strength.

From Figure 10, it is seen that the flow rate decreases with increasing electrolyte concentration after several-days of treatment. This can be attributed to the adsorption of the electrolyte ions onto the soil surface. According to Figure 5, the electrolyte (sodium acetate) appears to interact specifically with the soil surface, as indicated by an increase in pH_{zpc} upon an increase of electrolyte concentration. Figure 11 shows the solution pH at the various electrolyte concentrations. Based on data presented in Figures 5 and 11, it is seen that the absolute value of zeta potential decreases with increasing ionic strength at a given pH. According to Equation (16), the electrokinetic charge density

262

decreases as zeta potential decreases. Therefore, the EO flow decreases with increasing electrolyte concentration as expected.

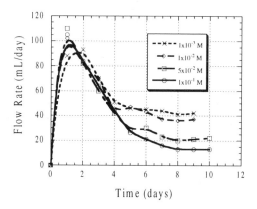

Figure 10. Effect of the electrolyte concentration on the electro-osmotic flow rate. Experimental conditions: voltage = 12 V; water content = 20% (w/w).

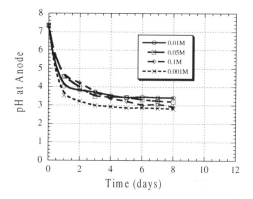

Figure 11. The anodic pH value as a function of time at various electrolyte concentrations. Experimental conditions: voltage = 12 V; water content = 20% (w/w).

Effect of Water Content

Figure 12 shows the EO flow rate as a function of time at various initial water contents. Regardless of the initial water content, the EO flow rate increases to a maximum level then decreases gradually to reach a steady state value. The initial surge in EO flow can be attributed to the flux of H⁺ ions into the soil column. When the H⁺ flux stops, the pH value of the soil water system decreases, approaching the pH_{zpc} value of the soil. As a result, the EO flow rate decreases.

Figure 13 presents the water content of the soil as a function of the normalized distance at the 6th-day which is the steady state condition. Results show that the final water content is different from the initial value. As the EO

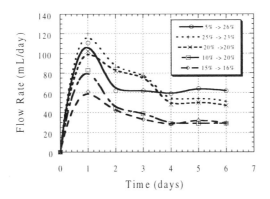

Figure 12. Effect of the water content on the electro-osmotic flow rate. Experimental conditions: electrolyte concentration = 10^{-2} M CH_3COONa; voltage = 12 V.

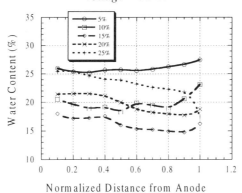

Figure 13. The final water content as a function of normalized distance at various initial water contents of the soil sample. Experimental conditions: electrolyte concentration = 10^{-2} M CH_3COONa; voltage = 12 V.

system reached a hydraulic steady state, the water content of soil varies from 5% to 26%, 10% to 20%, 15% to 16%, 20% to 20%, and 25% to 23%, respectively. The diversity between initial and final water content can be attributed to various hydraulic properties of EO systems. The use of different initial water contents could change the hydraulic properties of soil samples such as infiltrate rate, porosity, and hydraulic conductivity, which results in different final water contents. However, the hydraulic properties of soil samples can not be controlled, therefore, there is no relationship between initial and final water content.

Figure 14 presents the EO flow rate versus final soil water content. Equation (15) clearly predicts that the EO flow rate is proportional to the square of the water content (i.e., $Q \propto \omega^2$). Accordingly, by plotting logQ versus logω, one can have a straight line which slope is 2.0 and the intercept of y-axis is log ($K\sigma_eE_0/L$). With information, specifically, σe and E_0/L given from experimental conditions, the K value was calculated to be 57 $cm^3 \cdot \mu C^{-1} \cdot V^{-1}$. Figure 14 shows that experimental data could be fitted by a straight line as predicted. Furthermore, Gray and Mitchell (1967) reported a parabolic relationship between the EO flow and the water content in the soil-water system. Our observation was in agreement with what was reported by Gray and Mitchell.

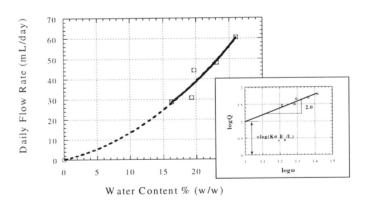

Figure 14. The electro-osmotic flow as a function of water content of the soil. Experimental conditions: electrolyte concentration = 10^{-2} M CH_3COONa; voltage = 12 V.

Conclusions

Based on experimental results, a semi-empirical equation for the EO flow rate in unsaturated soil-water system was established. A finite plate model was used to derive the EO flow rate in unsaturated soils. The derivation indicates that the EO flow rate is theoretically proportional to the extent of electrokinetic charge density, the electric field strength, and square of the soil water content in the soil-solution systems. Experimental results were in agreement with these predictions. For the effect of electrolyte concentration, results imply that there is no significant effect on the EO flow rate as long as the electrolyte ions do not interact specifically with the soil. The semi-empirical equation of EO flow was developed: $Q = K\sigma_e(E_0/L)\omega^2$. Where Q, K, σ_e, E_0/L, and ω, are electro-osmotic flow rate, characterized coefficient, electrokinetic charge density, potential gradient, and soil water content, respectively. The K value was 57 $cm^3 \cdot \mu C^{-1} \cdot V^{-1}$.

Acknowledgment

This work is supported by Department of Energy (Grant No. DE-FG07-96ER14716). Contents of this paper do not necessarily reflect the views of the funding agency.

References

1. Burgreen, D.; Nakache, F. R. J. Phy. Chem. **1964,** 68, 1084-1091.
2. Casagrande, I. L. Bautechnique **1949,** 1, 1-29,.
3. Dorn, E. Wied. Ann. **1880,** 10, 46.
4. Esrig, M. I.; Majtenyi, S. Highway Research Record, **1966,** 111, 31-41.
5. Gray, D. H.; Mitchell, J. K. J. Soil Mech. and Foun. Div., ASCE, **1967,** 93, 209- 236.
6. Helmholtz, H. Ann. Physik, **1879,** 7, 137.
7. Hunter, R. J. Zeta Potential in Colloid Science Principles and Applications; Academic Press: Orlando, FL, 1981.
8. Iwata, S.; Tabuchi, T.; Warkentin, B. P. Soil-Water Interactions: Mechanisms and Applications; Marcel Dekker: New York, NY, 1995.
9. Lyklema, J; Minor, M. Coll. Surfaces A: Physicochemical Eng. Aspects, **1998,** 140, 33-41.
10. Lyklema, J.; Rovillard, S.; Coninck, J. D. Langmuir **1998,** 14, 5659-5663.

11. Mitchell, J. K. Workshop on Electrokinetics Treatment and Its Application in Environmental-Geological Engineering for Hazardous Waste Site Remediation, Seattle, WA, Aug 4-5, 1986, p 20-31.
12. Mitchell, J. K. Fundamentals of soil behavior; 2nd Ed., John Wiley & Sons: New York, NY, 1993.
13. Neale, S. M.; Peters, R. H. Farad. Soc. Trans. **1946,** 42, 478.
14. Quincke, G. Ann. Physik. **1861,** 113, 513.
15. Reuss, F. Memories Soc. Imp. Naturalistes **1809,** 2, 327.
16. Schmid, G. Zhurnal fur Electrochemie **1950,** 54, 424.
17. Smoluchowski, von M. Handbuch der Elektrizitat und der Magnetismus II **1921,** 2, 366-428.
18. Stana-Kleinschek, K.; Ribitsch, V. Coll. Surfaces A: Physicochemical Eng. Aspects, **1998,** 140, 127-138.
19. Vane, L. M.; Zang, G. M. J. Hazardous Materials, **1997,** 55, 1-22.
20. Yeung, A. T.; Datla, S. Can. Geotech. J. **1995,** 32, 569-583.

Chapter 16

Reductive Dechlorination of Trichloroethylene and Carbon Tetrachloride at Iron Oxides and Basalt Minerals

Jani C. Ingram[1,*], Marnie M. Cortez[1], David L. Bates[2], and Michael O. McCurry[2]

[1]Idaho National Engineering and Environmental Laboratory, Idaho Falls, ID 83415–2208
[2]Department of Geology, Idaho State University, Pocatello, ID 83209

Abiotic degradation of trichloroethylene (TCE) and carbon tetrachloride (CCl_4) in aqueous solutions in the presence of iron oxides, basalt, and its minerals was investigated. The objective of this work is to gain an understanding of chemical interactions of chlorinated solvents with basalt in order to aid remediation efforts. The basalt samples used in these studies were collected near the Idaho National Engineering and Environmental Laboratory. Additionally, basaltic minerals (plagioclase, olivine, augite, magnetite, and ilmenite) and iron oxides (FeO, Fe_2O_3, Fe_3O_4, $FeTiO_3$, $FeOOH$) were investigated. Reactivity studies were conducted in which these materials were exposed to 16 ppm aqueous solutions of TCE and CCl_4 under anaerobic conditions. The reaction products were monitored over time using solid phase microextraction coupled with gas chromatography. The results indicate that reductive dechlorination of TCE occurred at FeO and $FeTiO_3$ only. In contrast, reductive dechlorination of CCl_4 was observed when reacted with all Fe(II)-bearing materials. Degradation of the CCl_4 was significantly faster than the TCE when reacted under the same conditions (hours compared to days or weeks for $FeTiO_3$ and FeO). Additionally, dechlorination of CCl_4 was observed to occur at plagioclase which was unexpected since plagioclase cannot support reductive dechlorination.

Introduction

The extensive use of chlorinated solvents in government and industrial processes and faulty management of wastes from these processes has resulted in chlorinated solvent contamination of subsurface environments. In this study, the specific areas of interest are the Department of Energy sites at the Idaho National Engineering and Environmental Laboratory (INEEL) and Hanford, Washington. It has been documented (1,2,3) that hundreds of thousands of kilograms of chlorinated solvents including carbon tetrachloride (CCl_4), trichloroethylene (TCE), tetrachloroethylene (PCE), have been disposed of as both organic waste and mixed waste. At both sites, these wastes have been detected in the ground water as well as in the vadose zone. In order to formulate remediation strategies for these contaminants, an understanding of their fate in the subsurface is required. Our work has been directed at gaining an understanding of the abiotic chemical interactions of chlorinated solvents with basalt as it is the main rock formation present at both Hanford and the INEEL. These geographical areas are dominated by diverse layers of basalt and sediments. Basalt is an inhomogeneous rock consisting of many mineral phases, which include, but are not limited to plagioclase, pyroxenes, olivines, and spinels (4). And there is added diversity within these classes of minerals, which arises from weathering, and the conditions present during their formation. This diversity leads to the reasonable expectation that a variety of contaminant-surface interactions are possible, which in turn may have a profound impact on contaminant fate. If the most significant interactions could be identified, then it might be possible to predict degradation or adsorption based on the presence or absence of the most reactive basalt phases.

Phases containing reducing metals, specifically Fe, are good candidates for high reactivity. It has been demonstrated that chlorinated organics (RCl) will oxidize Fe^0 to Fe^{2+}, and are reductively dechlorinated by hydrogenolylis in the process (5,6,7,8,9) as shown below:

$$Fe^0 + RCl + H^+ \rightarrow Fe^{2+} + RH + Cl^-.$$

Studies of the rates of these degradation reactions indicate that the dechlorination step is dependent on the direct interaction of the chlorinated organics with the Fe^0 located on the surface (6).

Additionally, Roberts and coworkers have shown that reductive elimination plays an important role in the dechlorination of chlorinated ethylene compounds by Fe^0(10), as shown below:

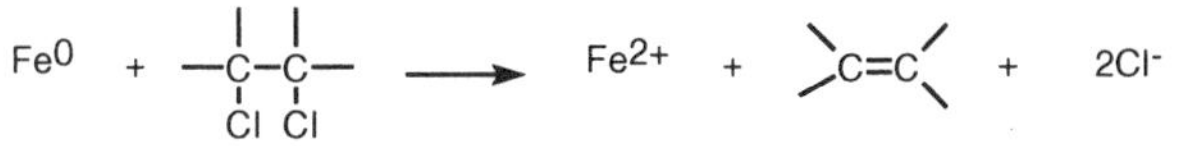

$$Fe^0 + \underset{\underset{Cl}{|}}{-}\overset{|}{C}-\underset{\underset{Cl}{|}}{\overset{|}{C}}- \longrightarrow Fe^{2+} + \quad C=C \quad + \quad 2Cl^-$$

It has been shown that surface-bound Fe(II) species are capable of reductive dechlorination. Adsorbed Fe(II) on iron oxides such as goethite, hematite, and ferrihydrite were shown to reductively dechlorinate CCl_4 (11,12) and TCE (13). Similarly, the importance of Fe(II)-bearing mineral surfaces for CCl_4 degradation was demonstrated by Kriegman-King and Reinhard (14,15).

The objective of these studies was to test the hypothesis that Fe(II)-containing mineral phases of basalt could abiotically dechlorinate TCE and/or CCl_4. The approach taken to test this hypothesis was to perform batch studies of 16 ppm aqueous TCE and CCl_4 with iron oxides, basalt minerals (plagioclase, olivine, magnetite, ilmenite, augite), and basalt taken from the field (Box Canyon, Idaho). Losses of the starting material and appearance of degradation products were monitored using gas chromatography (GC).

Experimental Description

Materials

The following iron oxides were purchased from Aldrich (>98% purity) and used with no additional clean-up in these experiments: Iron (II) oxide (FeO), Iron (III) oxide (Fe_2O_3), Iron (II) titanate ($FeTiO_3$), Iron (II, III) oxide (Fe_3O_4), and Iron (III) oxide, hydrated (FeOOH). The following minerals were purchased from Wards Natural Science Establishment, Inc. and used in these experiments: olivine (variation forsterite, Washington, USA), plagioclase (feldspar, Ontario, Canada), augite (pyroxene, Ontario, Canada), magnetite (spinel, Ontario, Canada), and ilmenite (Quebec, Canada) . The basalt characterized in these experiments was collected from a single olivine-tholeiite basalt flow from the Eastern Snake River Plain at an exposed basalt flow in Box Canyon, Idaho, approximately 2 km west of the INEEL. The basalt is considered to be representative of basalt flows which comprise the majority of the aquifer matrix in the Eastern Snake River Plain which the INEEL is located. The mineral composition of the basalt was determined by electron microprobe at the user facility at the University of Utah (Cameca SX-50). The average composition of the basalts were: apatite = 0.6%, plagioclase = 32.6%, olivine = 15.9%, magnetite/ilmenite = 7.8%, augite = 13.6%, and tachylite (a mixture of sideromelane and microcrystalline basaltic minerals) = 14.6%. The following chlorinated compounds were purchased from Supelco (>99% purity) and used as received: trichloroethylene (TCE), cis-dichloroethylene (cis-DCE), carbon tetrachloride (CCl_4), chloroform (CCl_3H), and methylene chloride (CCl_2H_2). 18MΩ water was used to make all aqueous solutions.

Instrumentation

Analyses were performed using a Hewlet Packard 5890 Series II Gas Chromatograph equipped with a flame ionization detector. The column used in these experiments was a 30 m x 0.32 mm ID x 1.8 μm df – Rt_x – 624. The analyses were run with a 12 psi head pressure, a spitless injection, He was used as the carrier gas at a rate of 10 ml/min through the column. The samples were injected directly from solid phase microextraction (SPME) fibers into the GC where the injector temperature was 250°C. The oven program was set to hold the temperature at 40°C for 8 minutes, then ramp the temperature at 70°C/min to 250°C.

Surface areas were determined by a Micromeritics Gemini 2360 instrument (based on the Brunauer, Emmet, and Teller (BET) method) using N_2 as the adsorption gas.

Procedures

Batch experiments were performed to determine the relative kinetics of the degradation of the TCE and CCl_4 in the presence of the iron oxides, basalt mineral, and field samples of basalt. These samples were prepared in an anaerobic chamber containing an atmosphere of 85% N_2, 10% CO_2, and 5% H_2. The samples were prepared using the following items (Supelco): 10 ml headspace glass vials, teflon faced butyl septa, and aluminum sealed (6ml/10ml/20ml). In the case of the iron oxide samples, 500 mg of the compound was weighed out in the vial. In the case of the minerals and the field samples, rock pieces were placed in the anaerobic chamber, then crushed with a heavy, stainless steel pestle which fit inside a steel chimney. The rock was crushed into pieces ranging from 0.0035 to 0.0937 in. The crushed material was used without sieving; thus a mix of sizes were used in these experiments. The purpose for crushing the rocks in the anaerobic chamber was to gain fresh surfaces and minimize oxidation. After being crushed, 500 mg of the powdered rock was weighed out and placed in the vial. The vials containing the material were spiked with 3 ml of 18 MΩ deaerated water (weighed after spiking), then the septa were crimped sealed onto the vials. Next, 100 µl of a 500 ppm solution of the chlorinated compound was added to the sealed vial using a syringe pump (weighed after spiking) to make a 16 ppm solution of the chlorinated analyte. The vials were placed in a box, then placed on a shaker at room temperature and shaken over the course of the experiment at 200 rpm. In order to minimize evaporation during the experiment, 1 vial was prepared for each data point collected which limited piercing of the septa.

The analyte concentration and degradation product concentrations were determined by sampling the head space of the reaction vials using a solid phase microextraction (SPME) fiber. The SPMEs used in these experiments were: TCE experiments used a 85 µm polyacrylate coated SPME (Supelco), and CCl$_4$ experiments used a 100 µm polydimethylsiloxane coated SPME (Supelco). In order to sample the head space of the vials, the SPME was injected in the vial, the fiber was exposed to the headspace for 6 minutes, it was retracted, and immediately injected directly into the GC. It has been demonstrated that the use of SPME technology to determine aqueous solution concentrations is more efficient than solvent extraction and purge and trap approaches (16,17,18).

The head space concentration of the analyte of interest were determined by comparing the response determined by the GC to a standard calibration curve for the analytes of interest using in-house prepared standards. The calibration curves were checked at least once a day using quality control (QC) standards (AccuStandard, Inc. – certified to ± 2%). The head space concentrations were converted to total concentrations (gas and aqueous phase) using gas partitioning constants (19).

Results

Trichloroethylene

The dechlorination of TCE at Fe oxides, basalt, and its minerals under aqueous, anaerobic conditions was investigated. It was expected that if reductive degradation of TCE occurred via hydrogenolysis, then cis-1,2-DCE would be detected as the first major degradation product (10). Alternatively, if reductive degradation of TCE occurred via reductive elimination, then chloroacetylene would be detected as the major degradation product (*10*).

The experiments were run over a number of weeks with the longest experiments lasting 12 weeks. Sampling for TCE and degradation products was performed either once a week or every two weeks. These experiments were repeated a minimum of 4 times.

It was found that under the conditions of these experiments, only the $FeTiO_3$ and FeO showed appreciable losses of TCE. Analyses for degradation products indicated the presence of cis-1,2-DCE, which was detected after only 2 days of exposure for the $FeTiO_3$ and 5 weeks of exposure for the FeO. None of the other materials used in these experiments produced cis-1,2-DCE or any other measureable quantities of degradation products. Additionally, post analysis of the supernatant liquid by ion chromatography showed that only the FeO and $FeTiO_3$ produced Cl⁻ after being reacted for 7 weeks. These results would suggest that hydrogenolysis is the major pathway by which dechlorination occurs. However, even though no acetylene compounds were detected, there is some question as to whether the SPME used in these experiments is capable of sorbing acetylene compounds at levels less than 1 ppm. We, therefore, suggest that although no evidence for reductive elimination was observed in these studies, there is a possibility that this pathway may also contribute to dechlorination of TCE by $FeTiO_3$ and/or FeO.

Some difficulties were encountered in performing these experiments mainly due to losses of TCE during the experiment which is attributed to either evaporation and/or absorption into the septa used to seal the sample vials. Attempts were made to minimize these losses which included using teflon coated septa, and preparing many similar vials for sampling such that the septa would be pierced only once during the experiment. However, even with these precautions, losses of TCE for the control samples (vials containing only aqueous TCE with no solid material present) near the end of the experiment (>10 weeks) were in the range of 25 to 40%. Others have reported comparable losses (*20*). As a result of the evaporative/absorptive loss of TCE in these experiments, quantitative measurements of the reductive degradation of TCE cannot be made from these results.

Carbon Tetrachloride

In a similar fashion to TCE, the dechlorination of CCl_4 at Fe oxides, basalt, and its minerals under aqueous, anaerobic conditions was investigated. It was expected that if reductive degradation of CCl_4 occurred, then CCl_3H would be detected as it is the first major degradation product reported for hydrogenolysis of CCl_4 at Fe^0 (*6*).

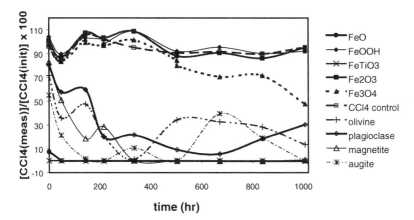

Figure 1. Loss of CCl₄ at iron oxides and basalt minerals.

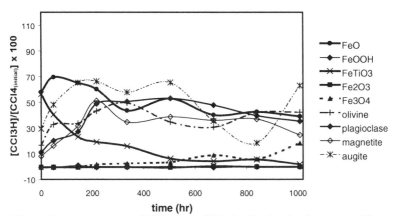

Figure 2. Appearance of CCl_3 from CCl_4 dechlorination by iron oxides and basalt minerals.

These experiments were run over an 8 week period and repeated a minimum of 2 times for the minerals and basalt, and 5 times for the iron oxides. Sampling for CCl_4 and its degradation products was performed either weekly or hourly, depending on the experiment.

It was found that under the experimental conditions used in this work, all Fe(II)-containing materials (FeO, $FeTiO_3$, Fe_3O_4, olivine, augite, magnetite, ilmenite, and the basalt samples) were capable of reductive degradation of CCl_4 as observed by the appearance of CCl_3H, and in some cases CCl_2H_2 in the latter parts of the experiments (Figures 1 through 4). Further, it was observed that plagioclase which is not an Fe-bearing mineral (an albite feldspar with composition $NaAlSi_3O_8$) also dechlorinated the CCl_4 producing CCl_3H (Figures 1 and 2).

274

In the case of FeO and FeTiO₃, the CCl₄ is very quickly degraded to CCl₃H; within the first few hours of exposure, CCl₃H is detected. By the end of the first day of exposure, all the CCl₄ is converted to CCl₃H. For the FeO, CCl₂H₂ is observed at 6 days into the experiment, although it may have been produced between day 2 and day 6 of the experiment as no analyses were done during that period. In the case of FeTiO₃, CCl₂H₂ is observed within the first hours of the experiment.

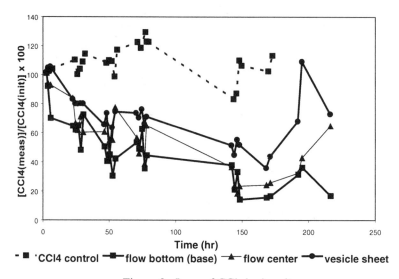

Figure 3. Loss of CCl₄ by basalt.

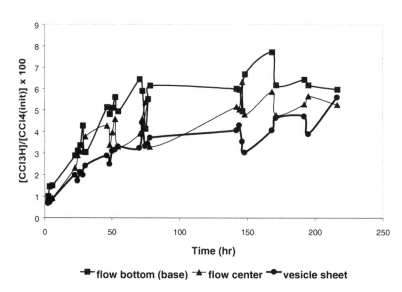

Figure 4. Appearance of CCl₃H from CCl₄ dechlorination by basalt.

The minerals and basalt are slower to degrade CCl_4, on the order of days (Figures 1 and 3). Although CCl_3H is observed within the first hours of the experiment, it is a few weeks into the experiment before approximately half of the CCl_4 is converted to CCl_3H. By the end of the experiments (8 weeks), the CCl_4 has not been fully converted to CCl_3H, and no CCl_2H_2 is observed for any of the minerals or basalt.

In contrast to the TCE experiments, CCl_4 losses due to evaporation and/or absorption into the septa were not observed. The CCl_4 control samples (vials containing aqueous CCl_4 only) remained relatively constant throughout the duration of the experiment (Figure 1 and 3). Thus, more quantitative measurements of the rates of dechlorination of CCl_4 and mass balances could be made.

Mass Balances

Mass balances were determined for the dechlorination of CCl_4. An example of those mass balances are shown in Figure 5 in which the difference between the initial number of μmoles of CCl_4 and the sum of measured μmoles of CCl_4 and its degradation products (CCl_3H and CCl_2H_2) for the basalt minerals (olivine, plagioclase, augite, magnetite, and ilmenite) are plotted versus time (hr). In the case of these minerals, with the exception of plagioclase, the total μmoles of measured chlorinated species (CCl_4, CCl_3H, and CCl_2H_2) accounted for the initial number of μmoles of CCl_4 to within experimental error (ca. 10%). Likewise, the mass balances of the Fe(II) oxides and basalt (not shown) account for the initial μmoles of CCl_4. In constrast, the plagioclase results show that at the 3 week mark, only 75% of the inital CCl_4 is accounted for in the sum of the measured CCl_4 and CCl_3H. The differences seen in the plagioclase data are discussed below.

Discussion

Reductive Dechlorination

The results reported above clearly show that dechlorination of CCl_4 in the presence of iron oxides, basalt, and its minerals readily occurs as compared to dechlorination of TCE under the same conditions. The reductive potentials of these species offer some insights into the differences observed. According to work reported by Vogel, Criddle, and McCarty (21), the E_o for the reductive

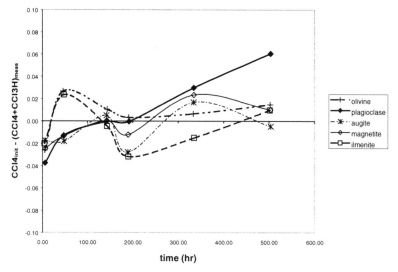

Figure 5. Mass balance plots (initial µmoles of CCl$_4$ − measured µmoles of (CCl$_4$ + CCl$_3$H)) for basalt minerals.

dechlorination of CCl_4 forming CCl_3H is approximately 1.1 V[i] whereas for the dechlorination of TCE forming cis-1,2-DCE the E_o is approximately 0.9 V. Thus, based solely on the reductive potentials, it would be predicted that CCl_4 would be more readily dechlorinated compared to TCE.

The results show that FeO and $FeTiO_3$ were the only Fe(II)-bearing materials which were strong enough reductants to dechlorinate the TCE. In contrast, all the Fe(II)-bearing materials were capable of mediating reduction reactions for CCl_4. The results show that FeO and $FeTiO_3$ are the strongest reductants as complete dechlorination of CCl_4 to CCl_3H occurs within the first day of the experiment. Subsequent dechlorination of the CCl_3H to CCl_2H_2 was observed to occur at both of theses iron oxides. Dechlorination of the CCl_4 at Fe_3O_4 and the Fe(II)-bearing minerals of basalt (olivine, augite, and magnetite) was observed to occur less rapidly compared to FeO and $FeTiO_3$ with complete conversion of CCl_4 to CCl_3H not observed over the 8 week period of the experiment. Additionally, CCl_2H_2 was not observed in these experiments.

The reduction potential for the oxidation of FeO in aqueous solution forming $Fe(OH)_3$ is reported to be between 0.2 and 0.3 V (22). Comparing the reductant strength of FeO to Fe^o (E_o=-0.44 V) and dissolved Fe^{+2} (E_o=0.77 V) (22), FeO is approximately half-way in between these two Fe-species. The rest potential (or equilibrium potential) for $FeTiO_3$ has been reported by White and Hochella (23) to be between 0.2 and 0.3 V suggesting similar reductant strength as compared to FeO. The surface area of the FeO used in these experiments was measured to be 0.07 m^2/g, and the $FeTiO_3$ surface areas was measured to be 0.7 m^2/g. This order of magnitude difference in surface area most likely accounts for the slight differences observed in the FeO and $FeTiO_3$ dechlorination data (particularly the TCE data).

The reductant strengths of the Fe(II)-basalt minerals olivine and augite as well as for basalt itself have not been studied. Based on the results observed in this study, the reductant strengths of these materials are speculated to be less than FeO or $FeTiO_3$. The surface area of olivine was measured to be 2 m^2/g, and augite was meaured to be 0.5 m^2/g. These values are similar or higher than the FeO and $FeTiO_3$ which does not explain decreased dechlorination activity. Based solely on the fact that the Fe content of olivine and augite is less than for FeO and $FeTiO_3$, one would predict olivine and augite would be less proficient at supplying electrons for reductive chemistry.

Interpretation of the magnetite results are not so straight forward. In Figures 1 and 3, dechlorination by magnetite (as the rock) is similar to the other basalt minerals (olivine and augite) and Fe_3O_4 (as the iron oxide powder) is the slowest Fe(II)-bearing material to dechlorinate CCl_4. The surface area of the magnetite rock was measured to be 0.09 m^2/g which is the lowest surface area of any of the

---

[i] All E_o values vs. normal hydrogen electrode

materials used in this work. The surface area of the Fe_3O_4 powdered was measured to be 2 m^2/g which is higher than either the FeO or $FeTiO_3$. Previous work by White and Hochella compared the redox chemistry of ilmenite ($FeTiO_3$) with that of magnetite (Fe_3O_4) for the reduction of Cr^{+6} to Cr^{+3} in aqueous solution (23). The rest potential reported for magnetite was similar to ilmenite (0.2 to 0.3 V), and the reductant strength of the two iron oxides were observed to be similar with respect to Cr^{+6} reduction. These results are in contrast to the results observed for the reductive dechlorination of TCE and CCl_4 reported here and suggest that thermodynamic factors do not thoroughly explain why CCl_4 dechlorinates more readily on some Fe(II)-bearing surfaces than others. Differences in adsorption of CCl_4 at the various Fe(II)-bearing surfaces may play a role in dictating the redox chemistry of these organic species.

Dechlorination at Plagioclase

In addition to reductive dechlorination of the CCl_4 at the Fe(II)-bearing materials, dechlorination of CCl_4 to form CCl_3H was observed for the samples containing plagioclase (surface area = 0.3 m^2/g). This was an interesting and unexpected result as it was believed that the plagioclase would not alter the CCl_4 since it is not capable of sustaining redox chemistry. However, the plots shown in Figures 1 and 3 clearly indicate that the CCl_4 is decreasing over time, and CCl_3H is being formed. The mass balance for the degradation of CCl_4 in the presence of plagioclase in which the unreacted CCl_4 and CCl_3H produced by dechlorination do not account for all the CCl_4 initially present. These data together with the fact that the plagioclase cannot support reductive dechlorination imply that an alternate mechanism for dechlorination to occur.

We suggest two possible explanations for these observations. First, the CCl_4 may be homolytically cleaved to form radical species. The formation of radicals at trace metal oxide sites such as V, W, and/or Mo in the plagioclase may be the responsible for the dechlorination of CCl_4 (24). It has been shown that these metals can partition into basalt minerals, including plagioclase (25). Bulk analysis of the plagioclase used in these experiments indicates Mo is present in trace amounts (ppb levels). The second explanation is that Fe is also present in the plagioclase at ppm levels which may be capable of reductive dechlorination. It should be noted that the bulk analysis was performed on plagioclase that was crushed using a ceramic mortar and pestle. In contrast, the plagioclase used in the reactivity studies was crushed using a stainless steel chimney which could also have contributed to the Fe content of the plagioclase.

Dechlorination at Basalt

The results indicate that abiotic dechlorination of aqueous CCl_4 in anaerobic conditions occurs when in the presence of basalt. Three different basalt samples taken from Box Canyon, ID were exposed to aqueous CCl_4. The samples chosen for the experiment included one from the lower quenched margin of the flow (flow bottom), one from the diktytaxitic center (flow center), and one from a horizontal vesicle sheet (vesicle sheet). These samples were chosen as they are representative of the three fundamental types of rock within the Box Canyon basalt flow. The surface areas of these samples were similar (ranging from 1.5 to 2.5 m^2/g). The results show that all three basalt samples provide similar degradation of the CCl_4 (Figures 3 and 4).

Comparing the conversion of CCl_4 to CCl_3H of the separate minerals of basalt (Figures 1 and 2) to the field samples of basalt (Figures 3 and 4), degradation occurs in relatively the same time frame. At approximately 8 days into the experiment (200 hr), >60% of the CCl_4 has been converted to CCl_3H for both the minerals and the basalt itself. As stated above, the mass balances of the basalt samples show that the initial μmoles CCl_4 are accounted for by summing the unreacted CCl_4 and the CCl_3H formed during the experiment. These results suggest that reductive dechlorination is the dominant degradation mechanism. It should be noted that plagioclase makes up the largest portion of the basalt (ca. 32%) which could result in alternate mechanisms associated with the degradation of CCl_4 at these mixed phase materials. The mass balance results are not indicative of mechanisms in which CCl_3H is not the dominant degradation product; however, additional investigation of alternate mechanisms is warranted.

Conclusions based on the work presented here suggest that subsurface basalt environments are capable of abiotic dechlorination of CCl_4 under anaerobic conditions. These results suggest that the first step of dechlorination (conversion of CCl_4 to CCl_3H) should be considered in the environmental modeling of the fate and transport of CCl_4 contamination of areas in which basalt is the major rock formation. Subsequent dechlorination (conversion of CCl_3H to CCl_2H_2) was not observed to occur on the time frame of the experiments performed in this work, but may also be possible over very long time frames. The TCE results suggest that basalt is not likely to abiotically dechlorinate TCE.

Currently, investigations of alternate dechlorination mechanisms for the degradation of CCl_4 in the presence of plagioclase are underway. Because of the aluminosilicate nature of the plagioclase, these investigations may have an important impact on understanding CCl_4 degradation in subsurface environments other those containing basalt.

Acknowledgments

The authors would like to thank Cathy Rae and Alana Oliver for their assistance in collecting the SPME/GC data, Byron White for collecting the ion chromatography data, Dr. Mason Harrup and the SIMS research group at the INEEL for technical discussions. Financial support for this work was provided by the Department of Energy, Office of Biological and Environmental Research through DOE Idaho Operations Office Contract DE-AC07-99ID13727 BBWI.

References

1. "A Comprehensive Inventory of Radiological and Nonradiological Contaminants in Waste Buried in the Subsurface Disposal Area of the INEL RWMC During the Years 1952-1983." Gov. Report EGG-WM-10903.
2. Kaminsky, J. F.; Wylie, A. H. *Ground Water Monitoring & Remediation* **1995**, *15*, 97-103.
3. "Groundwater/Vadose Zone Integration Project Background Information and State of Knowledge, Volume II", Gov. Report DOE/RL-98-48, June 1999.
4. Brown, G. M. In *The Poldervaart Treatise on Rocks of Basaltic Composition, Vol. 1*; Hess, H. H.; Poldervaart, A. Eds.; Interscience Publishers: New York, 1962; p. 103.
5. Wilson, E. K. *Chem. & Eng. News* **1995** (July 3), *73*, 19-22.
6. Gillham, R. W.; O'Hannesin, S. F. *Ground Water* **1994**, *32*, 958-967.
7. Burris, D. R.; Campbell, T. J.; Manoranjan, V. S. *Environ. Sci. Technol.* **1995**, *29*, 2850-2855.
8. Johnson, T. L.; Tratnyek, P. G. In *In-Situ Remediation: Scientific Basis for Current and Future Technologies*, (33rd Hanford Symposium on Health and the Environment, Pasco, WA, November 1994) Gee, G. W.; Wing, N. R. Eds.; Battelle Press, Columbus, OH, 1994, pp. 931-947.
9. Orth, S. W.; Gillham, R. W. Environ. Sci. Technol. **1996**, *30*, 66-71.
10. Roberts; A. L.; Totten, L. A.; Arnold, W. A.; Burris, D. R.; Campbell, T. J. *Environ. Sci. Technol.*, **1996**, *30*, 2654-2659.
11. Fredrickson, J. K.; Gorby, Y. A. *Curr. Opin. Biotechnol.*, **1996**, *7*, 287-294.
12. Gorby, Y. A., Amonette, J. E., Fruchter. J. S. In *In-Situ Remediation: Scientific Basis for Current and Future Technologies*, (33rd Hanford Symposium on Health and the Environment, Pasco, WA, November 1994) Gee, G. W.; Wing, N. R. Eds.; Battelle Press, Columbus, OH, 1994, pp. 233-248.

13. Sivavec, T. M., Horney, D. P., Baghel, S. S., In Emerging Techologies in Hazardous Waste management VII, Extended Abstracts for the Special Symposium. Am. Chem. Soc., Atlanta, GA, 1995, p. 42-45.
14. Kriegman-King, M.; Reinhard, M. "Abiotic Transformation of Carbon Tetrachloride at Mineral Surfaces", U. S. Environmental Protection Agency Report # EPA/600/SR-94-018, March 1994.
15. Kriegman-King, M. R.; Reinhard, M. *Environ. Sci. Technol.* **1994**, *28*, 692-700.
16. Arthur, C. L.; Potter, D. W.; Buchholz, K. D.; Motlagh, S.; Pawliszyn, J. *LC·GC* **1992**, *10*, 656-661.
17. Saraullo, A.; Martos, P. A.; Pawliszyn, J. *Anal. Chem.* **1997**, *69*, 1992-1998.
18. Ai, J. *Anal. Chem.* **1997**, *69*, 3260-3266.
19. Gossett, J. M. *Environ. Sci. Technol.* **1987**, *21*, 202-208.
20. Matheson, L. J.; Tratnyek, P. G. *Environ. Sci. Technol.* **1994**, *28*, 2045.
21. Vogel, T. M.; Criddle, C. S.; McCarty, P. L. *Environ. Sci. Technol.* **1987**, *21*, 722-736.
22. Pourbaix, M. Atlas of *Electrochemical Equilbria in Aqueous Solutions*, National Association of Corrosion Engineers, Houston, TX, 1966, pg. 307.
23. White, A. F.; Hochella, Jr., M. F. *Proc. 6th Int'l. Symp. Water-Rock Interaction*, 1989, pp. 765-768.
24. Sattari, D.; Hill, C. L. *J. Am. Chem. Soc.*, **1993**, *115*, 4649-4657.
25. Dunn, T.; Sen, C. *Geochim. Cosmochim. Acta*, **1994**, *58*, 717-733.

Physical and Radiation Chemistry

Chapter 17

Radiation-Induced Processes in Aqueous Suspensions of Nanoparticles and Nanoscale Water Films: Relevance to H$_2$ Production in Mixed Waste and Spent Nuclear Fuel

Thom M. Orlando[1] and Dan Meisel[2]

[1]Environmental Molecular Sciences Laboratory, Pacific Northwest National Laboratory, 908 Battelle Boulevard, Richland, WA 99352
[2]Radiaton Laboratory and Department of Chemistry and Biochemistry, University of Notre Dame, Notre Dame, IN 46556

The interaction of ionizing radiation with solids and particles generates electrons and holes with excess kinetic energies in the range of ~ 1-100 eV. For particles and solids in contact with the liquid phase, a portion of these non-thermal electrons may undergo inelastic scattering or trapping at the interface or be injected into solution prior to thermalization. Using pulse radiolysis, we follow those electrons that escape irradiated silica nanoparticles suspended in aqueous solution and appear as e$^-_{aq}$. We also probe the fate of holes produced in the silica by monitoring the generation of OH using oxidizable scavengers that remain in solution even at high silica loadings. For the 7 – 22 nm particle size range studied, essentially all electrons generated in the silica cross the interface and appear as e$^-_{aq}$, whereas, none of the holes escape. Modifying the particle surface with electron-acceptors can control interfacial charge transfer, but merely changing the surface potential seems to have little effect. We also report the yield of molecular hydrogen from these suspensions and suggest possible interfacial back reactions with trapped holes. In addition, we probe secondary electron interactions with nanoscale water films using ultrahigh vacuum surface science techniques and demonstrate the importance of exciton decay and H^- (D^-) reactive scattering in radiation-induced production of H_2 (D_2). These studies are relevant to the production of molecular hydrogen in mixed (radioactive/chemical) wastes and during storage of spent nuclear fuel.

The presence of interfaces, in particular the solid/liquid interface, introduces complications with regard to understanding radiation-induced chemical trans-formations in heterogeneous systems. Commonly, charge carriers (i.e. electron-hole pairs), are initially formed upon irradiation. The carriers may recombine releasing their excess energy to the surroundings or may be trapped at various sites in the material. In the presence of an interface, either one or both of the carriers may cross the interface. We attempt to quantitatively determine which carriers cross the interface, what material properties govern that process, and what chemical reactions these carriers initiate. We seek to determine whether interfacial electron or hole transfer depends upon the dimensions of each subphase and whether the presence of solid/liquid interfaces enhances or reduces the cross sections for water "radiolysis" and molecular hydrogen formation. Though there is some available literature on these issues (1-4), much of it is not quantitative and does not address the solid particle/water interface.

Many of the systems that encounter ionizing radiation in the real world are heterogeneous mixed phase systems. For example, nuclear materials are often stored, at least temporarily, as colloidal aqueous suspensions (5). Gases such as H_2, NH_3 and N_2O are also produced and released in some United States Department of Energy (DOE) mixed (radioactive/chemical) waste storage facilities. The detailed mechanisms governing the generation, retainment and release of these gases from the mixed wastes are not fully understood. The production of H_2 and O_2 from water radiolysis during the storage of spent nuclear fuel is also an important and potentially costly concern facing the DOE. Finally, the successful development of plasma or radiation-based remediation strategies relies upon understanding charge transfer and non-thermal carrier-induced processes at particle interfaces (6-8).

The storage of radionuclides and radioactive materials necessarily requires at least air/solid interfaces. These interfaces are usually in contact with a humid atmosphere. If the energy that is adsorbed by the colloidal particles or solid phases is transferred to the interface or released to the aqueous phase, the yields and chemical reactions may be dominated by interface reactions or aqueous radiation chemistry, respectively. In this paper, we present direct evidence that electrons produced in nanoparticles of silica appear in the aqueous phase. These e^-_{aq} electrons can contribute to the production of molecular hydrogen. We also demonstrate that the inelastic scattering of non-thermal electrons at water interfaces can produce molecular hydrogen via exciton decay and H^- reactive scattering. The latter channel is initiated via dissociative electron attachment (DEA) resonances and may be partially responsible for the non-scavangeable hydrogen produced via radiolysis of liquid water.

Experimental

We have carried out high-energy pulse-radiolysis of aqueous colloidal suspensions of silica and low-energy electron-beam bombardment of nanoscale water films. The pulse radiolysis experiments were performed using 2-3 ns pulses of 8 MeV electrons from the Notre Dame linear electron accelerator (TB-8/1S linac). The low-energy (5-100 eV), electron-beam irradiation studies were carried out at the Environmental Molecular Sciences Laboratory (EMSL) of Pacific Northwest National Laboratory (PNNL).

For the pulse-radiolysis studies, several principles guided our choice of particles. Colloidal solutions allow convenient spectrophotometric detection. Silica, in addition to being a common material in many applications, is also transparent across the near UV-visible range. Details of our experimental approach have been published (9-11). Briefly, we followed directly the absorption of e^-_{aq} at their red absorption . Holes (OH radicals) were rapidly scavenged using negatively charged scavengers (e.g. $Fe(CN)_6^{4-}$). All experiments were performed at the basic pH range [10-11], where silica is negatively charged and thus the scavengers remain in the water phase, removed from the interface. Silica particles were DuPont Ludox products ranging in size from 7 to 22 nm in diameter. High concentrations of particles are easy to obtain and the energy levels of the silica are conveniently posed relative to the redox levels of water.

The nanoscale water film experiments were carried out in an ultrahigh vacuum (UHV) chamber (base pressure $\sim 2 \times 10^{-10}$ torr) equipped with a pulsed low-energy electron gun, an effusive gas doser, a quadrupole mass spectrometer (QMS), a time-of-flight (TOF) detector and a Pt(111) crystal mounted on a liquid-nitrogen-cooled manipulator. Ice films were grown via vapor deposition of ultrapure D_2O and the thickness was determined by comparing D_2O temperature programmed desorption (TPD) spectra to published data (12), and has an estimated accuracy of ~ 20 %. D_2O was used to discriminate against background H_2O and H_2. Coverages are reported in terms of ice bilayers

($\sim 10^{15}$ molecules/cm^2). Film temperature was measured using a K-type thermocouple, and has an estimated accuracy of ± 5 K.

D$^-$ ions were generated by a monoenergetic electron beam, with an energy spread of ~ 0.3 eV full-width at half-maximum, and collected in the TOF spectrometer. The incident electron beam was pulsed at 100 Hz, with a pulse length of 1 μsec and an instantaneous current of $\sim 10^{-7}$ A in a spot size of ~ 0.1 cm^2, resulting in a time-averaged current density of $\sim 10^{-10}$ A / cm^2. The very low incident electron flux used in these experiments results in negligible charging or thermal heating of the substrate (13-15). For the temperature dependence measurements, the ice films were heated radiantly using a tungsten filament mounted behind the Pt crystal.

We have utilized (2+1) resonance enhanced multiphoton ionization (REMPI) spectroscopy via the E, F $^1\Sigma_g^+$ state to detect the D$_2$ products in the (X $^1\Sigma_g^+$) ground state (16). The laser wavelengths necessary for this detection scheme were generated by frequency tripling the output of a Nd:YAG pumped dye laser. Typical output powers ranged from 1-2 mJ/pulse and the detection efficiency is $\sim 10^6$ molecules/cm^3/quantum state. All laser experiments were done in the TOF mode in which the neutrals produced by the pulsed electron-beam traverse a 4 mm distance before they are resonantly ionized by the focused laser beam. Due to the small signal levels in these experiments, a relatively long ($\sim$20-25 μsec) electron beam pulse (20 Hz) was used which integrates a range of desorbate velocities. The ionizing laser pulse was delayed with respect to the electron beam pulse such that the D$_2$ molecules detected had translational energies < 0.5 eV.

Results and Discussion

I. Pulse-Radiolysis of Silica Nanoparticle Suspensions

A. Escape of Electrons

Figure 1 shows the relative yield of hydrated electrons observed in the silica suspensions. As the SiO$_2$ concentration increases, the density of the sample increases because the density of silica is 2.3-g cm^{-3}. Thus, the absorbed dose increases as shown by the upper solid curve that follows the density of the sample. Furthermore, as the SiO$_2$ concentration increases, the volume of water decreases and therefore less of the electron beam traverses water. This excluded volume effect further reduces the dose absorbed by the water. When the two effects are considered, the fraction of the dose absorbed by the silica equals the difference between the two solid curves.

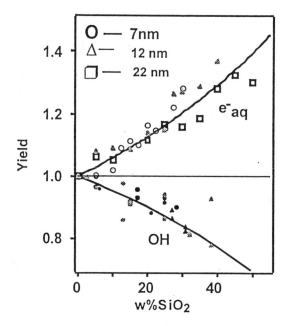

Figure 1. Relative concentration of e⁻ₐq and oxidized products from OH radicals upon increasing SiO₂ concentration. Upper curve is sample density and lower curve is volume fraction of water.

It is clear from the experimental results shown in Figure 1 that the concentration of e^-_{aq} *increases* upon addition of silica even though the dose absorbed by water *decreases*. Furthermore, the yield of electrons seems to follow the increase of total dose absorbed by the sample. (Note that all samples were irradiated with the same dose). This observation implies that regardless of the original energy deposition location, all the electrons become hydrated within ~100 ps after the pulse. This seems to occur with the same efficiency as if the electrons were originally generated in water. Though recombination processes on the 100 ps time scale seem to be as efficient in silica as in water, electron trapping in these small silica particles seems to be inefficient.

B. Trapping of holes and charge separation

The mobility of hot holes in silica is approximately two orders of magnitude slower than that of electrons. Furthermore, and contrary to electrons, many trapping sites for holes have been identified in silica. Thus, one may expect more efficient trapping of holes relative to electrons and consequently less escape into the aqueous phase. This expectation can be put to test by measuring the yield of OH radicals in the aqueous phase following radiolysis of a suspension of silica colloidal particles. Because of the technical difficulties in directly measuring OH radicals, we measured the yield of $Fe(CN)_6^{3-}$ (from ferrocyanide) and of $(SCN)_2^-$ (from SCN^-). These negatively charged scavengers are expected to reside exclusively in the aqueous phase, removed from the particle surface.

The result from suspensions of the two smaller particle sizes, 7 and 12 nm in diameter, are shown in the lower portion of Figure 1. Upon increasing the particles' concentration, the relative yield of OH radicals decreases. This in clear contrast with the results for e^-_{aq}, also shown in Figure 1. Note that the holes in silica do not generate OH radicals in water. Moreover, the results seem to follow the solid lower curve shown in Figure 1 which describes the fraction of energy absorbed by the aqueous phase. Thus, the only observed OH radicals seem to be those that were generated from radiolysis of the aqueous phase. No holes cross the interface to generate OH radicals in water even from the smallest particles.

It was shown above that electrons cross the interface from the solid particles to water whereas holes do not. Thus, we conclude that charge separation across the interface has occurred. The solid accumulates holes while electrons are trapped in water. Mechanisms to relieve this electrostatic asymmetry must intervene since electrons will not escape the continuously increasing attractive potential. Charge compensation by migration of ions from the solution towards the surface will undoubtedly occur and will relieve some of the electrostatic stress. The solution pH is the dominant parameter that determines the surface-potential for oxides. Therefore, protons or hydroxide ions are the most common ions to migrate to the surface in oxides. Similarly, they are also the ions that may be released from the solid particle. In the case at hand, protons may be released from the surface to reduce the electrostatic potential. The holes will initially be trapped at the various trapping sites in the material and eventually the valence band may become populated. At present, we have no estimate of the trapped hole energy levels, nor can we predict the extent charges accumulate.

C. Capture of electrons at the interface

Since e^-_{aq} contributes to H_2 generation in irradiated systems, the presence of oxide particles in mixed wastes may affect gas generation rates. We attempted to affect the escape of electrons across the interface by changing the surface potential and by modifying the particle surface to include deeper electron traps than the hydrated electron redox potential.

To change the surface potential of the silica particles we added Mg^{2+} ions to the suspension (up to 25 mM). At pH 10-11, these ions strongly adsorb at the surface. The experiments shown in Figure 1 for e^-_{aq} were then repeated. No effect of Mg^{2+} ions on the yield of e^-_{aq} could be observed, while the same effect of silica concentration shown in Figure 1 was reproduced. Thus, we conclude that under these experimental conditions, a significant reduction in the negative charge density at the surface has little effect on the yield of e^-_{aq}. Similar experiments to were repeated with the electron acceptor methyl-viologen, MV^{2+} added rather than Mg^{2+} ions and the results are presented in Figure 2.

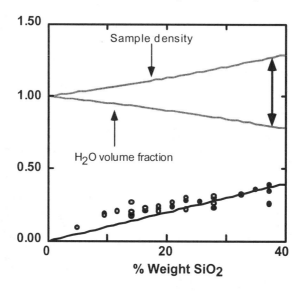

Figure 2. Fraction of MV^+ from capture of electrons escaping SiO_2 particles as a function of % weight SiO_2. Silica particles are 7 nm (o) or 12 nm (•) in diameter. The lowest solid line (and double-arrow) is the energy fraction absorbed by SiO_2. (Reproduced with permission from ref. 9. Copyright 1999.)

It was confirmed that all of the MV^{2+} acceptors are adsorbed on the silica particles. Because of the reduced concentration of MV^{2+} in the bulk of the aqueous phase, the rate of reaction of e^-_{aq} with the acceptor is very slow. Furthermore, from the ionic strength dependence of this rate constant, it was shown that under these conditions, e^-_{aq} reacts against the negative potential from the particle surface. Thus, the rate-determining step is the approach of e^-_{aq} to the negatively charged SiO_2 particle. To scavenge e^-_{aq} that originate from ionization of water, NO_3^- ions were added to the solution. As the concentration of the nitrate ion increases, the yield of MV^+ radical cations decreases until it reaches a plateau at $[NO_3^-]$ = 10 mM. At this concentration, all of e^-_{aq} reacts with nitrate and not with MV^{2+}. Figure 2 shows the fraction of captured e^-_{aq} expressed in terms of the fraction of MV$^+$. The solid line at the bottom of Figure 2 indicates the fraction of dose absorbed in silica. The double-arrow line shows the same parameter. The yield of captured electrons (i.e., the limiting yield of MV$^+$) tracks the fraction of dose absorbed by the silica particles. Since this is also the contribution of silica to the total generation of e$^-_{aq}$, one concludes that the acceptors at the surface eventually capture all the electrons that cross the interface. We cannot preclude the possibility that the acceptor's presence affects competing processes such as recombination, trapping, and escape.

D. Yield of H$_2$

The generation of molecular hydrogen resulting from irradiating suspensions of oxide particles is important and particularly relevant to gas production in DOE mixed wastes. The yields of hydrogen gas in irradiated oxide systems are related to the band-gap of the particles as well as the geometric and electronic structure of the adsorbate (water) substrate complex (17). There is also ample evidence to show that much of the H$_2$ produced in neat water involves reactions of hydrated electrons and their precursors (18). In spite of this, it is conceivable that new pathways may become available in the presence of silica particles. We, therefore, measured the yield of H$_2$ in silica suspensions at high SiO$_2$ loading. Figure 3 shows the yield of H$_2$ as a function of SiO$_2$ concentration at three particle sizes.

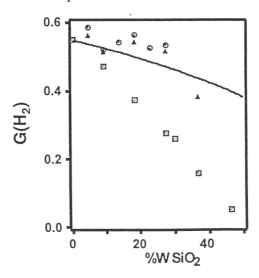

Figure 3: Yield of H$_2$ vs. SiO$_2$ loading for 7nm (circles), 12 nm (triangles), and 22 nm (squares) particles. Solutions contain 25 mM KNO$_2$.

The solid curve is the predicted yield, $G(H_2)$ in molecules per 100 eV of absorbed dose, if no interaction between the aqueous and solid phases occurs. This curve reflects the excluded volume effect described above. The two smaller sizes follow this curve reasonably well. The solutions in these experiments contained 25 mM NO_2^- ions to scavenge any OH radicals, H atoms and e_{aq}^- in the bulk aqueous phase. The latter two may otherwise produce H_2 from water radiolysis. Under these experimental conditions, no additional H_2 is produced in the suspensions containing the smaller SiO_2 particles and the larger particles seem to slightly inhibit H_2 production beyond the excluded volume effect. It should be emphasized that when molecular hydrogen was injected into the suspensions at concentrations similar to the radiolytic yield, essentially all of it was later recovered from the suspensions.

Our tentative rationalization for this result is a back-reaction of the trapped holes with molecular hydrogen. Indeed, when hole acceptors were adsorbed at the surface, the yield of H_2 recovered and the size dependence disappeared. As the size of the particle increases, the probability a given particle contains a trapped hole during irradiation also increases. Therefore, the probability that an H_2 molecule from the water will encounter a hole-containing particle is larger for the bigger particles. The probability that a single particle will contain two holes, required for oxidation of hydrogen to protons, is even more dependent on particle size. The rate at which the trapped hole can oxidize any reductant at the surface also decreases with increasing particle size.

II. Low-Energy Electron Bombardment of Nanoscale Water Films

A. Electron Emission and Dissociative Electron Attachment Resonances

It is well known that materials irradiated with high-energy electrons emit secondary electrons and that the maximum secondary electron emission yield for uncharged insulator surfaces, δ_m^o, is generally much greater than 1.0. The typical energy distribution of emitted secondary electrons is shown in Figure 4.

The energy distribution of secondary electrons generally peaks at a few eV, however, a considerable number have energies which extend out to > 50 eV. **These emitted secondaries must undergo multiple inelastic scattering events as they traverse the interfacial molecular layers prior to trapping or solvation.** The primary energy-loss channels for electrons with energies typical of the secondary distribution are ionization (electron-hole pair production in the condensed phase), direct electronic excitation and resonance scattering. Resonance scattering results in the formation of transient negative ion states, which decay via electron autodetachment and dissociative electron attachment (DEA). DEA usually involves multi-electron core-excited resonances, which consist of an excess electron temporarily bound by the positive electron affinity of an electronically excited target molecule. These are generally two-electron, one-hole configurations that are classified as either Feshbach or core-excited

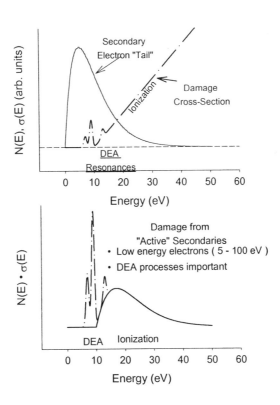

Figure 4. Top Frame:Typical energy distribution of secondary electrons emitted from keV electron-beam irradiated insulators. Also shown are the estimated relative cross-sections for some of the main inelastic energy-loss channels the secondaries undergo as they traverse the interface. Lower Frame: Estimate of the effective "damage" probability obtained by convoluting the energy-loss cross sections with the secondary electron energy distribution.

shape resonances. Since Feshbach resonance lifetimes are typically 10^{-12} - 10^{-14} seconds, dissociation into stable anion and neutral fragments may result if the resonance is dissociative in the Franck-Condon region and one of the fragments has a positive electron affinity. DEA has been observed at water interfaces, and many body interactions affect the energies, widths and lifetimes of the scattering resonances (13-15). The lower portion of Figure 4 estimates the effective "damage" probability of water due to ionization and DEA. Though exciton decay and direct excitation of dissociative excited states are not included in the convolution integral, Figure 4 demonstrates that DEA can be an important energy-loss channel.

The energy dependence of the D⁻ electron-stimulated desorption (ESD) yield from a 5-bilayer thick amorphous film of D_2O ice on Pt(111) is illustrated in Figure 5. The D⁻ yield rises significantly in the interval between 5 - 14 eV and the baseline, which is attributed to D- production via dipolar dissociation,

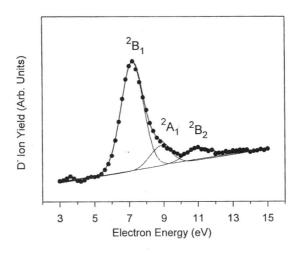

Figure 5. D⁻ signal vs. incident electron energy, collected at 120 K from a five-bilayer film of amorphous ice grown at 90 K. The solid lines are Gaussian fits to the data. (Reproduced with permission from ref. 15. Copyright 2000.)

gradually increases. The signal clearly peaks at ~ 7 eV, with a pronounced shoulder at ~ 9 eV and a slightly less intense feature at ~ 11 eV. As shown in the figure, a sum of three Gaussian peaks and a linearly increasing background fits this data well.

The ground state electronic structure of an isolated water molecule is $(1a_1)^2(2a_1)^2(1b_2)^2(3a_1)^2(1b_1)^2$, where the $1a_1$ orbital is essentially the O(1s) core level, the $2a_1$ and $1b_2$ orbitals are primarily involved in O-H bonding, and the $3a_1$ and $1b_1$ are non-bonding lone-pair orbitals. The lowest-lying unoccupied orbital is the $4a_1$. This strongly antibonding orbital mixes with the 3s Rydberg state, and hence is denoted as (3s:4a$_1$). The $^2B_1$, $^2A_1$ and $^2B_2$ DEA resonances, which correspond to states having two (3s:4a$_1$) electrons and a hole in the 1b$_1$, 3a$_1$, or 1b$_2$ orbital, respectively, have been observed in water vapor (19-21). Much of the Rydberg character of this level appears to be lost upon condensation due to short-range interactions between neighboring molecules, yet it retains its dissociative 4a$_1$ nature. The $^2B_1$ and $^2A_1$ resonances generate measurable ion yields in the ESD of condensed water films (13-15) and the first two peaks in Figure 5 are assigned as the $^2B_1$ and $^2A_1$ resonances, respectively. The third feature, at ~ 11 eV, is only resolvable at very low current densities and is assigned as the $^2B_2$ resonance (13-15).

B. D_2 Production and D⁻ Reactive Scattering

The temperature dependence of the electron-stimulated production and desorption of D$_2$ in the v=1, J=2 levels is presented in Figure 6. At 90 - 95 K, there is **no** D$_2$ signal above our detection sensitivity (~ 10^6 molecules-cm^{-3} quantum state^{-1}) in the v=1, J=2 levels (22). However, the yield of D$_2$ in the

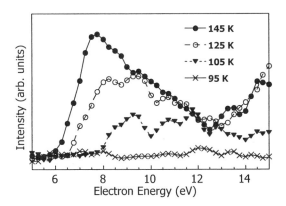

Figure 6. Temperature dependence of the D$_2$ (v=1, J=2) signal vs. incident electron energy from a sixty-bilayer film of amorphous ice. The films were grown at 90 K and data were collected with the films held at the indicated temperatures. (Reproduced with permission from ref. 15. Copyright 2000.)

v=1, J=2 levels clearly increases when the substrate is heated above 100 K and the threshold shifts to lower energy. This very reminiscent of the temperature and energy dependence of the D⁻ yield (13-15). The comparison for the yield of

D_2 in the vibrationally and rotationally relaxed levels ($v=0$, $J=0$) is not shown but is also similar (15). These temperature dependencies are attributed to thermally induced changes in the hydrogen bonding network, which changes the lifetimes of the predissociative states that lead to ESD and which also allows for reorientation of surface molecules. The exothermic ($\Delta H_g = -0.45$ eV) proton transfer reaction: D^- (H^-) + D_2O (H_2O) ---> D_2 (H_2) + OD^- (OH^-) has been studied in the gas-phase and is assumed to proceed via the formation of a proton-bound intermediate complex (23) which may resemble D-defects in ice. Evidence for this reaction has also been presented in studies of low-energy electron attachment to van der Waals clusters of D_2O (H_2O) (ref. 24). The D^- (H^-) + D_2O (H_2O) ---> D_2 (H_2) + OD^- (OH^-) reaction may be enhanced in the condensed phase due to the large number of collision partners and the increased exothermicity resulting from ion solvation. The data in Figure 6 suggests that the production of D_2 may involve reactive scattering of the D^- species, which is produced via DEA. To show this more clearly, we directly compare the D^- yield at 140 K with the yield of D_2 in the $v=1$, $J=2$ levels at elevated temperatures in Figure 7.

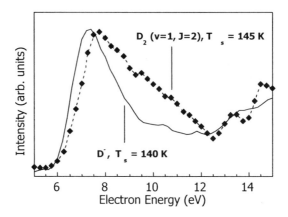

Figure 7. Comparison of D^- vs. electron energy with the D_2 ($v=1$, $J=2$) signal. (Reproduced with permission from ref. 15. Copyright 2000.)

These results indicate that low-energy electrons can lead to the production of vibrationally excited H_2 (D_2) via H^- (D^-) reactive scattering. The formation of these negative ions is primarily via the $^2B_1$ DEA resonance, however, the D_2 which is produced between 8-12 eV may involve the $^2A_1$ and/or $^2B_2$ resonance. The yield of vibrationally and rotationally relaxed H_2 (D_2) may also involve a similar reactive scattering event (15). In addition, we have previously shown that direct formation of H_2 (D_2) from water involves exciton decay and a molecular elimination step (26). Collectively, these channels can contribute to the formation of hydrogen at wet oxide interfaces, such as ZrO_2; an oxide component of some spent nuclear fuels.

Conclusions

In this report we focused our attention on irradiated suspensions of silica particles in water and low-energy electron interactions with water thin-films. We provide direct evidence that electrons produced in the silica rapidly appear in the water. As water is replaced by silica, more electrons are generated in water. However, our results show that the total yield of electrons (and therefore, holes as well) does not change significantly upon changing the composition of the suspension. Accordingly, we conclude that the competing recombination processes occur at the same efficiency in water and silica. Clearly, aqueous suspensions of small colloidal silica particles are a poor medium to store radionuclides. In addition, we have demonstrated the production of vibrationally excited H_2 (D_2) involves reactive scattering of H^- (D^-) ions which are produced primarily via the $^2B_1$ dissociative electron attachment resonance in water. The yield of vibrationally and rotationally relaxed H_2 (D_2) may also involve a similar reactive scattering event (15). This negative ion-molecule reaction and the direct formation of H_2 (D_2) via exciton decay, can contribute to the formation of hydrogen at wet oxide interfaces which may be present in mixed (radioactive/chemical) wastes and as components of spent nuclear fuels.

Acknowledgements:

This report summarizes the work of our colleagues, A. Cook (Brookhaven National Laboratory), N. Dimitrijevic (Argonne National Laboratory), A. Henglein (Notre Dame Radiation Laboratory-NDRL), H. Miyoshi (NDRL), T. Schatz (currently at NSF), W. C. Simpson (Pacific Northwest National Laboratory-PNNL) and G. A. Kimmel (PNNL). Their efforts and contributions are much appreciated. The work described here is supported by the US-Department of Energy (DOE), Office of Basic Energy Sciences and by the Environmental Management Science Program. This is contribution NDRL No. 4196 from the Notre Dame Radiation Laboratory. The Environmental Molecular Sciences Laboratory is a national scientific user facility sponsored by the DOE's Office of Biological and Environmental Research and located at Pacific Northwest National Laboratory (PNNL). PNNL is operated for DOE by Battelle under contract No. DE-AC06-76RLO 1830.

References:

1. Thomas, J. K., *Chem. Rev.* **1993**, *93*, 301-320.
2. Zhang, G.; Mao, Y.; Thomas, J. K., *J. Phys. Chem. B* **1997**, *101*, 7100-13.
3. Antonini, M.; Manara, A.; Lensi, P., *"The Physics of SiO$_2$ and Its Interfaces"*; Pantelides, S. T., Ed.; Pergamon Press: New York, **1978**, pp 316.
4. Seiler, H., *J. Appl. Phys.* **1983**, *54*, R1-18.
5. Lawler, A., *Science* **1997**, *275*, 1730.
6. Tonkyn, R. G., S. E. Barlow, and Orlando, T. M., *J. Appl. Phys.* **1996**, *80*, 4877-86.
7. Zacheis, G. A.; Gray, K. A.; Kamat, P. V., *J. Phys. Chem. B* **1999**, *103*, 2142.
8. Hilarides, R. J.; Gray, K. A.; Guzzetta, J.; Cortellucci, N.; Sommer, C., *Environ. Sci. & Technol.* **1994**, *28*, 2249.
9. Schatz, T.; Cook, A. R.; Meisel, D., *J. Phys. Chem. B*, **1999**, *103*, 10209-13.
10. Schatz, T.; Cook, A. R.; Meisel, D. *J. Phys. Chem. B* **1998**, *102*, 7225-30.
11. Dimitrijevic, N. M.; Henglein, A.; Meisel, D., *J. Phys. Chem. B*, **1999**, *103*, 7073-6.
12. Fisher, G. B., and Gland, J. L., *Surf. Sci.* **1980**, *94*, 446.
13. Simpson, W. C., Orlando, T. M., Parenteau, L., Nagesha, K., and Sanche, L, *J. Chem. Phys.* **1998**, *108*, 5027-34.
14. Simpson, W. C., Sieger, M. T., Orlando, T. M., Parenteau, L., Nagesha, K., and Sanche, L, *J. Chem. Phys.* **1997**, *107*, 8668-77.
15. Orlando, T. M., Kimmel, G. A., and Simpson, W. C., *Nucl. Instr. and Meth. in Phys. Res. B.* **1999**, *157*, 183-190.
16. K.-D. Rinnen et al., *J. Chem. Phys.* **1991**, *95*, 214.
17. Petrik, N. G., Alexandrov, A, Hall, A. I. , and Orlando, T. M., *Amer. Nucl. Soc. Trans.* **1999**, *81*, 101-103.
18. Pastina, B.; LaVerne, J. A.; Pimblott, S. M., *J. Phys. Chem. A* **1999**, *103*, 5841.
19. D. S. Belic, M. Landau, and R. I. Hall, J Phys. B: At. Mol. Phys. 1981, *14*, 175.
20. M. Jungen, J. Vogt, and V. Staemmler, Chem. Phys. **1979**, *37,* 49.
21. M. G. Curtis and I. C. Walker, J. Chem. Faraday Trans., 1992, *88*, 2805-10.
22. Kimmel, G. A. and Orlando, T. M., *Phys. Rev. Lett*, **1996**, *77*, 3983-86.
23. Betowski, D. et al., Chem. Phys. Lett. 1975, *31*, 321.
24. Knapp, M., Echt, O., Kreisle, and Recknagel, E., J. Phys. Chem. 1987, *91* 2601-07.
25. Klots, C. E., and R. N. Compton, *J. Chem. Phys.* **1978**, *69*, 1644-47.
26. Kimmel, G. A. and Orlando, T. M., *Phys. Rev. Lett*, **1995**, *75*, 2606-09..

Chapter 18

Thermochemical Kinetic Analysis of Mechanism for Thermal Oxidation of Organic Complexants in High-Level Wastes

Donald M. Camaioni and Tom Autrey

Pacific Northwest National Laboratory, 908 Battelle Boulevard, Richland, WA 99352

Complexants containing hydroxyl functionality, such as glycolate ion and hydroxyethylethylenediaminetriacetate (HEDTA), undergo aluminate-catalyzed thermally-activated oxidation by nitrite ion in alkaline high-level wastes (HLW) leading to the production of H_2 and nitrogenous gases. Aluminum species have been postulated to catalyze the formation of nitrite ester (RONO) intermediates that rapidly decompose to products at elevated temperatures. We tested this mechanism by measuring rate constants for RONO (for R = ethyl and $CH_2CO_2^-$) with and without sodium aluminate and found no effect on the rate of hydrolysis. The result suggests that different intermediates and mechanisms may be involved in aluminate-catalyzed oxidations.

Organic reactions induced by heat and radiation are studied to assess hazards of organic aging in HLW, understand methods for organic destruction, and predict rates of production of gases in the wastes. The major complexants used in Hanford nuclear materials processing were the sodium salts of hydroxyethylethylenediaminetriacetic acid (HEDTA), glycolic acid, ethylenediaminetetraacetic acid (EDTA), and citric acid (*1*). Of these, only HEDTA and glycolate ions are readily oxidized by thermally activated reactions to give flammable gases (*2-4*). EDTA, which is structurally quite similar to HEDTA, is relatively unreactive (*2,3*). Nitrite ion appears to supply the oxidizing equivalents for thermal oxidation while aluminate ions or other aluminum species are catalysts, and hydroxide ion promotes the oxidation (3-5). Work with waste simulants (2,3,5-7) has shown that the organic complexants containing hydroxyl groups (e.g., glycolate and HEDTA) decompose at a greater rate than the complexants without the hydroxylic functional groups (EDTA). Products from HEDTA include oxalate, formate, carbonate, ethylenediaminetriacetate, nitrogenous gases (N_2 and N_2O), and H_2 (3,4,7). Glycolate ion gives formate, oxalate, and carbonate, and the gases too (3,4,6).

Ashby and coworkers measured the kinetics of glycolate and HEDTA oxidations in waste simulants (3,6,7). For glycolate, the rate of oxidation is first order in glycolate, nitrite, and aluminate (3,6). For HEDTA, the kinetic behavior was complex, but clearly dependent on HEDTA concentration (7). To explain the observations, they proposed the following mechanism

$$Al(OH)_4^- + NO_2^- \rightleftarrows Al(OH)_3(ONO)^- + OH^- \tag{1}$$

$$Al(OH)_3(ONO)^- + HOCH_2R \rightarrow ONOCH_2R + Al(OH)_4^- \tag{2}$$

$$ONOCH_2R \rightarrow Products \tag{3}$$

in which $Al(OH_4)^-$ catalyzes nitrosation of glycolate ($R = CO_2^-$) and HEDTA ($R = (^-O_2CCH_2)_2NCH_2CH_2N(CH_2CO_2^-)CH_2$) (3,7). The intermediate, ONO-CH_2R, decomposes to NO^- and various organic moieties by competing pathways. Some of the pathways lead to formaldehyde and glyoxal. These products generate H_2 (8) and NO^- leads to N_2O, N_2 and NH_3. Note that by this mechanism, reaction 2 must be rate limiting for aluminate to appear in the rate law.

The sum of reactions 1 and 2 is the reverse of the well-known alkaline hydrolysis of alkyl nitrites (eq 4).

$$RCH_2OH + NO_2^- \rightleftarrows RCH_2ONO + OH^- \tag{4}$$

The equilibrium for eq 4 lies far to the left in alkaline solutions. The activation barriers for hydrolysis of nitrite esters (eq -4) are relatively large (9) such that even in molar hydroxide solutions, hydrolyses of alkyl nitrites occur slowly. Therefore, as shown in Figure 1, the mechanistic interpretation of Ashby and coworkers (3,7) is plausible provided aluminate actually does catalyze the nitrosation step.

As part of an effort to develop quantitative models for predicting reactions of organics in wastes, we recently measured the rates of alkaline hydrolysis of ethyl nitrite and nitritoacetate ion (nitrite ester of glycolate ion) in the presence and absence of dissolved $NaAl(OH)_4$. From Figure 1, we expected that if aluminum functions as a catalyst for nitrosation then the rate of the reverse reaction, hydrolysis of the nitrite ester, should also be catalyzed. The experiments seemed a promising way to obtain thermochemical kinetic information of the rate-controlling step. However, to our disappointment no catalysis was observed.

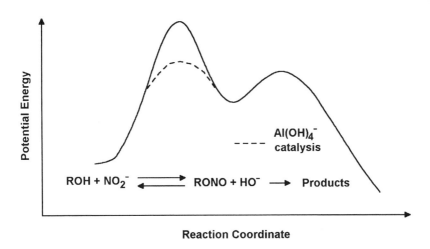

Figure 1. Potential energy diagram showing how Al(OH)₄⁻ may catalyze oxidation of glycolate and HEDTA (3).

Results

We present the results of our experimental work in three sections: 1) the reactions of ethyl nitrite and nitritoacetate in alkaline solutions, 2) nitrosyl exchange reactions between simple alkyl nitrite esters and glycolate ion, and 3) determination of nitritoacetate hydrolysis rate constants from kinetic measurements of the effect of glycolate ion on the hydrolysis rates of ethyl nitrite.

Hydrolysis Reactions of Ethyl Nitrite

We revisited the reactions of alkyl nitrites in sodium hydroxide solution as a preliminary step towards studying nitrite esters of complexants. Oae and coworkers (9) have provided thermochemical kinetic data on the base-catalyzed hydrolysis of a series of organic nitrite esters. They found that simple alkyl nitrite esters are more stable to base-catalyzed hydrolysis than their corresponding carboxylate esters.

Figure 2 shows the ultraviolet (UV) absorption spectra of ethyl nitrite in aqueous solution containing 1.5 M OH⁻ and 20 vol% dimethyl sulfoxide (DMSO) after mixing and after 4 h at room temperature.

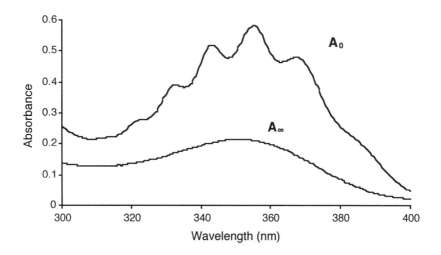

Figure 2. UV-visible absorption spectra of ethyl nitrite in 1.5 M OH⁻, aqueous 20 vol% DMSO. A_o: after mixing. A_∞: after 4 hours hydrolysis.

The absorbance at the beginning of the reaction shows the characteristic absorption features between 320 to 400 nm (bands separated by approximately 1000 cm⁻¹, $\varepsilon_{355} = 90$ M⁻¹ cm⁻¹), typical of the organic nitrites. At long reaction times, the absorption feature of the alkyl nitrite is replaced by the slightly blue-shifted weak broad absorbance of of sodium nitrite ($\lambda_{max} = 350$ nm, $\varepsilon_{350} = 25$ M⁻¹ cm⁻¹). Figure 3 shows that the absorbance at 355 nm decays exponentially over the course of the reaction.

Addition of aluminate ion has little effect on the observed rate of ethyl nitrite hydrolysis. Figure 4 compares the decay in absorbance at 355 nm in the presence and absence of 0.375 M aluminate (1.5 M OH⁻, aqueous 20 vol% DMSO). The rate constants for the two data sets are within 10% of the rate constant for the combined data.

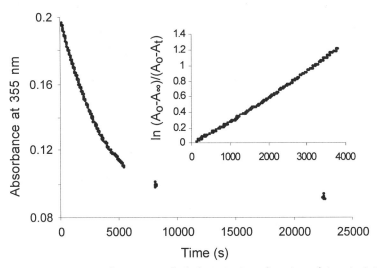

Figure 3. Decrease in absorption of ethyl nitrite as a function of time in 1.5 M OH⁻, aqueous 20 vol% DMSO. Insert: plot of ln[(Aₒ-A_∞)(Aₜ-A_∞)] verses time; Aₒ and A_∞ are initial and final absorbances.

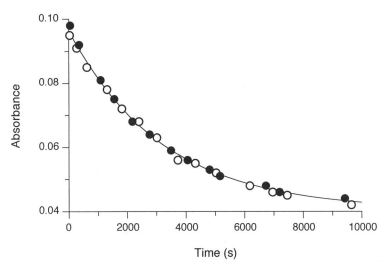

Figure 4. Decrease in absorbance at 355 nm of ethyl nitrite in 1.5 M OH⁻, aqueous 20 vol% DMSO with (O) 0.375 M sodium aluminate and (●) without sodium aluminate.

Hydrolysis of Nitritoacetate

We prepared nitritoacetate ion and nitritoacetic acid in situ by 2 independent routes: 1) reacting sodium nitrite with glycolic acid in the presence of a drying agent in DMSO and 2) reacting glycolic acid with t-BuONO in DMSO. The proton nuclear magnetic resonance (NMR) spectra showed a new methylene resonance (δ ~5.1 ppm) downfield from the methylene resonance (δ ~4 ppm) of glycolate ion/glycolic acid. Nitritoacetate ion hydrolyzed within seconds on mixing with water, aqueous sodium carbonate, or 0.1 M NaOH. The absorbance spectra recorded after mixing showed only the characteristic sodium nitrite ion band centered at 350 nm.

A comprehensive examination of the literature provided no evidence for the synthesis and isolation of the nitritoacetate ion. However, the synthesis and isolation of a related species, ethyl nitritoacetate, has been reported Kornblum and Eicher (*10*). They observed that hydrolysis of ethyl nitritoacetate occurred within a few seconds at room temperature to yield ethyl glycolate (not nitritoacetate) as shown in eq 5.

$$ONOCH_2CO_2Et + H_2O \rightarrow HOCH_2CO_2Et + HNO_2 \tag{5}$$

Therefore, to measure hydrolysis rates of nitritoacetate ion, we used an indirect kinetic method that is based on the tendency of nitrite esters to undergo rapid nitrosyl exchange reactions (Scheme 1).

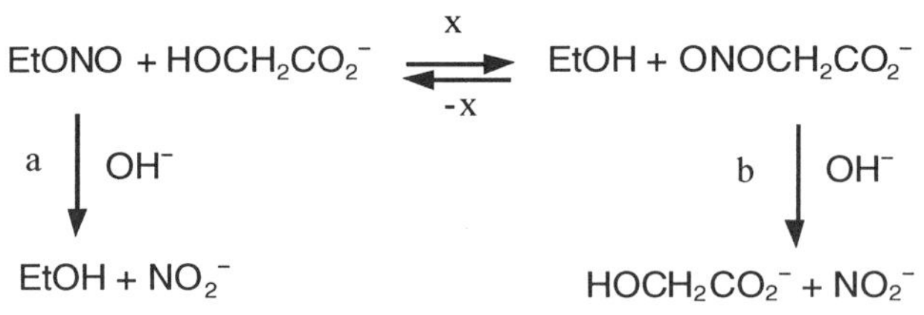

$$EtONO + HOCH_2CO_2^- \underset{-x}{\overset{x}{\rightleftharpoons}} EtOH + ONOCH_2CO_2^-$$

$$a \downarrow OH^- \qquad\qquad\qquad b \downarrow OH^-$$

$$EtOH + NO_2^- \qquad\qquad HOCH_2CO_2^- + NO_2^-$$

Scheme 1. Hydrolysis of ethyl nitrite in the presence of added glycolate ion.

If equilibrium is established rapidly enough, then glycolate ion should have the effect of accelerating the hydrolysis of ethyl nitrite. To obtain rate constant k_b, we need to know equilibrium constant for the nitrosyl exchange reaction. So next we report measurements of the equilibrium constant for nitrosyl exchange reactions with glycolate ion before presenting results of kinetic measurements on the system in Scheme 1.

Nitrosyl Exchange Reactions

Doyle and coworkers (*11*) have examined the nitrosyl exchange reactions of *t*-BuONO with a series of aliphatic alcohols as shown in eq 6.

$$ROH + t\text{-BuONO} \rightleftarrows RONO + t\text{-BuOH} \qquad (6)$$

The rate of exchange occurred within minutes at room temperature. They reported that the equilibrium constants for reactions of primary alcohols with *t*-BuONO are uniformly near 10 with the exception of a primary alcohol containing a β-electron withdrawing group, for example, 2-ethoxyethyl nitrite (EtOCH$_2$CH$_2$ONO), where it was observed that K_6 decreased to 5.8.

We measured K_6 for glycolate ion, methyl glycolate, and ethanol each reacting with *t*-BuONO. The measurements were made using ^{1}H-NMR spectroscopy. The protons attached to the carbon adjacent to the nitrosyl group are slightly broader and shifted downfield from the protons attached to the carbon adjacent to the hydroxyl group by approximately δ 1–1.1 ppm. The values of K_6 for EtOH and methyl glycolate were obtained in CDCl$_3$ solvent. The value of K_6 for glycolate ion was obtained in DMSO-d$_6$ solvent. The equilibrium constant is obtained by multiplying the normalized NMR area ratios for [RONO]/[ROH] and [*t*-BuOH]/[*t*-BuONO] such that the equilibrium expression, eq 7, is obtained.

$$K_6 = \frac{\left[t-\text{BuOH}\right]\left[\text{RONO}\right]}{\left[t-\text{BuONO}\right]\left[\text{ROH}\right]} \qquad (7)$$

We obtained $K_6 = 9.8$ for EtOH that agrees well with the value of 10.6 reported by Doyle et al. (*11*). For *t*-BuONO reacting with methyl glycolate in CDCl$_3$ and reacting with sodium glycolate in DMSO, we obtained a $K_6 = 0.094$ and 0.1, respectively. The observation that $K_6 < 1$ (ROH = glycolate) shows that nitritoacetate ion is less stable than primary alkyl nitrite esters. From the measured values, we estimate the $K_x \sim 0.01$ for exchange between EtONO with glycolate ion (Scheme 1).

Effect of Added Glycolate Ion on the Rate of Ethyl Nitrite Hydrolysis: the Hydrolysis Rate Constant for Nitritoacetate

This approach offers three distinct advantages to obtaining the rate constant for alkaline hydrolysis of nitritoacetate: 1) nitrosyl exchange is rapid and the equilibrium concentration of both ethyl nitrite and nitritoacetate can be calculated from our NMR experiments, 2) the small equilibrium constant assures low steady-state concentration of nitritoacetate and allows the kinetics to be followed by simple mix and observe techniques, and 3) aluminate does not affect the rate of hydrolysis of ethyl nitrite, permitting a direct measure on the effect of nitritoacetate hydrolysis with and without aluminate.

Rate Law and Kinetic Results: Scheme 1

From results shown in Figure 4, the lifetime, τ, $(1/k_{obs})$ of EtONO in 1.5 M OH$^-$ is 4150 ± 250 seconds ($k_a = (1.6\pm0.1)\times10^{-4}$ s^{-1}). With glycolate ion present at 0.375 M, we observed the lifetime of EtONO to decrease significantly ($\tau = 1210\pm300$ seconds). When [glycolate], [OH$^-$] and [EtOH] are all $>>$[EtONO], the system (Scheme 1) will exhibit first order kinetic behavior: rate = k_{obs}[EtONO] with

$$k_{obs} = \left(\frac{\frac{k_x k_b}{k_{-x}}[\text{Glycolate}]}{\frac{k_b[\text{OH}^-]}{k_{-x}} + [\text{EtOH}]} + k_a \right) [\text{OH}^-] \tag{8}$$

The expression simplifies further depending on whether the nitrosyl exchange equilibrium is rapidly established. The concentration of ethanol is a controlling factor. At ethanol concentrations high enough such that [EtOH] $>>$ k_d[OH$^-$]/k_b, the equilibrium is established and eq 8 simplifies to:

$$k_{obs} = k_b K_x([\text{Glycolate}]/[\text{EtOH}])[\text{OH}^-] + k_a[\text{OH}^-] \tag{9}$$

with $K_x = k_x/k_{-x}$. If instead k_b[OH$^-$] $>>$ k_{-x}[EtOH], then eq 8 simplifies to

$$k_{obs} = k_x[\text{Glycolate}] + k_a[\text{OH}^-] \tag{10}$$

This distinction is important for our purpose because if we are to determine the effect of aluminate ion on the hydrolysis rate of nitritoacetate, then hydrolysis and not exchange must be rate limiting in Scheme 1.

We varied the concentration of ethanol and observed behavior consistent with Scheme 1 and eqs 9 and 10. Figure 5 shows the effect of ethanol on the hydrolysis of ethyl nitrite in alkaline aqueous solutions (20 vol% DMSO) containing 0.375 M glycolate ion. At 1 M ethanol, the observed lifetime is ~3 times longer than the EtONO lifetime in the absence of added ethanol. Since a dependence on EtOH is observed, we suggest k_{obs} is defined by eq 9 at least when concentrations of ethanol are ≥0.1 M.

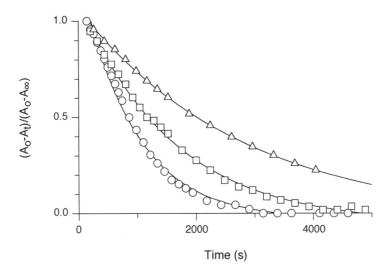

Figure 5. Decrease of ethyl nitrite as a function of ethanol concentration in 1.5 M OH⁻, 0.375 M sodium glycolate, aqueous 20% DMSO: (○) 0.0 M ethanol, (□) 0.1 M ethanol, (△)1.0 M ethanol.

Effect of Aluminate Ion on the Alkaline Hydrolysis of Nitritoacetate.

We repeated the experiments shown in Figure 5, but with 0.375 M aluminate also present. If aluminate catalyzes the hydrolysis of nitritoacetate, then we should expect shorter lifetimes for ethyl nitrite when it is present, and if catalysis by aluminate is particularly effective, then the rate-dependence on ethanol (Figure 5) might even be expected to disappear since the exchange step would become rate limiting. Figure 6 shows time-dependent absorbances at 355 nm of aqueous alkaline EtONO solutions (0.375 M glycolate, 1.5 M HO⁻, aqueous 20 vol% DMSO) in the presence and absence of aluminate (0.375 M) for ethanol concentrations of 0.1 and 1 M EtOH. As can be seen, added aluminate ion had negligible effect in either case (Figure 6A and Figure 6B, respectively).

Discussion

We have shown that due to facile exchange of nitrosyl between ethyl nitrite and glycolate, the rate constant for hydrolysis of nitritoacetate is measurable from the effect of glycolate on the hydrolysis rate of ethyl nitrite. The magnitude of the effect is dependent on glycolate, ethanol and hydroxide ion concentration in accord with Scheme 1. However, aluminate had negligible effect on the hydrolysis rates of ethyl nitrite and nitritoacetate. Results of our kinetic measurements are summarized in Table I.

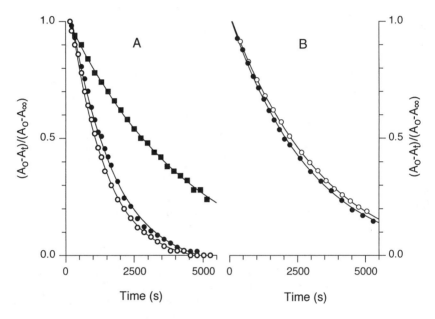

Figure 6. Effects of ethanol, glycolate ion, and aluminate ion on hydrolysis of ethyl nitrite in alkaline aqueous solutions (1.5 M OH⁻, 20 vol% DMSO). Normalized absorbances at 355 nm: A) 0.1 M ethanol with (○) 0.375 M glycolate and 0.375 M aluminate, (●) 0.375 M glycolate and no aluminate, or (■) no glycolate and no aluminate; B) 1 M ethanol with (○) 0.375 M glycolate and 0.375 M aluminate, or (●) 0.375 M glycolate and no aluminate.

Table I. Summary of Rate Data for Alkaline Hydrolysis of Ethyl Nitrite and Nitritoacetate in the Presence and Absence of Sodium Aluminate

No.	EtOH M	Glycolate M	Al(OH)$_4^-$ M	OH$^-$ M	$k_{obs}/10^{-4}$ s^{-1}	$k_a/10^{-4}$ M^{-1} s^{-1}	$k_b/10^{-2}$ M^{-1} s^{-1}
1	0.01	0	0	2	3.45	1.7	-
2	0.01	0	0.5	2	2.73	1.4	-
3	0.00	0	0	1.5	2.56	1.7	-
4	0.10	0	0	1.5	2.27	1.5	-
5	0.10	0	0	0.75	5.12	1.7	-
6	0.09	0.65	0	2	27.4	1.6	1.7
7	0	0.375	0	1.5	12.2	-	-
8	0.10	0.375	0	1.5	6.67	1.6	0.76
9	0.10	0.375	0.375	1.5	8.27	1.6	1.0
10	1.0	0.375	0	1.5	3.36	1.6	1.7
11	1.0	0.375	0.375	1.5	2.95	1.6	0.98

NOTE: All solutions are aqueous 20 vol% DMSO. NaAl(OH)$_4$ solutions prepared from 3:1 NaOH:Al(NO$_3$)$_3$. Entry no. 6 is average of 4 experiments (standard error ±3%).

Comparing the two 0.1 M EtOH experiments, no. 8 with no. 9 (Table I), shows that the rate constant of the EtONO increases slightly in the presence of aluminate. On the other hand, comparing the 1 M EtOH experiments, no. 10 with no. 11, shows that the observed rate constant of RONO decreases in the presence of aluminate ion. We consider these small differences to be within the experimental error.

Thermochemical Kinetic Analysis of Scheme 1

Figure 7 shows a free energy diagram for Scheme 1. We estimate from thermochemical data: $\Delta G°_r$ ~18 kcal/mol for alkaline hydrolysis of ethyl nitrite. Oae determined $k_{(BuONO + OH^-)}$ = 5.6×10^{-5} M^{-1} s^{-1} and $\Delta G^{\ddagger}_{(BuONO + OH^-)}$ = 24 kcal/mol (9). Assuming similar frequency factors, $\Delta G_r^{\ddagger}$ = ~23 kcal/mol for ethyl nitrite. We found that hydrolysis of nitritoacetate is 100 times faster that ethyl nitrite. This corresponds to a barrier for nitritoacetate hydrolysis that is ~3 kcal/mol lower, or ~20 kcal/mol. Since the equilibrium constant for nitrosyl exchange, EtONO + glycolate $\rightleftarrows$ nitritoacetate + EtOH, is ~0.01, the products are ~3 kcal/mol less stable than the reactants. So it seems that the transition states for ethyl nitrite hydrolysis and nitritoacetate hydrolysis are similar in energy and the difference in reactivities between ethyl nitrite and nitritoacetate is mainly due to the difference in their ground state stabilities and not their transition states. Therefore, it is no easier for glycolate to be nitrosated in alkaline solution by nitrite ion than it is for ethyl nitrite: $\Delta G^{\ddagger}_r \approx 41$ kcal/mol for eq 4.

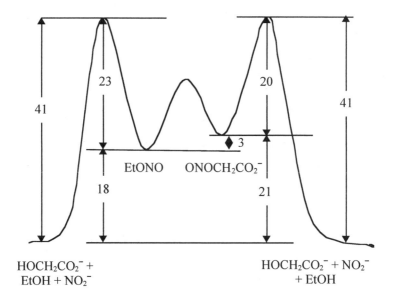

Figure 7. Free-Energy Diagram for Scheme 1 (kcal/mol)

Aluminate Catalysis

The results confirm the presence of a large endergonic barrier for the formation of nitritoacetate from sodium nitrite and glycolate ion (reaction 4). Therefore, if nitritoacetate is a key intermediate in the thermal aging of glycolate in Hanford waste, then a catalyst is required and the enhancement must be substantial. It has been proposed that aluminate catalyzes this reaction (3). However, we find no evidence for it under conditions used here. Ashby et al. reported that glycolate decomposes in homogeneous waste simulants containing aluminate, nitrite and hydroxide ions with a rate constant of 2×10^{-4} M^{-1} h^{-1} at 120 °C (3). From our rate data and above thermochemical analysis, we estimate that nitrosation of glycolate by nitrite ion will have a rate constant of $\sim4\times10^{-9}$ M^{-1} h^{-1} at 120°C. The glycolate degrades $\sim10^{5}$ times faster in the simulant. Transition state theory predicts an order of magnitude rate enhancement for every 1.8 kcal/mol decrease in the barrier at 120°C. A change of five orders of magnitude in rate corresponds to $\Delta\Delta G_r^{\ddagger}$ ~9 kcal/mol. Therefore, if aluminate catalyzes nitrosation, then the barrier to hydrolysis should decrease to 11 kcal/mol (see Figure 1 and Figure 7) since hydrolysis is the reverse of nitrosation. The hydrolysis rate constant should increase $>10^{5}$ fold at room temperature!

Whether the full effect would be observable for nitritoacetate in scheme 1 is a point to consider. If k_d were to become very large, the observed rate for ethyl nitrite hydrolysis by the process in Scheme 1 would become limited by exchange, k_x, such that k_{obs} would take the form of eq 10, provided k_x is unaffected by aluminate catalysis. The observed rate enhancement would be smaller, but at least >4 for conditions of experiment nos. 8/9 and 10/11 in Table I.

The speciation of aluminum in alkaline aluminate solutions is complex (*12*). While it is possible that catalytically-active species do not form under conditions of our experiments, we also now consider that alternative mechanisms and reactive intermediates may account for the thermally-activated aging reactions. Stock (*13*) has proposed that the mechanism may involve a cyclic transition state in which aluminum ion coordinates nitrite and substrate to foster "intramolecular" transfer of H from carbon to nitrite with expulsion of HNO.

The mechanism, which is analogous to the Oppenhauer oxidation (*14*), provides a pathway for transformation of alcohol functional groups in complexants to carbonyl groups. It explains the observed kinetics and major product, oxalate. However, it does not explain how glycolate converts to formate ion, and HEDTA to ethylenediaminetriacetate and formate ion. One possibility

$$R = CO_2^- \text{ for glycolate ion}$$
$$R = (^-O_2CCH_2)NCH_2CH_2(^-O_2CCH_2)NCH_2 \text{ for HEDTA}$$

is that electron transfer from coordinated ROH to NO_2^- is concerted with cleavage to fragments. The types of intermediates depend on substrates, e.g., glycolate goes to CO_2, formaldehyde and NO^-; HEDTA to iminium ion, formaldehyde and NO⁻. While we show intermediates resulting from 2 electron transfer processes, one electron transfer processes generating radical intermediates may be possible (i.e., R• and NO).

Conclusions

In summary, we measured the rate of alkaline hydrolysis of ethyl nitrite and nitritoacetate in the presence and absence of aluminate. We determined that alkaline hydrolysis of nitritoacetate occurs at room temperature with a second order rate constant of 0.017 M^{-1} s^{-1} ($\tau \sim 60$ s in 1 M NaOH). This rate is approximately two orders of magnitude faster than ethyl nitrite and other primary alkyl nitrites (9). However, as is clear in Figure 7, the faster rate is largely due to differences in ground state energies of ethyl nitrite and nitritoacetate. The data in Figures 4 and 6 and Table I provide no evidence for aluminate catalysis of eq 4. Perhaps the conditions of our experiments do not favor the catalytically-active forms of aluminum. Alternatively, mechanisms different from reactions 1-3 and Figure 4 govern aluminate-catalyzed organic aging reactions.

Acknowledgement

We benefit from the efforts of many of our colleagues. Their contributions are much appreciated. Support from the Environmental Management Science Program is acknowledged. Our interactions with Richland Operations and the Hanford site provided the impetus to this study. Pacific Northwest National Laboratory is operated by Battelle Memorial Institute for the U.S. Department of Energy under Contract DE-AC06-76RLO 1830.

References

1. Meacham, J.E.; Cowley W.L.; Webb, A.B.; Kirch, N. W.; Lechelt, J. A.; Reynolds, D. A.; Stauffer, L. A.; Bechtold, D. B.; Camaioni, D. M.; Gao, F.; Hallen, R. T.; Heasler, P. G.; Huckaby, J. L.; Scheele, R. D.; Simmons, C. S.; Toth, J. J.; Stock, L. M. *Organic Complexant Topical Report,* HNF-3588, 1998; publicly available.
2. Camaioni, D. M.; Samuels, W. D.; Linehan, J. C.; Sharma, A. K.; Autrey, S. T.; Lilga, M. A.; Hogan, M. O.; Clauss, S. A.; Wahl, K. L.; Campbell,

J. A., *Organic Tanks Safety Program Waste Aging Studies Final Report,* PNNL-11909 Rev. 1, 1998, ; Avail. INIS.

3. Ashby, E.; Barefield, E.; Liotta, C.; Neumann, H.; Doctorovich, F.; Konda, A.; Zhang, K.; Hurley, J.; Boatright, D.; Annis, A.; Pasino, G.; Dawson, M.; Juliano, M. in *Emerging Technologies for Hazardous Waste Management,* ACS Sym. Ser., Vol. 554, Tedder, D. and Pohland, F., Ed.; American Chemical Society: Washington, DC; 1994, pp 247-283.

4. Delegard, H. C., Identities of HEDTA and Glycolate Degradation Product in Simulated Hanford High-Level Waste, RHO-RE-ST-55P, 1987; publicly available.

5. Delegard, C. *Laboratory Studies of Complexed Waste Slurry Volume Growth in Tank 241-SY-101,* RHO-LD-124, 1980; Avail. NTIS.

6. (a) Ashby E.C.; Annis, A.; Barefield, E. K.; Boatright, D.; Doctorovich, F.; Liotta, C. L.; Neumann, H. M.; Konda, A.; Yao, C.-F.; Zhang, K.; Duffie; N.G. *Synthetic Waste Chemical Mechanism Studies.* WHC-EP-0823, 1994, Avail. NTIS. (b) Barefield, E. K.; Boatright, D.; Deshpande, A.; Doctorovich, F.; Liotta, C. L.; Neumann, H. M.; Seymore, S. *Mechanisms of Gas Generation from Simulated SY Tank Farm Wastes: FY 1994 Progress Report,* PNL-10822, 1995, Avail. NTIS.

7. Barefield, E. K.; Boatright, D.; Deshpande, A.; Doctorovich, F.; Liotta, C. L.; Neumann, H. M.; Seymore, S. *Mechanisms of Gas Generation from Simulated ST Tank Farm Wastes: FY 1995 Progress Report* PNNL-11247, 1996; Avail. NTIS.

8. (a) Ashby, E. C.; Doctorovich, F.; Liotta, C. L.; Neumann, H. M.; Barefield, E. K.; Konda, A.; Zhang, K.; Hurley, J.; Siemer, D. D. *J. Am. Chem. Soc.* **1993,** *115*, 1171-73. (b) Kapoor, S.; Barnabas, F.A.; Sauer, Jr., M. C.; Meisel, D.; Jonah, C. A. *J. Phys. Chem.* **1995,** *99,* 6857-6863

9. Oae, S.; Asai, N.; Fujimori, K. *JCS Perkin II* **1978**, 571-77.

10. Kornblum, N.; Eicher, J. H. *J. Am. Chem. Soc.* **1956**, 78, 1494-7.

11. Doyle, M. P.; Terpstra, J. W.; Pickering, R. A.; LePoire, D. M. *J. Org. Chem.* **1983**, *48*, 3379-82.

12. Akitt. J. W.; Gessner, W. *J. Chem. Soc. Dalton Trans.* **1984**, 147, 148.

13. Stock, L. M.; Pederson, L. R. *Chemical Pathways for the Formation of Ammonia in Hanford Wastes*, 1997, PNNL-11702 Rev.1; Avail. NTIS.

14. March, J. Advanced Organic Chemistry: Reactions, Mechanisms, and Structure; 4th ed.; John Wiley and Sons: New York, 1992; pp. 917, 1169.

Chapter 19

Kinetics of Conversion of High-Level Waste to Glass

P. Izak, P. Hrma, and M. J. Schweiger

Pacific Northwest National Laboratory, 908 Battelle Boulevard,
Mail stop K6–24, Richland, WA 99352

The kinetics of the conversion of high-level waste (HLW) feed to glass controls the rate of HLW processing. Simulated HLW feed and low silica–high sodium (LSHS) feed with co-precipitated Fe, Ni, Cr, and Mn hydroxides (to simulate the chemical and physical makeup of these components in the melter feed) were heated at constant temperature-increase rates (0.4, 4, and 14°C/min), quenched at different stages of conversion, and analyzed with optical microscope, scanning electron microscope, and x-ray diffraction (XRD). Quartz, sodium nitrate, carnegieite [$Na_8Al_4Si_4O_{18}$], sodalite [$Na_8(AlSiO_4)_6(NO_2)_2$], and spinel were identified in the samples. Mass fractions of these phases were determined as functions of the temperature and the heating rate. The fractions of nitrates and quartz decreased with increasing temperature, starting above 550°C and dropping to zero at 850°C. Spinel was present in the feed within the temperature interval from 350°C to 1050°C, peaking between 550°C and 700°C. Sodalite (in HLW feed) and carnegieite (in LSHS feed) formed at temperatures above 600°C and then began to dissolve. Thermogravimetric analysis (TGA) and differential scanning calorimetry (DSC) were used to determine the mass loss and the conversion heat as functions of temperature and heating rate and were compared with the reaction progress reached in quenched samples.

Introduction

Waste loading, melting efficiency, and low-risk operation of the melter determine the effectiveness of the high-level waste (HLW) vitrification process. Waste loading controls the amount of waste glass to be produced *(1, 2)*. Melting efficiency affects the rate at which the glass is made. The need to minimize risk in operation limits the waste loading *(3)* for those HLW streams that contain significant portions of Fe_2O_3, NiO, Cr_2O_3, MnO, FeO, and other spinel-forming oxides (Savannah River HLW streams and most of the HLW streams at Hanford). If spinel precipitates in the melter, it may plug the melter spout, obstruct the drain, and interfere with the flow of glass within the melter. The fraction of spinel that precipitates from HLW glass generally increases with the waste loading.

The kinetics of conversion of HLW feed to glass affects the rate at which the feed is processed in the melter *(4–6)*. Spinel is an intermediate product of this conversion. One objective of this study is to obtain information regarding the kinetics of feed-to-glass conversion in terms of melting reactions between the starting and intermediate phases. This information is needed to mathematically model the velocity and temperature fields that control spinel behavior in the melter *(7)*. A second objective is to find out whether spinel is nucleated during the feed-to-glass conversion or at some later stage of the process. The nucleation rate determines the crystal size and number density, which in turn determine the settling rate *(8)*.

Experimental

We used a simplified version of a typical HLW glass, an 11-component glass with the liquidus temperature (T_L) = 1080°C. The glass composition for this feed (denoted as MS-7 throughout this paper) was Al_2O_3 (0.080), B_2O_3 (0.070), Cr_2O_3 (0.003), Fe_2O_3 (0.115), Li_2O (0.0451), MgO (0.006), MnO (0.0035), Na_2O (0.153), NiO (0.0095), SiO_2 (0.454), and ZrO_2 (0.060) (in mass fractions). The second feed was used to check the impact of a significant change in feed composition on the reaction path. The glass composition of low silica–high sodium feed (LSHS) was Al_2O_3 (0.047), B_2O_3 (0.041), Cr_2O_3 (0.006), Fe_2O_3 (0.230), Li_2O (0.027), MgO (0.0035), MnO (0.0058), Na_2O (0.317), NiO (0.0196), SiO_2 (0.268), and ZrO_2 (0.035).

We prepared simulated feeds using a modified standard procedure *(9)*. Spinel-forming components Fe, Ni, Cr, and Mn in the form of nitrates were dissolved in deionized water and co-precipitated by mixing the solution, at 40°C, with enough 10 M NaOH to supply the total amount of Na_2O for the glass. Other feed components, $ZrO(NO_3)_2 \cdot H_2O$, $Mg(NO_3)_2 \cdot 6H_2O$, Li_2CO_3, SiO_2 sand (≤ 0.19 mm),

Al_2O_3 ($\leq$100 μm), and H_3BO_3 were stirred into the slurry with co-precipitate. All work with nitrates was performed in the fume hood. The resulting slurry was dried in an oven at 90°C for 2 days and heat-treated at 200°C for 2 days. The dry feed was weighed to assure the absence of crystalline water in it. Samples of the feed were subjected to thermo-gravimetric analysis (TGA) and differential scanning calorimetry (DSC).

Approximately 1.3-g samples were heat-treated in platinum crucibles at 0.4, 4, and 14°C/min, starting from 200°C. These heating rates correspond to those estimated for feed melting in Joule-heated melters *(10)*. Crucibles were removed from the furnace at different temperatures and quenched. Samples were analyzed using X-ray diffraction (XRD), optical microscopy, and scanning electron microscopy (SEM) with energy-dispersive spectroscopy (EDS). To obtain mass fractions of crystalline phases, XRD was calibrated by mixing spinel, carnegieite, sodium nitrate, and quartz (one at a time) with crystal-free MS-7 glass and with dried feed. An example of a calibration line is shown in Figure 1.

Results and Discussion

The TGA curves (Figure 2) show that mass loss began at 300°C at a slow rate. The mass-loss rate increased at about 400°C, peaked, and then stopped at temperatures listed in Table I. A faster heating rate shifted the reaction to higher temperatures. The measured total mass loss was 26% for MS-7 and from 34% to 37% for LSHS. The calculated mass loss from carbonate decomposition was 3.7% for MS-7 and 2.3% for LSHS, and the loss from nitrate decomposition was 22.2% for MS-7 and 33.7% for LSHS. Thus, the measured and calculated total mass losses are in good agreement.

The DSC curves (Figure 3) show two endothermic peaks for each feed and heating rate. A sharp peak (MS-7 at 270°C; LSHS at 285°C) corresponds to the melting of sodium nitrate. The melting heat per feed unit mass is 18 J/g to 20 J/g for MS-7 and 47 J/g to 49 J/g for LSHS. This corresponds to approximately 83 J/g $NaNO_3$ in MS-7 feed and 123 J/g $NaNO_3$ in LSHS feed. The literature value is 176.5 J/g $NaNO_3$ *(11)*. The difference is probably cause by interaction between nitrate and borate melts in the feeds. A broad peak above 500 C marks mainly the decomposition of nitrates and dissolution of quartz. The reaction heat per feed unit mass is 277 J/g to 288 J/g for MS-7 and 825 J/g to 890 J/g for LSHS. At a faster heating rate, the peaks shifted to higher temperatures (Table I).

The XRD data are displayed in Figures 4, 5, 6 and 10, 11, 12. Sodium nitrate began to decompose at temperatures above 500°C and disappeared completely by 750°C (Figure 4). The initial fractions of $NaNO_3$ in the feeds (0.28 in MS-7 and 0.40 in LSHS) are in good agreement with the $NaNO_3$ content measured by XRD

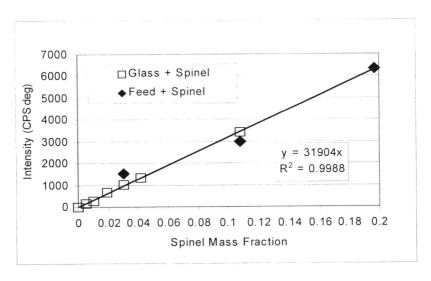

Figure 1. Calibration curve for spinel: Main peak area versus mass fraction.

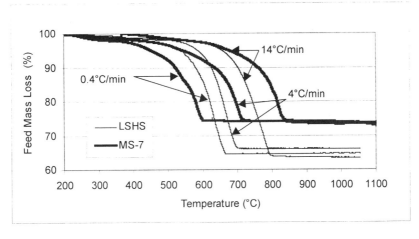

Figure 2. Feed mass loss versus temperature and heating rate.

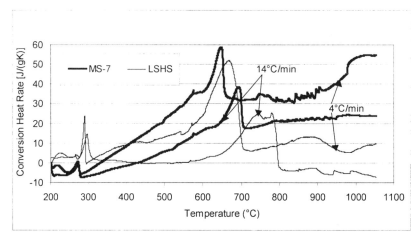

Figure 3. Conversion heat of the feeds versus temperature and heating rate.

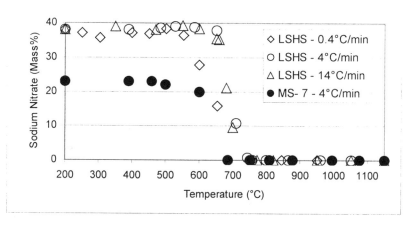

Figure 4. Sodium nitrate mass fraction in MS-7 and LSHS feeds versus temperature and heating rate.

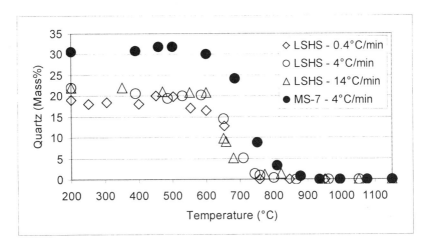

Figure 5. Quartz mass fraction in the feed versus temperature and heating rate.

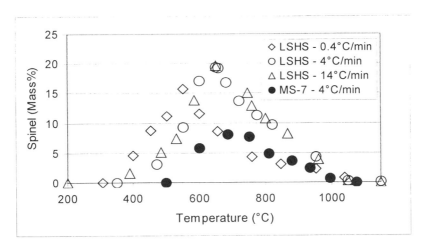

Figure 6. Spinel mass fraction in the feeds versus temperature and heating rate.

(0.23 in MS-7 and 0.39 in LSHS). Table I summarizes the commencing, peak, and ending temperatures of the nitrate decomposition reactions. Also included in Table I are preliminary data for evolved gas analyses (EGA). Thermoanalytical (TGA and DSC) reaction temperatures generally increase with increasing heating rate whereas the corresponding temperatures from XRD analysis are little affected by heating rate. The effects of the rate of heating on the reaction kinetics are not as pronounced with larger samples, probably as a result of temperature gradients. Generally, sample volume affects the kinetics of feed reactions. Extremely small amounts (45 mg) used for TGA and DSC expose the reactants to the atmosphere and possible effects of crucible wall. The ambient atmosphere decreases the pressure of the gaseous reaction products and thus accelerates decomposition reactions *(12, 13, and 14)*.

Table I. the commencement (C), peak (P), and end (E) of nitrate decomposition in MS-7 and LSHS feeds by different methods (in °C)

	dT/dt	DSC			TGA		
Feed	°C/min	C	P	E	C	P	E
MS-7	0.4	-	-	-	300	440	600
	4	600	660	680	410	585	710
	14	640	700	715	480	785	830
LSHS	0.4	-	-	-	410	625	690
	4	590	680	705	540	660	720
	14	650	770	800	540	750	800

	dT/dt	XRD			EGA		
Feed	°C/min	C	P	E	C	P	E
MS-7	4	500	650	690	450	680	790
LSHS	0.4	530	630	750	-	-	-
	4	610	685	750	-	-	-
	14	610	685	750	-	-	-

As Figure 5 shows, silica sand (quartz) began to dissolve at 550°C, and its dissolution was completed at between 760°C and 880°C. The initial fractions of SiO_2 in the feeds (0.32 in MS-7 and 0.16 in LSHS) are in reasonable agreement with those measured by XRD (0.31 in MS-7 and 0.20 in LSHS). The silica-dissolution kinetics is remarkably similar to the nitrate decomposition kinetics, suggesting a reaction of sodium nitrate with silica. Another possible reaction is that between sodium nitrate and boric acid. The reaction products, sodium silicates and sodium borates, generate glass-forming melt. Quartz dissolution temperatures do not appear to be affected by heating rate.

Spinel began to form above 350°C (Figure 6) soon after the melting of nitrates. The hydroxide products of co-precipitation dissolved in the nitrate melt, from which spinel precipitated until the nitrates began to decompose. Once the glass-forming melt appeared, spinel began to dissolve in it and disappeared at 1050°C. The fraction of spinel peaked (between 650°C and 700°C) shortly after the decomposition of nitrates commenced.

The spinel XRD pattern (Figure 7, measured d-spacing = 25.135 nm) corresponded to that of trevorite ($NiFe_2O_4$; d-spacing = 25.140 nm) or lithium-nickel-iron oxide (d-spacing = 25.137 nm). Trevorite with lithium was reported by M. Greenblatt *(15)*.

Spinel crystals were isolated from partially converted HLW feeds (MS-7 and LSHS) by first dissolving the feed in 16 M nitric acid for 1 h at 60°C and then magnetically separating the spinel from the resulting sludge. Most of the spinel crystals found in the feed were of submicron size—Figure 8; while the size of spinel crystals found in glass were typically 2–20 μm—Figure 9. Spinel composition was determined by SEM EDS (Table II). According to Table II, spinel is a solid solution of $NiFe_2O_4$, $MnFe_2O_4$, and Fe_3O_4. However, EDS is not capable of measuring Li

Table II. SEM EDS analyses of spinel composition (in atomic %)

Feed	T(°C)	SEM - EDS Analyses		
		Mn	Fe	Ni
LSHS	960	9.5	60.1	30.4
	878	7.0	87.9	5.1
MS-7	992	4.4	86.8	8.8
	1040	2.5	68.6	28.9

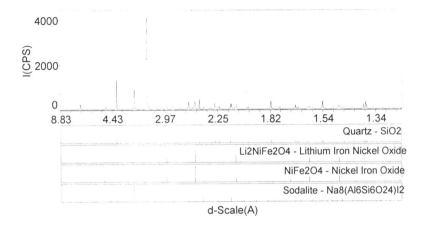

Figure 7. XRD pattern of MS-7 feed at 685°C, heating rate 4°C/min. Intensity (Counts per second) versus d-spacing (10⁻¹⁰ m).

Figure 8. Spinel crystals isolated from a HLW feed (MS-7 and LSHS at 650°C-SEM).

Figure 9. Spinel crystals isolated from a HLW glass (SEM).

Unlike spinel, sodalite [$Na_8(AlSiO_4)_6(NO_2)_2$], (in MS-7 feed) and carnegieite [$Na_8Al_4Si_4O_{18}$], (in LSHS feed) were observed only when sufficient fractions of glass-forming melt accumulated (Figure 10) in the reacting mixture. Carnegieite and sodalite *(16)* are intermediate solid-reaction products of the dissolution of corundum, the source of Al_2O_3 in the feed.

The evolution of all crystalline phases detected by XRD from MS-7 and LSHS is shown in Figures 11 and 12.

The SEM (Figure 13) found large particles (40 μm) of zirconium silicate in the heat-treated LSHS feed. An approximately 5-μm-thick layer of sodium zirconium silicate covered these particles. These sparse particles, not detected by XRD, are a product of a reaction between of zirconium hydroxide and silica.

Several visual observations were made during the heat treatments, such as the segregation of chromate melt in the Cr-rich LSHS feed or feed volume expansion. None of these phenomena have been characterized quantitatively. A segregated yellow-green chromate melt *(17, 18)* was observed in the LSHS feed samples quenched from 700°C to 900°C. In larger feed samples, chromate is usually reduced to chromium oxide at temperatures above 850°C. From our small samples, the chromate phase spread on crucible walls and evaporated.

Cr_2O_3 did not participate in spinel formation in MS-7 and LSHS feeds (Table II). This is probably caused by the lack of Cr(III) in the nitrate melt, in which most of the Cr was dissolved in the form of chromate. The absence of Cr_2O_3 in spinel has a significant consequence for spinel dissolution in the glass forming melt as the feed temperature increases. The liquidus temperature of glasses with spinel primary phase sharply increases as the Cr_2O_3 content increases *(19, 20)*. This is because the Cr-containing spinel forms in molten glass at a higher temperature than trevorite and magnetite. Consequently, the trevorite-magnetite spinel from the feed likely dissolves in the glass below its liquidus temperature.

One of the questions this study aimed to resolve was whether the spinel from feed can survive in the melt and thus control the crystal number density and the crystal size, the quantities important for the settling rate. The results from this study indicate that the spinel from the feed may not survive in the melt. However, the mechanism of nucleation of the high-temperature spinel that contains Cr_2O_3 is not yet clear. It remains uncertain whether any interaction exists between the Cr-containing spinel and the dissolving spinel from the feed.

We also observed feed-volume expansion at temperatures above 700°C. This primary foam *(21)* collapsed by the time the temperature reached 1000°C. Similar volume expansion and gas evolution was observed in simulated Hanford low-activity aqueous waste feed at 10°C/min *(22)*. The continuous growth and collapse of bubbles accelerates the homogenization of melt, but bubbles slow the heat transfer rate from molten glass to the feed in the melter *(4, 5)*.

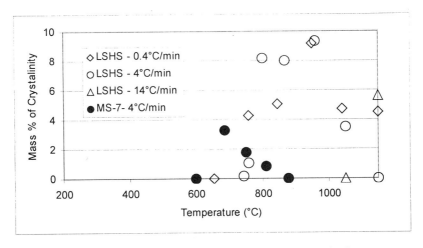

Figure 10. Sodalite and carnegieite mass fraction in the feeds versus temperature and heating rate.

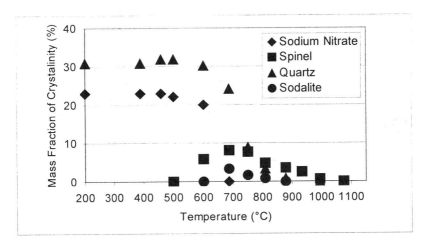

Figure 11. Mass fractions of crystals in MS-7 feed versus temperature at 4°C/min.

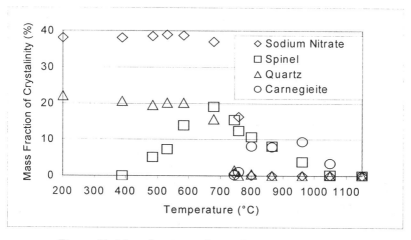

Figure 12. Mass fractions of crystals in LSHS feed versus
temperature at 4°C/min.

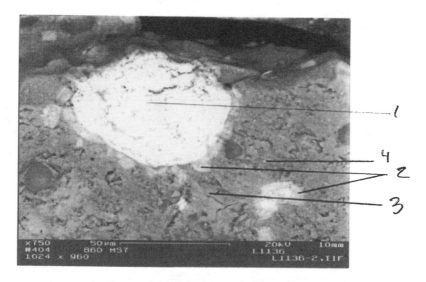

Figure 13. A Zr-containing particle from LSHS feed (4°C/min, 860°C) (SEM);
1-Zirconium silicate, 2-Sodium zirconium silicate, 3- Matrix,
4- Carnegieite.

Conclusions

Nitrates and quartz, the major HLW feed components, are converted to glass-forming melt by 880°C. The conversion heats for this reaction are 277 J/g feed to 288 J/g feed for MS-7 and 825 J/g feed to 890 J/g feed for LSHS. Spinel forms in the nitrate melt starting from 350°C and eventually dissolves in the borosilicate melt by 1050°C. Sodalite and carnegieite are intermediate products of corundum dissolution. They appear (sodalite in the HLW feed and carnegieite in high sodium–low silica feed) above 600°C and eventually dissolve in glass. The effect of the spinel produced in the feed on spinel formation in molten glass remains unclear.

Acknowledgments

The authors thank John Vienna, Jim Young, Jaroslav Klouzek and Jarrod Crum for laboratory support and insightful advice. Pavel Izak is grateful to Associated Western Universities for the fellowship appointment at Pacific Northwest National Laboratory operated by Battelle Memorial Institute for U.S. Department of Energy under contract 1830 provided funding for this study.

References

1. Hrma, P.; Smith, P. A. The Effect of Vitrification Technology on Waste Loading. *Proc. Int. Top. Meeting Nucl. Hazard. Waste Manag. Spectrum* **1994**, 2, 862-867.
2. Hrma, P. Toward Optimization of Nuclear Waste Glasses Constraints, Property Models, and Waste Loading. *Ceram. Trans.* **1994**, 45, 391-401.
3. Hrma, P.; Vienna, J. D.; Schweiger, M. J. Liquidus Temperature Limited Waste Loading Maximization for Vitrified HLW. *Ceram. Trans.* **1996,** 72, 449-456.
4. Hrma P. Melting of Foaming Batches: Nuclear Waste Glass. *Glastech. Ber.* **1990**, 63K, 360-369.
5. Kim, D. S.; Hrma P. Laboratory Studies for Estimation of Melting Rate in Nuclear Waste Glass Melters. *Ceram. Trans.* **1994**, 45, 409-419.
6. Smith, P. A.; Vienna J. D.; Hrma P. The Effects of melting Reactions on Laboratory-scale Waste Vitrification. *J. Mat. Res.* **1995**, 10, 2137-2149.

7. Schill, P. *Batch Melting in Mathematical Simulation of Glass Furnaces,* 3rd Int. Seminar on Math. Simulation in Glass Melting Proc.: Horni Becva, Czech Rep., 1995; p.97.

8. LaMont, M. J.; Hrma P. A. Crucible Study of Spinel Settling in a High-Level Waste Glass. *Ceram. Trans.* **1998**, 87, 343-348.

9. Russell, R. L.; Smith, H. D. Simulation and Characterization of a Hanford High-Level Waste Slurry. *PNNL–11293*, Pacific Northwest National Laboratory: Richland, WA, 1996.

10. Hrma, P. Thermodynamics of Batch Melting. *Glastechn. Ber.* **1982**, 55, 138-150.

11. *Handbook of Chemistry and Physics*; Lide, D. R.: London, UK, **1994**, Vol.6, pp. 126.

12. Bader, E. On the Effect of Grain Size on Melting Reactions in Soda-Lime-Quartz and Soda-Dolomite-Quartz Batch. *Silikattechnik,* **1983**, 34.01, 7-11.

13. Hrma, P. *Complexities of Batch Melting*, Adv. in Fusion of Glass: Am. Ceram. Soc., Westerville, OH, 1988, p.10.

14. Hrma P.; Paul *A. Batch Melting Reactions in Chemistry of Glass*, Prentice Hall: London, UK, 1990, p.157.

15. Chen C. J.; Greenblatt M. Lithium Insertion into Spinel Ferrites, *Solid State Ionics,* **1986**, 18, 838-846.

16. Klingenberg R.; Felsche J. Interstitial Cristobalite-type Couponds $(Na_2O)_{0.33}Na[AlSiO_4]$, *J. Solid State Chem.* **1986**, 61, 40-46.

17. Maruskova, K.; Had, J.; Maryska, M.; Sanda, L.; Hlavac, J. High-Temperature Reactions of Sodium Chromate with Silica, *Ceram – Silikaty.* **1997**, 41, (3), 105-111.

18. Li, H.; Hrma, P.; Langowski, M. H.; Hlavac, J.; Vojtech, O.; Krestan, V.; Exnar, P. Vitrification and Chemical Durability of High-Level Nuclear Waste Glasses with High Concentration of Cr_2O_3 and Al_2O_3, *Ceram. Trans.* **1996**, 72, 299-306.

19. Mika, M.; Schweiger M. J.; Hrma, P. Liquidus Temperature of Spinel Precipitating High-Level Waste Glass, *Scientific Basis for Nuclear Waste Management* **1997**, 465, 71-78.

20. Hrma P.; Vienna, J. D.; Crum J. V.; Piepel J. F. Liquidus Temperature of High-Level Waste Borosilicate Glasses with Spinel Primary Phase, *Mat. Res. Soc. Proc.*, **2000**, in press.

21. Kim, D. S.; Hrma P. Volume Changes During Batch to Glass Conversion, *Ceram. Bull.* **1990**, 69.06, 1039-1043.

22. Darab, J. G.; Meiers E. M.; Smith A. P.; Behaviour of Simulated Hanford Slurries during Conversion to Glass, *Mat. Res. Soc. Symp. Proc.* **1999**, 556, 215-222.

Chapter 20

The Nature of the Self-Luminescence Observed from Borosilicate Glass Doped with Es

Zerihun Assefa and Richard G. Haire

Chemical and Analytical Sciences Division, Oak Ridge National Laboratory, Bethel Valley Road, MS 6375, Oak Ridge, TN 37831

We have been conducting spectroscopic investigations of transuranium elements in silicate matrices. Glasses doped with Es show a blue self-luminescence (SL) band having a maximum at ~ 455 nm, and minor bands at 650 and 730 nm. With time, the intensity of the 455 nm band decreases, whereas the 650 and 730 nm bands increase. These SL bands are believed to arise from defect centers that are produced by alpha-particles and recoiling atoms. The band at 455 nm is assigned to originate from oxygen defect centers within the Si-O-Si network, whereas the bands at 650 and 730 nm are believed to originate from non-bonding oxygen hole centers (NBOHC), and Si-micro cluster sites, respectively. The latter two bands (650 and 730 nm) were also observed in the photo-luminescence(PL) spectra obtained with a 488 nm excitation. However, the PL profile showed a dependence on the power of the excitation flux. The 730 nm band dominates at low power, while at higher levels this band's intensity was "quenched" relative to the 650 nm emission. A new band also emerges, concomitantly, around 770 nm on freshly prepared samples and becomes dominant at laser powers greater than 1W. Details of these investigations are discussed in this paper.

Introduction

The effects of radiation in solid waste materials are complex, and a fundamental understanding is often limited. As radioactive decay in a solid matrix can alter physical and chemical behaviors, fundamental understanding of its effects is an essential factor in evaluating long term effectiveness of the immobilization matrices, as well as radionuclides release to the biosphere (1). Borosilicate glasses are one of the primary materials being considered for long term isolation of radionuclides (1, 2). Studies involving radiation effects in actinide-doped glasses have been conducted to ascertain volume and micro-structural changes, as well as radiolytic decomposition effects (3 -6).

Radiation can affect a solid matrix by several processes, and may include the transfer of energy to electrons and/or nuclei. Deposition of nuclear energy into electronic processes results primarily in ionization and/or electronic excitation, where as energy transferred to atomic nuclei, involves elastic collision and atomic displacements (1). An alpha decay process normally results in the formation of a high-energy (4 - 6 MeV) α-particle and a recoiling nucleus having an energy of ~100 KeV (2). It is known that, while the alpha particle loses most of its energy by ionization processes, the recoiling nucleus produces mainly atomic displacements (7 - 9). On the other hand, beta decay produces very little atomic displacement directly, although localized electronic excitations generated by ionization processes may produce selective atomic displacements (10). Thus, the main effect of beta radiation is ionization. The overall consequences of these and other radiation effects on a given matrix are complex and require thorough investigations.

Many studies involving radiation effects have been conducted by using external radiation sources. For example, electron irradiation studies on thin specimens or bulk gamma irradiations using Co-60 sources have been used to simulate beta radiation effects (11 - 15). The long term consequences of both displacement and ionization damage can be best simulated using short-lived actinide dopants (1). In this study we have conducted spectroscopic investigations of borosilicate glasses doped with einsteinium, with one goal being to monitor the consequences of the high-energy alpha radiation on the glass matrix. Es-253, with an alpha decay of 6.6 MeV and a short half-life of 20.5 days, is convenient to simulate both displacement and ionization effects in boro-silicate matrices. We have applied spectroscopic techniques to follow these effects. Self-luminescence, as defined previously (*16 - 17*), refers to luminescence which is excited under the effect of alpha and/or beta decay of actinide isotope under study.

Experimental

Materials.

The glass matrix used in this study had the following composition: SiO_2 (53 %); B_2O_3 (19%); Na_2O (25%); CaO (3%). A glass stock was prepared by grinding thoroughly the components and heating until dissolution was complete (850 °C). The molten material was slowly cooled to room temperature to provide a transparent glass. Einsteinium was incorporated into this base matrix by remelting the glass together with the oxide (~100 μg in 10 mg glass) compound in a small platinum heater of local design. Incorporation of the Es in the matrix provided a clear glass initially but it darkened with time.

Spectroscopy

Both self-emission and absorption spectra were collected using an Instrument SA's optical system, which consisted of a model 1000 M monochromator equipped with CCD, PMT and IR detectors. For the absorption studies a 400 W Xe lamp was used as the light source. The sample, doubly contained in quartz capillary tubes, was placed under a microscope objective and analyzed with the light from the Xe lamp delivered via a fiber optic setup. The transmitted light, collected with a second fiber, was directed into the monochromator.

The self emitted light was collected with fiber optics and directed to the monochromator. The photo-luminescence studies were conducted using argon ion lasers (Models 306 and 90, Coherent) as the excitation sources and a double meter-spectrophotometer (Raman Model HG.2S, Jobin-Yvon) using the procedure described previously (*18*). Spectramax for Windows software (Instruments SA) was used for data acquisition, while Grams 32 software (Galactic Industries) was used for data analysis.

Results and Discussion

When Es is doped in borosilicate matrix, a clear glass is obtained initially, although the sample darkens within a few hours. The darkened glass is easily "bleached" and a clear glass again attained, when visible light is focused on the glass for several minutes. For example, the darkening is removed when the glass is radiated with the 514 nm argon-line at 10 mW of laser power for about 10 minutes. All of the spectra discussed here were collected on clear glasses following radiating with the 514 nm laser line for ~ 20 minutes.

As a secondary issue, the oxidation state of Es in the glass matrix was determined to be trivalent. Shown in Figure 1 is the absorption spectrum, where the bands at ~485, 492, 555, and 598 are assignable (*19*) to f - f transitions in Es(III). Evidence for Es(II) has not been obtained in this study.

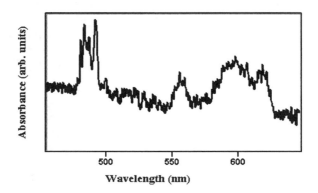

Figure 1. Absorption spectrum of Es in a boro-silicate glass. The absorption bands correspond to Es(III)

Self-luminescence spectra

A feature associated with these Es doped glasses is the observance of a blue self emission visible in the dark. The self-luminescence (SL) spectra of such a sample are shown in Figures 2 and 3. Initially a broad band (FWHM ~ 66 nm) that maximizes at ~455 nm dominates the spectrum with minor features at 544, 650, and 730 nm. As depicted in Figure 3, the SL spectra collected from older samples indicate a significant reduction in the intensity at 455 nm. A slight blue shift to

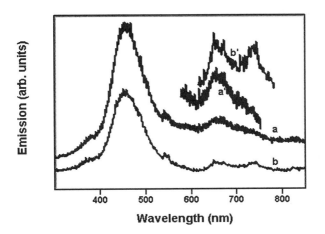

Figure 2. Radiation induced luminescence from a boro-silicate glass doped with Es. Spectra collected following a)3; b) 15 days of sample preparation. Expansions of the low-energy regions are shown in Figures a' and b'.

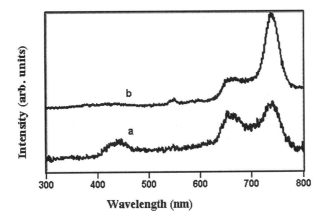

Figure 3. Radiation induced luminescence from borosilicate glass doped with Es. Spectra collected after, a) 50; b) 90 days from preparation.

~445 nm is also evident from older samples. In contrast the intensities of the low-energy bands at 650 and 730 nm increases, concomitantly, while there was no change in the intensity of the 544 nm feature. As shown in Figure 3, the ratio of the 730 to 650 nm band increases and the former band is dominant.

The blue-emission band.

As shown in previous studies (*20 - 22*), one of the consequences of radiation in silica materials is the creation of defect centers. Some of the known effects include creation of oxygen vacancies, free O-atoms, and free electron and hole centers. Basic intrinsic defects found in silica matrices include the E'_1 center; non-bridging oxygen-hole center (NBOHC), $\equiv$Si-O•; peroxy radical, $\equiv$SiO-O•, and oxygen deficient centers (*20*). The E'_1 center is one of the best studied oxygen-vacancy-related defects in silica matrix. It has been observed in all forms of silica including crystalline and glassy materials. As a radiation induced defect, the E'_1 center was first studied in 1956 (*23*) and it's importance for detailed analysis of ageing effects and radiation degradations have been demonstrated since then (*24 -26*).

A characteristic feature of the E'_1 center is that the silicon atom is bonded to just three oxygens and contains an unpaired spin (*27*) in a dangling tetrahedral orbital, $\equiv$Si•. Most commonly, E'_1 defects are induced in silica glasses by energetic radiations including neutron and/or γ radiation.

Being one of the few optically active sites, the E'_1 center is characterized by absorption bands in the far UV region and emission bands in the UV and blue spectral regions. Numerous spectroscopic studies on irradiated silica glasses have indicated that the E'_1 center displays a UV emission band at 4.4 eV and a blue luminescence at 2.7 eV (~450 nm) when excited with a 7.6 eV (163 nm) photon (*28*).

It is important to note that Es-doped glasses exhibit a strong self-luminescence band at 455 nm (2.7 eV) indicating that the alpha radiation is responsible for the displacement of an O⁻ within the Si-O-Si matrix and the creation of the E'_1 center. In addition to defect formation, the alpha radiation from Es is also responsible for the electronic excitation of the site and, thus, the observance of the 2.7 eV self emission of the matrix.

Emission bands in the red-spectral region

In addition to the blue emission band at 455 nm, the SL spectrum of Es-doped glasses exhibit major features in the red spectral region. Emission bands in the red region have been reported in several silica glasses exposed to x-ray, gamma, and/or neutron radiations (*29, 30*). The reported emission energies vary slightly from sample to sample, but generally lie within the 1.7 - 2.2 eV range. Although, a luminescence band around 1.9 eV (650 nm) is thought to signify intrinsic defects in silica glasses (*31*), the exact origin of this emission has been controversial for some time. Cumulative literature data indicate that two major defect centers could provide emission in the 1.7 - 2.2 eV region.

The NBOHC is one of the defect sites believed to provide emission bands in the 1.8 - 2.0 eV range (*31 - 33*), although in few instances these bands have been assigned to an interstitial ozone molecule. The non-bridging center is a localized defect consisting half of the permanently broken Si-O bonds. Such defects are formed as a result of permanent bond breaking under ionizing radiation, such as gamma, neutron, and/or beta. Although, the recoiling nucleus accompanying alpha decay is mainly involved in atomic displacements (*2,7,8,9*), and hence in the creation of vacancies, the alpha particle loses most of its energy by ionization processes. Due to the high alpha activity of einsteinium, creation of non-bridging oxygen defect sites is, thus, expected.

Other luminescent centers are also known to provide emission in the red spectral region. One such center involves a Si-micro cluster defect site created due to a high oxygen deficiency in the matrix. For example, a recent study (*34, 35*) indicated that a silica matrix exposed to a high-dose of γ-irradiation exhibits photo-luminescence at ~1.8 eV(~690 nm) when excited with a 496 nm radiation. The band's intensity was reported to show an increase, albeit with a slight red-shift, when the partial pressure of O_2 decreases. A large dose of γ-irradiation produces a high concentration of oxygen-deficient centers (*35*) that ultimately lead to micro-clustering of the Si atoms. Thus, Si-micro cluster centers are believed to be the source for red emission in silica glasses. Support for this conclusion has been derived from a recent Si-implantation study (*36, 37*), where samples annealed at 1100 °C (a temperature at which the Si precipitates) exhibited photo-luminescence at 1.7 eV. The band has been assigned to a defect site at the interface between crystalline Si-nanoparticles and SiO_2.

Implantation of Si ions into silica glasses provided two photo-luminescence bands, at 2.0 and 1.7 eV, when excited with a 488 nm Ar-ion laser (*38*). Both emissions were thought to originate from Si-rich defect sites (*38, 39*). It is interesting to note that silicate glasses doped with Es also display two SL bands in the red spectral region. We have undertaken photo-luminescence (PL) studies in order to facilitate the assignments of the observed bands.

Photo-luminescence study

The photo-luminescence spectra of the silicate matrix doped with Es are shown in Figure 4 . The spectra were recorded using the argon laser's 488 nm line as the excitation source. Two major PL bands are observed at 650 and 730 nm. The bands are situated at similar positions to that found in the SL spectrum shown in Figure 3.

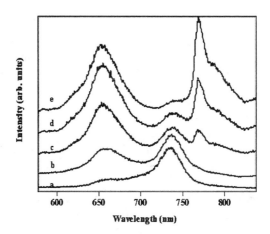

Figure 4. PL spectra of a boro-silicate glass doped with Es. The 488 nm laser line was used for excitation. The spectra are recorded at laser powers of: a)100; b) 310; c) 600; d) 800; e) 1000 mW. The spectra are separated for clarity

The overall spectral profile showed a dependence on the power of the excitation flux. As shown in Figure 4a, the 730 nm band dominates at low power, while at higher levels the 730 nm emission band is "quenched" relative to the 650 nm band. The overall dependence of the 730 nm band on laser power is depicted in Figure 5. Although, an initial increase is evident between 70 and 200 mW, the intensity of the 730 nm band decreases drastically as the laser power is increased further. Also, a new band emerges, concomitantly, around 770 nm and becomes dominant at laser powers greater than 1W. In contrast, the band at 650 nm (1.9 eV) remained linear with an excitation power of 70 to 500 mW, although saturation was evident above 500 mW (Figure 6).

The PL signal exhibited by the sample is much stronger than the SL signal. Hence, at the experimental conditions under which the PL data were collected (example, significantly reduced slit width), the SL signal simply contributed to the noise level. As discussed above, the PL and SL spectra have similar profile in the low-energy spectral region suggesting that both photon and nuclear decay processes provided similar electronic excitations of the emitting sites.

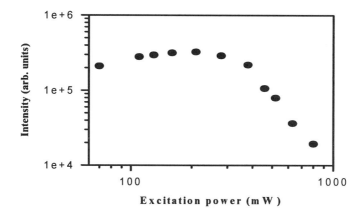

Figure 5. A log-log plot showing the dependence of the 730 nm emission band on the laser power.

As noted earlier, and shown in Figures 5 and 6, the dependence of the two emission bands on the laser power is complex. On one hand, the intensity of the 650 nm band increases linearly up to a 500 mW power and then reaches saturation. On the other hand, the intensity of the 730 nm band, although it slightly increases initially, is drastically "quenched" at higher powers. The spectral profile behaves as if two different species operate independently, suggesting that the two emission bands originate from different centers in the matrix.

In terms of energy position, power dependence, and FWHM, the 650 nm band shows similar profile to that found in silica glasses studied previously (*29 - 33*). Thus, consistent with previous assignments, the 650 nm band is assigned to a NBOHC defect site. Moreover, comparison of the PL profile with time (not shown) indicated that emission from Es(III) also contributed in this spectral region. The details of this phenomenon will be published elsewhere (*40*).

In contrast, the 730 nm band can not be assigned to the non-bridging oxygen centers, since the PL data suggested different origins for the two emissions. One center likely responsible for this emission is a Si-micro cluster unit. Emission in the region of 1.7 eV has previously been reported (*36*) on silica glasses irradiated with very high gamma radiation (~10 Gy), and the band has been attributed to highly oxygen deficient centers. Similarly, Es with its highly energetic (6.6 MeV) alpha particle should also facilitate the creation of oxygen deficient centers that can ultimately lead to Si-micro clustering in the silica matrix. We suggest that the 730

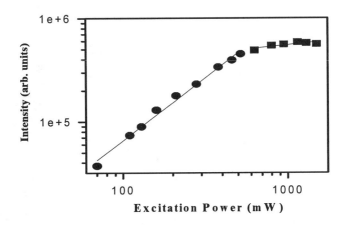

Figure 6. A log-log curve showing the dependence of the 650 nm (1.9 eV) emission band on laser power. Note that saturation occurs after ~800 mW of laser power.

nm band (1.7 eV) originates from centers involving Si-micro clustering.

The unusual power dependence exhibited by the 770 nm emission (Figure 3) makes assignment of this band difficult. The 770 nm emission, which is absent both in the SL and low power PL spectra, appears only under high power excitation. As is the case in a photon avalanche phenomenon, a threshold power appears necessary to observe this emission. The appearance and increase in the intensity of the 770 nm band is also accompanied by a concomitant decrease in the 730 nm emission band. Moreover, a non-linear dependence has been noted between the emission intensity and the excitation flux. These observations are indicative of energy transfer between the Si-micro cluster units and the center responsible for the 770 nm emission. However, the exact nature of the defect center responsible for this emission is not fully established. Although numerous studies have been conducted on radiation-induced defects, to the best of our knowledge, the spectral profile exhibited by the 770 nm band is observed for the first time, and may be unique to alpha radiation effects. Further study on this system is underway in order to better understand the phenomenon.

Summary

We have conducted fundamental optical studies on defects formed by alpha radiation from within boro-silicate glasses doped with einsteinium. It is thought that these studies would help in establishing the influence of defects on the physical and chemical durability of the matrix. The self-luminescence profile of the glass shows bands at 455, 650 and 730 nm, indicating that both the creation of defect centers and their electronic excitations are accomplished by the alpha radiation produced during the decay of Es-253. Based on previous studies, the 455 nm band is assigned to the well known E'_1 defect center. The bands at 650 and 730 nm are believed to originate from non-bonding oxygen hole centers, and Si-micro cluster sites, respectively. The photo-luminescence data showed similar spectral profile indicating that both photon and nuclear processes provided similar electronic excitations to the sites responsible for the emissions. The overall PL profile, however, showed a dependence on the power of the excitation flux. The 730 nm band dominates at low power, while at higher levels the band's intensity is "quenched"relative to the 650 nm emission. Also, a new band emerges concomitantly around 770 nm and becomes dominant at laser powers greater than 1 W. Presence of excited state energy transfer between the Si-micro cluster sites and the unidentified defect centers has been inferred from the PL data.

Acknowledgment

This research was sponsored by the Division of Chemical Sciences, Office of Basic Energy Sciences, US Department of Energy, under contract DE-AC05-96OR22464 with Lockheed Martin Energy Research Corporation. The authors are indebted for the use of Es-253 to the Office of BES, distributed through the trans-plutonium element production program at the Oak Ridge National Laboratory.

References

1. Lutze, W. In *Radioactive Waste forms for the future*; Lutze, W.; Ewing R. C., Eds.; North Holland, Amsterdam, 1988; pp 1- 159.
2. Weber, W. J. *JOM*, **1991**, 43, 35.
3. Weber, W. J.; Roberts, F. P. *Nucl. Tech.* **1983**, 60, 178.
4. Weber, W. J. *Nucl. Instr. Meth.* **1988**, B32,471.

5. Bibler, N. E. In *Scientific Basis for Nuclear Waste Management*; Topp, S. V. Ed.; Materials Research Society Symposium Proceedings; Elsevier Science Publishing Company: New York, 1982; Vol. 6, pp 681.

6. Spilman D. B.; Hench L.L.; Clark, D.E. *Nuclear and Chemical Waste Management*, **1986**, 6, 107.

7. Grambow, B. *Mater. Res. Soc. Bull.* **1994**. XIX (12), 20.

8. Ewing, R. C.; Weber, W. J.; Clinard, F. W. Jr. *Prog. Nucl. Energy.* **1995**, 29, 63.

9. Weber, W. J. et. al., *J. Mater. Res.* **1998**, 13, 1434.

10. Clinard, F. W.; Hobbs, L. W. In *Physics of Radiation effects in Crystals*; Johnson, R. A.; Orlov, A. N., Eds.; **1986**, pp 387.

11. Hosono, H.; Kawazoe, H. *Nucl. Instr. Meth.* **1994**, B91, 395.

12. Mori, H.; Suzuki, Y.; Hirai, M. *Nucl. Instr. Meth.* **1994**, B91, 391.

13. Sigel, G. H. *J. Non-Cryst. Solids*, **1973**, 13, 372.

14. Friebele, E. J.; Griscom, D. L.; Marrone, M. J. *J. Non-Cryst. Solids*, **1985**, 71, 133.

15. Nagasawa, K.; Hoshi, Y.; Ohki, Y.; Yahagi, K. *Jpn. J. Appl. Phys.* **1986**, 25, 464.

16. Finch, C. B.; Young, J. P. *J. Inorg. Nucl. Chem.* **1976**, 38, 45.

17. Barbanel, Y. A.; Chudnouskaya, G. P.; Gaurish, Y. I.; Dushin, R. B.; Kolin, V. V.; Kotlin, V. P. *J. Radioanal. Nucl. Chem.* **1990**, 143, 113.

18. Assefa, Z.; Haire, R. G.; Stump, N. *J. Alloys & Compds.,* **1998**, 271/272, 872.

19. Carnall, W. T.; Cohen, D.; Fields, P. R.; Sjoblom, R. K.; Barnes, R. F. *J. Chem. Phys.* **1973**, 59, 1785.

20. Griscom, D.L. *J. Ceram. Soc. Jpn.* **1991**, 99, 923.

21. Feigl, F. J.; Fowler, W. B.; Yip, K. *Solid-State Commun.* **1974**, 14, 225.

22. Yip, K. L.; Fowler, W. B. *Phys. Rev.* **1975**, 11, 2327.

23. Weeks, R. A. *J. Appl. Phys.* **1956**, 27, 1376.

24. Devine, R. A. B. *Nucl. Instr. Meth.* **1994**, B91, 378

25. Grove, A. S. In *Physics and Technology of Semiconductor Devices;* 1967, John Wiley, New York.

26. Amossov, A. V.; Rybaltovsky, A. V. *J. Non-Cryst. Solids,* **1994**, 179, 226.

27. Weeks, R. A. *J. Non-Cryst. Solids,* **1994**, 179, 1.

28. Trukhin, A. N.; Fitting, H. J. *J. Non-Cryst. Solids,* **1999**, 248, 49.

29. Skuja, L. N.; Silin, A. R. *Phys. Status Solidi,* A, **1979**, 56, K11.

30. Skuja, L. N.; Silin, A. R.; Boganov, A. G. *J. Non-Cryst. Solids*, **1984**, 63, 431.

31. Skuja, L. N.; Suzuki, T.; Tanimura, K. *Phys. Rev.* **1995**, B 52, 208.

32. Skuja, L. N. *J. Non-Cryst. Solids*, **1994**, 179, 51.

33. Skuja, L. N.; Tanimura, K.; Itoh, N. *J. Appl. Phys.* **1996**, 80, 3518.

34. Sakurai, Y.; Nagasawa, K.;Nishikawa, H.; Ohki, Y. *J. Appl. Phys.* **1999**, 88, 370.

35. Nishikawa, H.; Watanabe, E.; Ito, D.; Sakurai, Y.; Nagasawa, K.; Ohki, Y. *J. Appl. Phys.* **1996**, 80, 3513.
36. Nishikawa, H.; Miyake, Y.; Watanabe, E.; Ito, D.; Seol, K. S.; Ohki, Y.; Ishii, K.; Sakurai, Y.; Nagasawa, K. *J. Non-Cryst. Solids*, **1997**, 222, 221.
37. Kaschieva, S.; Yourukov, I. *Solid-State Electron.* **1998**, 42, 1835.
38. Iwayama, S. T.; Ohshima, M.; Niimi, T.; Nakao, S.; Saitoh, K.; Fujita, T.; Itoh, N. *J. Phys. Condens. Matter,* **1993,** 5, L375.
39. Iwayama, S. T.; Fujita, K.; Nakao, S.; Saitoh, K.; Fujita, T.; Itoh, N. *J. Appl. Phys.* **1994**, 75, 7779.
40. Assefa, Z.; Haire, R. G. In Preparation.

Chapter 21

Radiation and Chemistry in Nuclear Waste: The NO$_x$ System and Organic Aging

Dan Meisel[1], Donald M. Camaioni[2], and Thom M. Orlando[3]

[1]Notre Dame Radiation Laboratory, and Department of Chemistry
and Biochemistry, University of Notre Dame, Notre Dame, IN 46556
[2]Energy and Health Sciences Division, Pacific Northwest National Laboratory,
Richland, WA 99352
[3]Environmental Molecular Sciences Laboratory, Pacific Northwest National
Laboratory, Richland, WA 99352

We describe results that advance the understanding of radiation effects in high level waste (HLW) stored at DOE sites. The scientific issues on .which we focus include: a) reactions of primary radicals (e$^-$, OH, and H) of water radiolysis with NO$_3^-$/ NO$_2^-$, b) redox chemistry of NO$_x$ radicals and ions, c) degradation mechanisms and kinetics of organic components of HLW, and d) interfacial radiolysis effects in aqueous suspensions and at crystalline NaNO$_3$ interfaces. Understanding these effects and the chemical reactions they induce have contributed to resolving safety issues and setting waste management guidelines at Hanford.

Several safety issues concerning stored high level waste (HLW) at the Hanford Site have been identified prior to the start of the EMSP. The aging of organics in the waste had been shown to lead to several safety concerns such as flammable and noxious gases generation (H$_2$, N$_2$O, NH$_3$, volatile organic chemicals). Possible runaway reactions of organics (largely chelates) mixed with nitrate salts or organic liquids pooled on the surface of the waste were another significant concern in HLW tanks. We embarked on a coordinated effort to understand the processes that are initiated by radiation and lead to aging of HLW. Our main objective in this project is to assist the safety programs at Hanford and other DOE sites in resolving outstanding safety issues.

HLW invariably contain high concentrations, often above saturation levels, of nitrate and nitrite salts. Because of the high efficiency of scavenging of the primary radicals from water radiolysis by NO_3^-/NO_2^-, a majority of the radiolytically generated radicals are of the NO_x family, reactions 1-4 (*1, 2*).

$$OH + NO_2^- \rightarrow NO_2 + OH^- \quad k_1 = 1.0 \times 10^{10} \text{ M}^{-1}\text{s}^{-1} \qquad (1)$$

and at high pH

$$O^- + NO_2^- (+H_2O) \rightarrow NO_2 + OH^- \quad k_{1a} = 3.1 \times 10^8 \text{ M}^{-1}\text{s}^{-1} \qquad (1a)$$

$$e_{aq}^- + NO_3^- \rightarrow NO_3^{2-} \quad k_2 = 9.7 \times 10^9 \text{ M}^{-1}\text{s}^{-1} \qquad (2)$$

$$NO_3^{2-} (+ H_2O) \rightarrow NO_2 + 2OH^- \quad k_3 = 5.5 \times 10^4 \text{ M}^{-1}\text{s}^{-1} \qquad (3)$$

At high nitrite concentrations, H is primarily converted to NO, reaction 4.

$$H + NO_2^- \rightarrow NO + OH^- \quad k_4 = 7.1 \times 10^8 \text{ M}^{-1}\text{s}^{-1} \qquad (4)$$

The initial reduction product of nitrate, NO_3^{2-}, is expected to be a strong reductant, but NO_2 and NO are oxidants ($E°(NO_2/NO_2^-) = 1.04$ V) (*3*). Therefore, the early events that follow the absorption of radiation, and the subsequent redox chemistry of NO_x radicals and ions will determine the redox environment in HLW. Thus, they are highly relevant to understanding radiolytic aging of wastes.

Results and Discussion

In this report we focus on the following processes: a) reactions of primary radicals (e⁻, OH, and H) of water radiolysis with NO_3^-/NO_2^-, b) redox chemistry of NO_x radicals and ions, c) degradation reactions of organic components, and d) interfacial radiolysis effects in aqueous suspensions and at crystalline $NaNO_3$ interfaces. We used pulse radiolysis in conjunction with time resolved spectrophotometric, EPR, and conductivity detection techniques. Continuous γ radiolysis and NO_2 gas contacting of HLW simulants, with subsequent product analyses by ion chromatography and NMR techniques. Computation of the structure and energies of intermediates using *ab initio* theory and solvation models was used in support of the experimental effort. Ultra-high vacuum surface analysis techniques were applied to measure electron beam interfacial and bulk damage mechanisms of $NaNO_3$ crystallites. In a separate chapter of this volume we detail our studies on the effects of interfaces of oxide powders and colloidal suspensions in HLW radiolysis.

Formation, Reactions and Properties NO_3^{2-}

Reaction 3 is a bottleneck in the transformation from a reducing environment to an oxidizing one. To our knowledge, e_{aq}^- is the only radical that reduces nitrate ion. Mezyk and Bartels showed that H atoms do react with NO_3^-, but they concluded that the product is directly NO_2 (4). As long as the NO_3^{2-} radical exists in solution, the potential for the generation of H_2 from water cannot be excluded. Therefore, we studied the relevant chemistry of NO_3^{2-}.

The absorption spectrum of NO_3^{2-} peaks at $\lambda_{max}=260$ nm with $\varepsilon=2.0 \times 10^3$ M^{-1} cm^{-1}. It decays within a few μs to generate a weakly absorbing species $\lambda_{max}=390$ nm attributed to NO_2. A small fraction of NO_2 dimerized on the same time window to N_2O_4 (5), which leads to a band at $\lambda_{max}=270$ nm.

The charge of the initial product was verified to be -2 by measuring the ionic strength effect (controlled by varying the concentration of $NaNO_3$) on the rate of electron transfer from NO_3^{2-} to methyl viologen (MV^{2+}), reaction 5.

$$NO_3^{2-} + MV^{2+} \rightarrow NO_3^- + MV^+ \qquad (5)$$

A least squares fit to the Brönsted-Bjerrum modified Debye-Huckel equation, Eq. 6, yields: the rate constant at zero ionic strength, $k_0=(1.3 \pm 0.1) \times 10^{10}$ M^{-1} s^{-1}, and an effective reaction distance $d=2.48 \pm 0.2$ Å. The product of the charges, $Z_a Z_b = -4.2 \pm 0.2$, confirms the proposition that the radical is indeed NO_3^{2-}.

$$\log \frac{k}{k_0} = \frac{1.018 Z_a Z_b \sqrt{\mu}}{1 + 0.329 d \sqrt{\mu}} \qquad (6)$$

Reactions of the Precursor of Hydrated Electrons

The nitrate ion is one of the most efficient scavengers for the precursor to the hydrated electron e_{th}^-. The concentration at which only 37% of the initially produced thermal electrons escape scavenging, is $C_{37} = 0.45$ M (6, 7).

$$e_{th}^- + NO_3^- \rightarrow P \qquad (2a)$$

Because of the low C_{37} and because the concentration of nitrate in HLW may be extremely high (1-5 M), the identity of the product, P in reaction 2a, is of considerable significance. To test that question we measured the yield of MV^+ at increasing $[NaNO_3]$ and at several doses per pulse. Up to 30% decrease in the yield of MV^+ was observed at 5 M of $NaNO_3$. Some of the radiation is absorbed directly by the nitrate salt at such high concentrations. This "direct effect" produces directly NO_2, and perhaps NO, but not the reducing NO_3^{2-} (8). Thus, it may be concluded that the thermalized precursor to e^-_{aq}, labeled e^-_{th} in reaction 2a, reduces nitrate to NO_3^{2-}.

Acid-Base Chemistry of NO_3^{2-}

The lifetime of NO_3^{2-} was determined across a broad pH range by following its characteristic absorption at 270 nm. The lifetime decreases with pH and with total buffer concentration as previously reported. This trend has been ascribed to acid-base equilibria of reactions 7 and 8 with the protonated forms assumed to have shorter lifetimes than NO_3^{2-} (1, 9).

$$H_2NO_3 \rightleftarrows H^+ + HNO_3^- \quad (pK_{a1} = 4.8) \tag{7}$$

$$HNO_3^- \rightleftarrows H^+ + NO_3^{2-} \quad (pK_{a2} = 7.5) \tag{8}$$

$$NO_3^{2-} + HA \rightleftarrows HNO_3^- + A^- \tag{9}$$

$$HNO_3^- \rightarrow NO_2 + OH^- \quad (k_{10} = 2.3 \times 10^5 \text{ s}^{-1}) \tag{10}$$

$$H_2NO_3 \rightarrow H_2O + NO_2 \quad (k_{11} = 7.0 \times 10^5 \text{ s}^{-1}) \tag{11}$$

Table I lists rate constants for the reactions of various "acids" with NO_3^{2-}. All were obtained from the dependence of the rate of decay of the radical dianion on the concentration of the acid.

Details of the kinetic analysis are given elsewhere (10), but they indicate that the protonated intermediate is either very short-lived or a transition state structure for reactions 3 and 10, or 11. This conclusion concurs with that of Mezyk and Bartels (4). Addition of H atom in reaction 12 would lead to the same intermediate discussed above (reactions 7, 8).

$$H + NO_3^- \rightarrow NO_2 + OH^- \tag{12}$$

Furthermore, the fast reaction of the various acids (e.g., boric acid, which is an OH^- acceptor and not H^+ donor) with the radical dianion suggests that the reaction is not simply a proton transfer. It seems better described as an O^{2-} transfer, even in the case of water, reaction 3.

Table I. Rate Constants For the Reaction of Various "Acids" with NO_3^{2-}

HA	$\dfrac{k}{M^{-1}s^{-1}}$	pK_a (of parent)	$\dfrac{\text{Lit. } k}{M^{-1}s^{-1}}$	Ref.
H^+	$(4.5\pm0.5)\times10^{10}$		$>2.0\times10^{10}$	11
$H_2PO_4^-$	$(5.3\pm0.5)\times10^8$	7.21	5.0×10^8	1
NH_4^+	$(3.6\pm0.4)\times10^8$	9.25	2.0×10^8	11
$B(OH)_3$	$(8.4\pm0.8)\times10^7$	9.24		
HCO_3^-	2.8×10^7	10.3		12
$CAPSH^+$	$(1.1\pm0.1)\times10^7$	10.4		
NH_3	$(3.4\pm0.6)\times10^0$			
$MeOH$	$(1.2\pm0.4)\times10^4$	15.0		
$t\text{-BuOH}$	$\leq2\times10^4$	16.9		
H_2O	$(1.2\pm0.1)\times10^3$ $(6.8\pm0.8\times10^4\,s^{-1})$	15.7	1.00×10^3	1 9

Redox Potential of NO_3^-/NO_3^{2-}

The redox potential of the nitrate/NO_3^{2-} couple was determined using pulse radiolysis via equilibrating the radical dianion with redox couples of know potentials (reaction 13).

$$NO_3^- + Q \rightleftarrows NO_3^{2-} + Q^+ \qquad (13)$$

Equilibrium was established with the bridged bipyridinium salt, 2,2'-tetra-methylenebipyridium, V^{2+}. Using the equilibrium constants determined experimentally we calculate a mid-point potential of $E_m = -0.89\pm0.02$ V for the NO_3^-/NO_3^{2-} couple at pH 9.2 and $\mu = 0$. This value is also close to predicted estimates from theoretical calculations (-1.04±0.1 V) (10). At a more commonly encountered ionic strength of $\mu = 1.0$, and at that same pH, we estimate $E_m = -0.79\pm0.02$ V.

Possible Electron Transfer from NO_3^{2-} to NO_2^-

Of particular interest in the context of HLW chemistry is the possible reaction of NO_3^{2-} with nitrite (reaction 14).

$$NO_3^{2-} + NO_2^- \rightleftarrows NO_3^- + NO_2^{2-} \qquad (14)$$

Because of the high concentration of nitrite in waste tanks, even relatively small rate constant may imply an efficient conversion of NO_3^{2-} to NO_2^{2-} and eventually to NO (reaction 15).

$$NO_2^{2-} + H_2O \rightarrow NO + 2OH^- \tag{15}$$

The redox potential of the NO_2^-/NO_2^{2-} couple was determined in a similar method to that described above to be $E_m = -0.47 \pm 0.02$ V under similar conditions. Thus, reaction 14 is thermodynamically feasible.

Attempts were made to measure the rate constant of reaction 14 using competition between the bipyridinium ions and nitrite for NO_3^{2-}. Only an upper limit for the rate constant, $k_{14} \leq 5 \times 10^4$ $M^{-1}s^{-1}$, could be estimated. Whereas this is a relatively slow rate constant, we cannot rule out the possibility that the reduction equivalents are partially converted to NO via reaction 14, 15 in highly concentrated nitrite solutions.

Other Routes to NO_3^{2-}

From the one-electron redox potential $E°(NO_2/NO_2^-) = 1.04$ V and the two-electron redox potential of nitrate, $E°(NO_3^-/NO_2^-) = 0.01$ V, one estimates for Reaction 3: $K_3 \approx 1 \times 10^{-2}$ M. Thus, it is conceivable that at high OH^- concentrations the oxide ion, O^{2-}, may react with NO_2 to produce NO_3^{2-}. Such a proposition is unexpected. If oxide ions add to NO_2, a strongly oxidizing environment is converted to a strongly reducing one. Another previously unreported route to the production of NO_3^{2-} is the addition of O^- to NO_2^-. Using pulse radiolysis with time-resolved ESR detection, we obtained recently unequivocal evidence that O^- adds, at least partially, to nitrite ions, reaction 16.

$$NO_2^- + O^- \rightarrow NO_3^{2-} \tag{16}$$

As can be seen in Figure 1, the ESR signal of NO_3^{2-} can be produced from solutions that contained only nitrite (and N_2O) at high base concentrations. The same ESR signal is obtained in nitrate containing neutral solutions via reaction 2. Also shown if Figure 1 are the results of simulations of the known processes that occur in that system, summarized in Table II. In order to obtain reasonable fit to experimental results, both reaction 16 and the reverse of reaction 3 are necessary. However, the evidence for reaction -3 is weak because adjustments to other rate constants could eliminate it from the scheme.

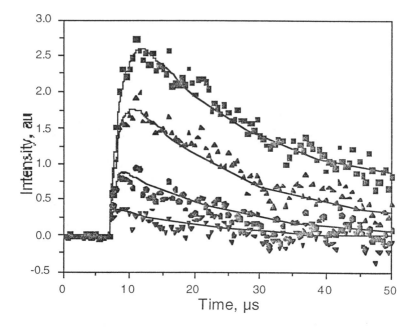

Figure 1: ESR signal of NO$_3^{2-}$ produced from solutions that contained only nitrite, N$_2$O, and increasing [OH]=0.1, 0.3, 1,2 M from bottom up.

To conclude our studies of the NO$_3^{2-}$ radical, it seems clear that it is more prevalent in irradiated solutions of nitrate and nitrite than initially thought. It is a strong reductant that interconverts with a strong oxidant, the NO$_2$ radical. However, its importance in HLW is still questionable. Because of its relatively short lifetime and because of the fact that it is a single electron redox reagent, its participation in gas generation processes is probably minimal. On the other hand, its successor, the NO$_2$ radical is relatively long lived and participates in oxidative degradation of organic compounds in HLW tanks.

Reactions of Model Organic Complexants with NO$_2$

The radiolysis of homogeneous aqueous HLW simulants generate primarily OH/O$^-$ and NO$_2$. The reactions of NO$_2$ with organic chelates and carboxylate salts were studied in order to develop a quantitative model for their degradation. In these experiments we contacted aqueous alkaline solutions of organic species with N$_2$ containing 10-100 ppm NO$_2$. The latter dissolves in the solution, it dimerizes and hydrolytically disproportionates to nitrite and nitrate. In the presence of the organic solute NO$_2$ may be reduced directly to nitrite. We use the competition between these two pathways to estimate the rate of the organic oxidation. Below we discuss the results from several of these organic solutes.

Table II. Kinetic Model For Simulating the Pulse Radiolysis of Alkaline Nitrite Solutions

No.	Reaction	Forward (Back) Rate Constants	Notes/Refs.
1	$e^-_{aq} + N_2O \rightarrow N_2 + \bullet O^-$	9.1×10^9 $M^{-1} s^{-1}$	2
2	$e^-_{aq} + NO_2^- (+ H_2O) \rightarrow NO + OH^-$	3.5×10^9 $M^{-1} s^{-1}$	13
3	$OH + NO_2^- \rightarrow NO_2 + OH^-$	1×10^{10} $M^{-1} s^{-1}$	2
4	$OH + OH^- \rightleftarrows \bullet O^- + H_2O$	1.3×10^{10} $M^{-1} s^{-1}$ $(9.4 \times 10^7 \ s^{-1})$	14
5	$H + OH^- \rightarrow e^-_{aq} + H_2O$	2.0×10^7 $M^{-1} s^{-1}$	15
6	$O^- + NO_2^- (+ H_2O) \rightarrow NO_2 + 2OH^-$	4.7×10^7 $M^{-1} s^{-1}$ [a]	a
7	$O^- + NO_2^- \rightarrow NO_3^{2-} {}^-$	1.8×10^7 $M^{-1} s^{-1}$	a
8	$NO_3^{2-} + H_2O \rightleftarrows NO_2 + 2OH^-$	1×10^3 $M^{-1} s^{-1}$ $(500 \ M^{-2} s^{-1})$	a, forward rate: 10, 1
9	$2NO_2 \rightleftarrows N_2O_4$	5×10^8 $M^{-1} s^{-1}$ $(6800 \ s^{-1})$	16, 5
10	$N_2O_4 (+ H_2O) \rightarrow NO_3^- + NO_2^- + 2H^+$	$1000 \ s^{-1}$	16, 5
11	$NO + NO_2 (+ 2OH^-) \rightarrow 2NO_2^-$	1×10^9 $M^{-1} s^{-1}$	1, 5, 16

[a]Rate constants for Reactions 6 and -8 were adjusted to give the best fit to the curves in Figure 1.

Formate Ion

Formate ion degrades cleanly to carbonate ion upon reaction with NO_2 (reactions 18, 19).

$$NO_2 + HCO_2^- \rightarrow NO_2^- + CO_2^- + H^+ \tag{18}$$

$$NO_2 + CO_2^- + H_2O \rightarrow NO_2^- + CO_3^{2-} + 2H^+ \tag{19}$$

Figure 2 shows results from contacting 25 and 100 ppm NO_2 in N_2 with a 0.1M formate in 1M NaOH solution. Both $[NO_2^-]$ and $[NO_3^-]$ increase linearly with time, but $[NO_2^-]$ grows faster indicating that oxidation of formate ion occurs competitively with NO_2 hydrolysis (Reaction 20) (*17*, *18*)

$$2NO_2 + H_2O \rightarrow HNO_2 + HNO_3 \tag{20}$$

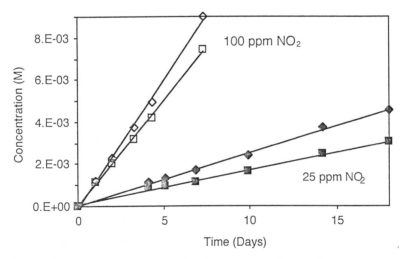

Figure 2. Concentrations of NO_3^- and NO_2^- in solution vs. time upon contact with NO_2/N_2 gas. NO_2^- grows faster than NO_3^- due to reduction by formate. Diamonds: nitrite, Squares: nitrate.

The stoichiometry of formate oxidation was verified using ^{13}C-labeled formate. Conversion to ^{13}C-carbonate after the contact experiments follows the stoichiometry shown in reactions 18, 19. A model to describe the competition kinetics of the contact experiments was developed and is described elsewhere.

Glycolate, Glycine, Iminodiacetictate (IDA), and Nitrilotriacetate (NTA)

We utilized ion chromatography and ^{1}H and ^{13}C NMR spectroscopies in analysis of products from these compounds following the reaction with NO_2. Glycolate degrades mainly to formate and oxalate. NTA and IDA degrade via stepwise loss of $-CH_2CO_2^-$ groups, producing mainly formate, carbonate and

$$N(CH_2CO_2^-)_3 \xrightarrow{\ NO_2\ } HN(CH_2CO_2^-)_2 \xrightarrow{\ NO_2\ }$$

$$H_2NCH_2CO_2^- \xrightarrow{\ NO_2\ } HCO_2^- \qquad (21)$$

oxalate. Formate and carbonate dominate over oxalate indicating that oxidative decarboxylation is the dominate reaction mechanism. Note that in 1 M NaOH solutions used for these studies, formaldehyde and glyoxal convert to formate and oxalate, respectively, producing H_2 gas (reaction 25) (*19, 20*).

$$R_2NCH_2CO_2^- + NO_2 \rightarrow R_2NCH_2\bullet + CO_2 + NO_2^- \qquad (22)$$

$$R_2NCH_2\bullet + NO_2 \rightarrow R_2N^+{=}CH_2 + NO_2^- \qquad (23)$$

$$R_2N^+{=}CH_2 + HO^- + H_2O \rightarrow R_2NH + H_2C(OH)_2 \qquad (24)$$

$$H_2C(OH)_2 + OH^- \rightarrow HCO_2^-\ H_2 + H_2O \qquad (25)$$

Oxalate results from oxidation at the methylene group.

$$R_2NCH_2CO_2^- + NO_2 \rightarrow R_2NCH(\bullet)CO_2^- + NO_2^- + H^+ \qquad (22a)$$

Studies with 1- and 2-^{13}C-labeled glycine show it to degrade to formate, oxalate and carbonate ions. The branching ratio is ~60/40 formate/oxalate. Therefore, glycine is functionally similar to IDA and NTA, and glycolate except that it tends to give higher fractions of oxalate.

The details of NO_2 attack on these compounds are unknown. In the case of aminocarboxylate ions, we consider the possibility that reactions occur by electron transfer from N followed by, or concerted with, decarboxylation of the radical. A precedent for such a path is found in the autoxidation of trialkylamines in alkaline aqueous solutions (*21*), the decarboxylation of amino acid anions by hydroxyl radical (*22, 23*), and decarboxylation of anilinoacetate radical cations (*24*). Even the pathway that yields oxalate may originate from attack at N. For example, branching reactions may follow electron transfer from the amino group (reaction 26).

$$\text{NO}_2^\bullet + \text{H}_2\text{N-CH}_2\text{-CO}_2^- \longrightarrow \left[\text{H}_2\overset{+\bullet}{\text{N}} \overset{a}{\frown} \text{CH}_2\text{-CO}_2^- \underset{b}{\searrow} \text{H}_2\text{O} \right] \overset{a}{\underset{b}{\longrightarrow}} \begin{array}{l} \text{H}_2\text{N-CH}_2^\bullet + \text{NO}_2^- + \text{C(O)(O)} \\[4pt] \text{H}_3\text{O}^+ + \text{NO}_2^- + \text{HN}^\bullet\text{-CH}_2\text{-CO}_2^- \end{array} \qquad (26)$$

The aminomethyl radical, H_2NCH_2 (path a, reaction 26), and the nitrogen-centered radical, $\text{NHCH}_2\text{CO}_2^-$ (path b, reaction 26), may abstract H from another glycine molecule to give carbon-centered radical, $\text{H}_2\text{NCHCO}_2^-$ (*23*). Alternatively, the $\text{NHCH}_2\text{CO}_2^-$ may undergo water assisted 1,2 H-shift to form $\text{H}_2\text{NCHCO}_2^-$. For glycolate ion, analogous paths following electron transfer oxidation at the OH group are possible. Alternatively, the $\text{NHCH}_2\text{CO}_2^-$ may undergo water assisted 1,2 H-shift to form $\text{H}_2\text{NCHCO}_2^-$. For glycolate ion, analogous paths following electron transfer oxidation at the OH group are possible.

The dimer, N_2O_4, may also reacts with the organic substrates. Challis and coworkers reported that secondary amines (*25*) and glycylglycine (*26*) are nitrosated on contact with NO_2 in alkaline aqueous solutions. They suggest nucleophilic attack on N_2O_4, reaction 27.

$$\text{R}_2\text{NH} + \text{N}_2\text{O}_4 \rightarrow \text{R}_2\text{NNO} + \text{NO}_3^- \qquad (27)$$

Depending on conditions, other adducts may also form during oxidation of glycine and glycolate. We identified hydroxyaspartate ion on oxidation of glycine by NO_2 and by γ radiolysis of alkaline nitrate solutions that contained glycine. However, when 1 M NO_2^- is present in the solution, production of hydroxyaspartate (and other adducts) is substantially reduced and formate, oxalate and carbonate ions amount to ~90% of the products. Thus, it seems that the presence of high nitrite concentrations may prevent the radiolytic formation of higher molecular weight products in tank wastes.

Relative reactivities of organic substrates with NO_2 were determined by competition with ^{13}C-formate ion. Conversions of the unlabeled substrate to ^{12}C-formate and labeled formate to [13]CO_3^{2-} were measured by NMR and used to calculate the rates. The trend obtained is formate:glycolate:glycine:IDA:NTA $\approx$ 1:19:7:11:19. Because the absolute rate for NO_2 reaction with formate is ~1 M^{-1} s^{-1}, the relative reactivities may also be equated to the absolute rates.

γ-Radiolysis Aging of Organic Solutes

Waste aging studies were performed at the Hanford Site to explain waste characterization data and determine the effect of organic aging on combustion hazards of HLW (*27*). These studies simulated waste aging by γ-irradiating non-radioactive waste simulants. The relative rates of complexants disappearance during irradiation are summarized in Table III. The compounds are the main chelates used in Hanford processes and some degradation intermediates. Where direct comparisons can be made, the relative reactivities are similar to those obtained from the NO_2 contact experiments. Significantly, formate is substantially less reactive than aminocarboxylate ions and glycolate ions. This selectivity is counter to that expected for OH/O^- (*2*).

Table III. Relative Rates of Disappearance of Organic Compounds in Waste Simulants upon γ Radiolysis[a]

Substrate[b]	k_{rel}
u-Ethylenediaminediacetic (EDDA)	13
s-EDDA	13
IDA	12
NTA	10
HEDTA	14
Glycine	7
Glycolate	5
EDTA	6
Formate	1[c]
Citrate	0.7
Acetate	0.7

[a]Reproduced from Ref. *27*. [b]In 3.75 M $NaNO_3$, 1.25 M $NaNO_2$, 2 M NaOH and at 20°C. [c]Defined value.

For the Hanford waste aging studies (*27*) we also measured the radiolytic yield of carbonate from formate in the simulants. The yield increases with the concentration of formate reaching a maximum value of ~2 molecules/100 eV (see Table IV). Since the necessary rate constants were measured we were able to model the degradation processes. Results are compared with experimental observations in Table IV.

Table IV. Yields and Fractions of Carbonate Ion Produced from H, OH, O⁻,
and NO₂ During γ-Radiolysis of HLW Simulants[a] Containing Formate Ion

Concentration M		Dose Rate	$G(CO_3^{2-})$		% Distribution of CO_3^{2-}			
OH^-	HCO_2^-	Rad/h	Model	Expt	OH	O^-	NO_2	H
2	1.0	4.6×10^5	1.9	2.0^b	57	13	13	5
2	0.1	4.6×10^5	0.8	0.8^b	7	87	5	1
2	0.03	9.3×10^4	2.5	2.2^c	3	96	0	2
0.1	0.1	4.6×10^5	0.2		42	32	22	5
2	0.1	4.6×10^2	1.5		3	44	52	1

[a]Simulant: 1.25 M NaNO₂, 3.75 M NaNO₃, 2 M NaOH. [b]Ref. *27*. [c]0.03 M NaNO₂, 0.1 M NaNO₃ ref. *28*.

Table IV also details the fraction of carbonate resulting from attack by OH, H, O⁻, and NO₂. The model shows that O⁻/OH is the dominant oxidant of formate in the high radiation fields used in the experiments. However, dose rates in the tank wastes are much smaller; they range between 10^2-10^4 Rad/h (*29*). The last line in Table IV shows that reducing the dose rate increases the yield of carbonate and the fraction of oxidation by NO₂. This behavior is expected since NO₂ oxidation of formate competes with second-order radical recombination reactions.

Role of NO in Flammable Gas Generation

In radiolysis of formate, NO mainly terminates with NO₂. Radiolysis of simulants containing formate as the only organic solute does not generate H₂ and N₂O gases. However, the complexants do (*30*). If NO couples with organic radicals from these compounds, RNO compounds are formed. These are expected to undergo solvolytic and ionic reactions to form carbonyl compounds and hydroxylamine. The latter ultimately produce H₂, N₂O, N₂, and NH₃ in waste simulants (*19, 20,* 31). However, simulations show that as long as nitrite reacts with the organic radicals with rate constants of $>10^4$ M^{-1} s^{-1}, then termination with NO cannot compete. The organic radicals can react with nitrite by electron transfer and/or addition reactions. Electron transfer will produce NO and aldehydes that liberate H₂ in strongly alkaline solutions, but electron transfer will not lead to N₂O, N₂ and NH₃. The latter path produces metastable radical anions (RNO₂⁻) that may live long enough to terminate with NO. Accordingly, RNO and gases may arise by this indirect but kinetically favorable path.

Radiation Effects on Crystalline NaNO₃

Sodium nitrate is a wide band-gap ionic material with the calcite-type rhombohedral lattice. It is a major component of HLW wastes currently in underground storage tanks. As part of our study on radiation-induced processes within porous solids and at interfaces, we have performed a detailed study of electron interactions with $NaNO_3$ single crystals using the tools and techniques of modern surface science (*32*).

Solution-grown $NaNO_3$ single crystals were annealed for 2-3 hours at 430 K in an ultrahigh vacuum (1×10^{-10} Torr) chamber. The sample temperature was then varied from 100-450 K. A Kimbal Physics ELG-2 low-energy electron gun was utilized as the electron-beam source. The incident energy was variable from 5-100 eV. The pulsed beam supplied an electron fluence of ~ 10^9 electrons per cm^2 per pulse; the continuous irradiation was applied at the current density of ~10 $\mu A/cm^2$ and a maximal dose of ~10^{16} electron/cm^2. Yields of neutral desorbates were obtained utilizing quadrupole mass-spectrometer (QMS), which was programmed to detect 10 masses. The time-of-flight (TOF) and thus the velocity distributions of the neutral desorbates were measured by phase-locking the QMS to the electron-beam frequency. The AES spectra and SEEM images of the damaged $NaNO_3$ single crystals were taken using a Physical Electronics 680 Auger Nanoprobe.

Figure 3 is a SEEM image of the $NaNO_3$ surface showing the effects of electron irradiation. The square area (center portion) irradiated with 10-keV electrons is topologically rough with a large amount of disorder and deep cone-like damage features. The AES spectrum of the damaged area demonstrates the dominance of the oxygen and sodium peaks. This clearly indicates a sizable depletion in both nitrogen and oxygen, pointing to presence of Na_2O as one of the main surface damage products.

The Electron-Stimulated Desorption (ESD) of all these molecular products is thermally activated in the temperature region between 110-440 K and the most significant temperature effect can be observed at temperature above 300 K. The normalized data for NO and O_2 ESD yields have essentially identical temperature dependencies. Recent studies report NO and O (^3P_J) fragments as primary desorption products under low-dose nanosecond-pulse electron- and UV-photon irradiation (*33, 34*). The molecular products released from the $NaNO_3$ single crystal surface under microsecond pulses of the 100 eV electron-beam irradiation are NO and O_2 with a small amount of NO_2. The temperature dependence of the ESD yields of these molecules is presented in Arrhenius coordinates in Figure 4.

The data can be approximated by a sum of two Maxwell-Boltzmann-type equations of the form: $I = I_{01} \exp(-E_1/kT) + I_{02} \exp(-E_2/kT)$ where E_1 and E_2 are activation energies, 0.16 ± 0.03 and 0.010 ± 0.004 eV, respectively; I_{01} and I_{02} are constants. The solid line in Figure 4 is a fit using this equation. The similar temperature dependencies for NO and O_2 ESD, and the constant yield ratio,

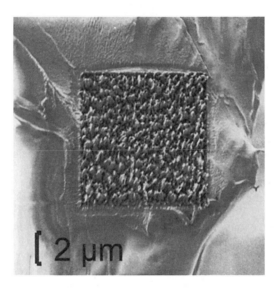

Figure 3. SEEM image of the NaNO₃ surface, whose central area has been irradiated with 10 - keV electron-beam. (Reproduced from Ref. 32 with permission from the American Chemical Society)

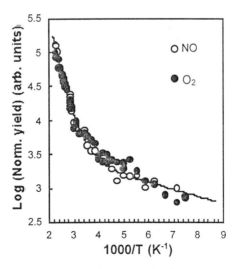

Figure 4. Temperature dependence of ESD yields of NO (open circles) and O₂ (filled circles) in Arrhenius coordinates. The yields are normalized for the sake of comparison. (Reproduced from Ref. 32 with permission from the American Chemical Society)

indicate that these molecules are produced via the same process and likely involve a common precursor.

The NO_2, O_2 and NO TOF distributions resulting from pulsed (1 µs, 20 Hz, 100 eV) electron-beam irradiation of $NaNO_3$ crystals held at 423 K are shown in Figure 5.

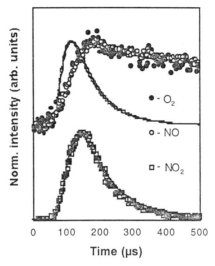

Figure 5. Velocity distributions for NO (open circles), O_2 (solid circles) and NO_2 (squares) following 100 eV electron irradiation of $NaNO_3$ surfaces at 423 K. The intensities are normalized. (Reproduced from Ref. 32 with permission from the American Chemical Society)

The lines are the calculated 423 K Maxwell-Boltzmann distributions for O_2, NO and NO_2 respectively assuming instantaneous desorption during the electron pulse. To facilitate comparisons of the leading edges, all intensities are normalized. The experimental NO_2 velocity distribution fits the Maxwell-Boltzmann approximation, but the emission of the NO and O_2 is extended for several milliseconds after the electron pulse and the delayed NO and O_2 emission kinetics are similar. We suggest that the observed NO and O_2 ESD temperature dependence is due to the temperature dependence of NO_2^- production. The NO_2^- is a source of holes, which migrate to the surface and recombine with NO_3^- to produce NO_3 precursors. The $^2E'$ excited state of NO_3 can undergo unimolecular dissociation yielding NO ($^2\Pi$) and O_2 ($^3\Sigma_g^-, ^1\Delta_g$) molecules due to vibronic coupling to the NO_3 ($^2A_2'$) ground state. This excited state is long-lived, dissociating on the time scale of hundreds of microseconds in the gas phase. This proposed mechanism does not need a fundamental ionization event and can explain delayed NO and O_2 emission induced by 5 and 6.4 eV photons. Therefore, thermal annealing of irradiated $NaNO_3$ crystals, which activates the decay of bulk radiation defects, produces oxygen that diffuses to the surface.

With respect to HLW, the results suggest that radiolysis of the bulk and interfacial $NaNO_3$ crystalline components may be a source of O_2 in the wastes. Allowing for continuous dissolution-reprecipitation of $NaNO_3$ due to thermal gradients and seasonal temperature swings of the wastes, O_2 and nitrite ions formed in the bulk crystals may be released to the solution. Assuming that surface radiolysis is analogous to that at aqueous-crystalline interfaces, the interfacial radiolytic process nay be a direct source of NO.

Conclusion

With respect to issues of relevance to DOE's Environment Management operations, we conclude that:

- The probability that H_2 can be formed from NO_3^{2-} is rather small. Experimentally, the yield of H_2 decreases upon increasing NO_3^- concentrations, even though NO_3^{2-} is efficiently produced in these solutions. The generation of H_2 from water requires two reduction equivalents, but in the absence of catalysts NO_3^{2-} is able to provide only one equivalent before it dissociates. The protonated form is much shorter lived, if it exists at all. Thus, it is converted to NO_2 even faster in the presence of Lewis acids, reducing the probability of fuel generating reduction processes.

- NO_2 contact experiments corroborate the proposition that NO_2 is generated during radiolysis of waste simulants. Thus, it contributes to oxidative degradation of organic solutes in HLW.

- NO radicals, which are produced by reduction of nitrite (by H atoms and organic radicals), or by direct effect on crystalline $NaNO_3$, may accumulate to relatively high steady state levels. They can then terminate with various organic radicals. The products RNO compounds lead to H_2, N_2O and NH_3.

- Nitrite ions scavenge organic radicals thus promoting organic oxidation. However, they also facilitate gas generation as described in the previous item. They may also convert NO_3^{2-} to NO, thereby reducing the efficiency of oxidatively degrading organic waste components.

These insights and some of the quantitative rate determinations have been incorporated into safety analyses at the Hanford site. In that context they contributed to the resolution of several of the safety issues at the site, including the organic tanks and gas generation issues.

Acknowledgement

This is a summary of the efforts of many of our colleagues. Their contributions are much appreciated. This report describes coordinated efforts by three projects of the Environmental Management Science Program ("The NO_x System in Nuclear Waste," "Mechanisms and Kinetics of Organic Aging in High Level Nuclear Wastes," and "Interfacial Radiolysis Effects in Tanks Waste"). Support from the EMSP of these projects is acknowledged. The insight developed here would not be possible without many years of support by the Office of Basic Energy Sciences – Chemical Sciences Division. Our interactions with Richland Operations and the Hanford and Savannah River sites during those years provided the impetus to these studies. This is document NDRL No. 4193 from the Notre Dame Radiation Laboratory. Pacific Northwest National Laboratory is operated by Battelle Memorial Institute for the U.S. Department of Energy under Contract DE-AC06-76RLO 1830.

References

1. Gratzel, M.; Henglein, A.; Taniguchi, S. *Ber. Bunsenges. Phys. Chem.* **1970**, *74*, 292, (1970).
2. Buxton, G. V; Greenstock, C. L.; Helman, W. P.; Ross, A. B. *J. Phys. Chem. Ref. Data* **1988**, *17*, 513.
3. Stanbury, D. M. *Reduction Potentials Involving Inorganic Free Radicals in Aqueous Solution*; Sykes, A. G., Ed.; Academic Press, Inc.: San Diego, 1989; Vol. 33, pp. 69-138.
4. Mezyk, S. P.; Bartels, D. M. *J. Phys. Chem. A* **1997**, *101*, 6233.
5. Gratzel, M.; Henglein, A.; Lilie, J.; Beck, G. *Ber. Bunsenges. Phys. Chem.* **1969**, *73*, 646.
6. Wolff, R. K.; Bronskill, M. J.; Hunt, J.W. *J. Chem. Phys.* **1970**, *53*, 4211.
7. Lam, K. Y.; Hunt, J. W. *Int. J. Radiat. Phys. Chem.* **1975**, *7*, 317.
8. (a) Daniels, M. *J. Phys. Chem.* **1966**, *70*, 3022. (b) Daniels, M. *Adv. Chem. Ser.* **1968**, *81*, 153. (c) Daniels, M. *J. Phys. Chem.* **1969**, *73*, 3710. (c) Daniels, M.; Wigg, E. E. *J. Phys. Chem.* **1969**, *73*, 3703. (d) Kiwi, J. T.; Daniels, M. *J. Inorg. Nucl. Chem.* **1978**, *40*, 576.
9. Logager, T.; Sehested, K. *J. Phys. Chem.* **1993**, *97*, 10047.
10. Cook, A. R.; Dimitrijevic, N.; Dreyfus, B. W.; Curtiss, L. A.; Camaioni, D. M., *Submitted for publication*.
11. Benderskii, V. A;.Krivenko, A. G.; Ponomarev, E. A.; Federovich, N. V. *Sov. Electrochem.* **1987**, *23*, 1343.
12. Alfassi, Z. B.; Dhanasekaran, T.; Huie, R. E.; Neta, P. *Radiat. Phys. Chem.* **1999**, *56*, 475.

13. Elliot, A. J. M., D. R.; Buxton, G. V.; Wood, N. D., *J. Chem. Soc., Faraday Trans.* **1990**, *86*, 1539.
14. Zehavi, D.; Rabani, J. *J. Phys. Chem.* **1971**, *75*, 1738
15. Han, P.; Bartels, D. M., *J Phys Chem* 1990, *94*, 7294
16. Treinin, A. H., E., *J. Am. Chem. Soc.* **1970**, *92*, 5821.
17. Park, J.-Y.; Lee, Y.-N. *J. Phys. Chem.* **1988**, *92*, 6294-6302.
18. Lee, Y.-N. and Schwarz. *J. Phys. Chem.* **1981**, *85*, 840-848.
19. Kapoor, S.; Barnabas, F. A.; Sauer, Jr., M. C.; Meisel, D.; Jonah, C. A. *J. Phys. Chem.* **1995**, *99*, 6857-6863.
20. Ashby, E. C.; Annis, A.; Barefield, E. K.; Boatright, D.; Doctorovich, F.; Liotta, C. L.; Neumann, H. M.; Konda, A.; Yao, C.-F.; Zhang, K. *J. Am. Chem. Soc.* **1993**, *115*, 1171.
21. Chen, M. J.; Linehan, J. C.; Rathke, J. W. *J. Org. Chem.* **1990**, *55*, 3233.
22. Mönig, J.; Chapman, R.; Asmus, K.-D. *J. Phys. Chem.* **1985**, *89*, 3139-3144.
23. (a) Bonifacic, M.; Stefanic, I.; Hug, G.L.; Armstrong, D.A. Asmus, K.D. *J. Am. Chem. Soc.* **1998**, *120*, 9930-9940. (b) Bonifacic, M.; Armstrong, D. A.; Carmichael, I.; Asmus, K.-D. *J. Phys. Chem. B* **2000**, *104*, 643-649.
24. Su, Z.; Falvey, D. E.; Yoon, U. C.; Mariano, P. S. *J. Am. Chem. Soc.* **1997**, *119*, 5261.
25. (a) Challis, B. C.; Kyrtopoulos, S. A. *J. Chem. Soc., Perkin Trans. 2* **1978**, 1296-302. (b) Challis, B. C.; Kyrtopoulos, S. A. *J. Chem. Soc., Perkin Trans. 1* **1979**, 299-304.
26. Challis, B. C.; Carman, N.; Fernandes, M. H. R.; Glover, B. R.; Latif, F.; Patel, P.; Sandhu, J. S.; Shuja, S. in *Nitrosamines and Related N-Nitroso Compounds. Chemistry and Biochemistry*; Loeppky, R. N. Michejda, C. J.; ACS Sym. Ser. Vol. 553; American Chemical Society, Washington DC, 1994; pp 74-92.
27. Camaioni D. M.; Samuels, W. D.; Linehan, J. C.; Clauss, S. A.; Sharma, A. K.; Wahl, K. L.; Campbell; J. A. *Organic Tanks Safety Program Waste Aging Studies Final Report,* PNNL-11909 Rev 1, Pacific Northwest Laboratory, Richland, Washington.
28. Lilga, M. A.; Hallen, R. T.; Alderson, E. V.; Hogan, M. O.; Hubler, T. L.; Jones, G. L.; Kowalski, M. R.; Lumetta, W. F.; Schiefelbein, G. F.; Telander, M. R., *Ferrocyanide Safety Project, Ferrocyanide Aging Studies. Final Report,* PNL-11211, Pacific Northwest National Laboratory, Richland, Washington.
29. Stauffer, L. A. *Temperature and Radiation Dose History of Single-Shell Tanks,* HNF-SD-WM-TI-815, Lockheed Martin Hanford Corporation, Richland, Washington.
30. Meisel, D.; Diamond, H.; Horwitz, E. P.; Matheson, M. S.; Sauer, Jr., M. C.; Sullivan, J. C.; Barnabas, F.; Cerny, E.; Cheng, Y. D. *Radiolytic*

Generation of Gases from Synthetic Waste, ANL-91/41, Argonne National Laboratory, Argonne, Illinois.

31. Barefield, E. K.; Boatright, D.; Desphande, A.; Doctorovich, R.; Liotta, C. L.; Neumann, H. M.; Seymore, S. *Mechanisms of Gas Generation from Simulated SY Tank Farm Wastes: FY 1995 Progress Report*, PNNL-11247, Pacific Northwest National Laboratory, Richland, Washington.

32. Petrik, N. G.; Knutsen, K.; Paparazzo, E.; Lea, S.; Camaioni, D. M.; Orlando, T. M. *J. Phys. Chem. B* **2000**, *104*, 1563-1571.

33. Knutsen, K.; Orlando, T. M. *Phys. Rev. B* **1997**, *55*, 13246.

34. Knutsen, K.; Orlando, T. M. *Appl. Surf. Sci.* **1998**, *127 - 129*, 1.

Analytical Chemistry

Chapter 22

Novel Spectroelectrochemical Sensor for Ferrocyanide in Hanford Waste Simulant

Mila Maizels[1], Michael Stegemiller[1], Susan Ross[1], Andrew Slaterbeck[1], Yining Shi[1], Thomas H. Ridgway[1], William R. Heineman[1], Carl J. Seliskar[1,*], and Samuel A. Bryan[2]

[1]Chemistry Department, University of Cincinnati, Cincinnati, OH 45221–0172
[2]Pacific Northwest National Laboratory, 908 Battelle Boulevard, Richland, WA 99352

A new type of spectroelectrochemical sensor that embodies two modes of instrumental selectivity (electrochemical and spectroscopic) in addition to selective partitioning into an applied film barrier is described. The sensor consists of a planar optical substrate/electrode coated with a chemically-selective film. Sensing is based on the change in the attenuation of light passing through the guided wave optic which accompanies a chemical reaction of an analyte induced by electromodulation. Threefold selectivity for a chosen analyte relative to other environmental components is obtained by the choice of coating material, the electrolysis potential, and the wavelength for optical monitoring. The sensor concept is demonstrated with an indium tin oxide coated glass guided wave device that has been over-coated with a sol-gel derived charge-selective thin film. This device is then shown to be able to sense ferrocyanide in Hanford waste tank simulant solution. The ongoing development of a small portable sensor unit including a virtual interface, control electronics and optics is also described.

Introduction

The required remediation of over 300 underground nuclear waste storage tanks at USDOE sites together with the associated needs to characterize and monitor the chemical compositions of the tanks themselves presents a major scientific challenge (*1-6*). In addition to the previously identified chemical complexity of such storage materials, the added dimensions of limited tank access and harsh chemical and radiological environment preclude the straightforward application of well-established laboratory-based chemical analysis techniques to this national problem. Beyond any solution to this immediate problem also lies a present and long-term need to monitor low-level subsurface contaminations associated with such storage facilities. While an approach to removing high-level nuclear wastes has been identified (vitrification), the chemical analysis technology available in-hand to assist in this important task is inadequate.

The general aim of the work embodied in this project is to design and implement a new sensor technology which offers the unprecedented levels of specificity needed for analysis of the complex chemical mixtures found at USDOE sites nationwide. The new sensor concept combines the elements of electrochemistry, spectroscopy and selective partitioning into a single device that provides three levels of selectivity.

This type of sensor has many potential applications at DOE sites. As an example, the enhanced specificity embodied in this new sensor design is well-suited to the unique analytical problem posed by the addition of ferrocyanide to radioactive tank wastes at the USDOE Hanford Site (*7*). Various radioactive wastes from defense operations have accumulated at the Hanford Site in underground waste tanks since the early 1940s. During the 1950s, additional tank storage space was required to support the defense mission. Hanford Site scientists developed two procedures to obtain this additional storage volume within a short time period without constructing additional storage tanks. One procedure involved the use of evaporators to concentrate the waste by removing water. The second procedure involved developing precipitation processes for scavenging radio-cesium and other soluble radionuclides from tank waste liquids. The scavenging processes used sodium and potassium ferrocyanide and nickel sulfate to precipitate radioactive cesium from solutions containing nitrates and nitrites. Radioactive strontium and cobalt were scavenged from some of the solutions using calcium or strontium nitrate and sodium sulfide, respectively. After allowing the radioactive precipitates to settle, the decontaminated solutions were pumped to disposal cribs, thereby providing additional tank storage volume. Later, some of the tanks were found to be leaking; pumpable liquids were removed from these tanks, leaving behind a wet solid (sludge) residue containing the ferrocyanide precipitates (7). In implementing this process, approximately 140 metric tons of ferrocyanide, [calculated as $Fe(CN)_6^{-4}$], were added to waste that was later routed to 18 large (750,000 to 1,000,000 gallon) underground single shell tanks (SSTs). Since ferrocyanide precipitated with nickel is potentially explosive in the dry state at elevated temperatures, a ferrocyanide sensor could help characterize *solubilized* tank waste before and during any disposal process should the need arise. It is to be noted, however, that ferrocyanide added to Hanford tanks has apparently degraded sufficiently that it is no longer an issue of tank safety (8).

A chemical sensor is a device that responds to varying concentrations of a single chemical (the analyte) or a specific class of chemicals. Interest in the development of new sensors for numerous chemical species in a wide range of applications has soared in the past few years (9-12). The significant problem in chemical sensor development is achieving adequate selectivity for determination of an analyte in a real sample where interferences often confuse the sensor measurement. Selectivity in presently available sensors is usually achieved in one of several ways. Electrochemical sensors based on electrolysis typically have two modes of selectivity as illustrated in **Figure 1**.

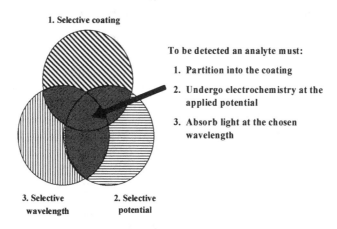

To be detected an analyte must:

1. Partition into the coating

2. Undergo electrochemistry at the applied potential

3. Absorb light at the chosen wavelength

Figure 1. A triad of sensor selectivity is shown to illustrate the interrelationships among sensor selectivities (electrochemical, optical and chem-selective film). The spectroelectrochemical sensor concept joins these three levels of selectivity and is denoted by the central area of common overlap of the three circles.

The first level of selectivity is based on the specificity which the applied potential provides. This alone is usually inadequate. The second mode of selectivity is typically provided by modification of the electrode surface with a selective coating or membrane. Whereas in fiber optic chemical sensors, the first level of selectivity employs light of a wavelength specific to an analyte. This first level is similar to that in electrochemical sensors in that it is permissive. Thus, the level of specificity supplied by wavelength choice is rarely adequate to meet all the challenges which these sensors experience in real samples where interferences of similar optical properties might be present. As a result, an additional level of selectivity is often supplied much in the same way that it has been done in electrochemical sensors, that is, with a selective coating or membrane.

Recently, it has been our motivation to increase the selectivity in chemical sensors by amalgamating the capabilities of electrochemical and optical sensors into a single device. In general, such a strategy for chemical sensing would require three characteristics of an analyte: that it specifically interact with an applied sensor coating, that it be electrolyzed at the chosen potential, and that it have an optical change at the wavelength chosen for sensing. This is indicated in Fig. 1 by the overlap of the three circles of sensor selectivity.

This sensor concept is further illustrated in **Figure 2** where the diagram of the critical interfacial region of such a device is shown. One can think of the sensor structure as one based on an optically transparent electrode (OTE), in other words, a guided wave optic with an electrode. The OTE is coated by a thin chemically-selective film that serves to enhance detection limit by preconcentrating the analyte at the OTE surface. The evanescent field at the points of internal reflection within the guided wave optic penetrates the film so that electrochemical events within the film can be monitored optically. In its operation, a potential excitation signal is applied to the OTE to cause electrolysis of analyte that has partitioned into the film and the change in the light propagated by attenuated total internal reflection due to disappearance of analyte by its electrolysis or appearance of an electrolysis product is monitored. Thus, sensor transmittance or absorbance changes measured in concert with an electrochemical modulation becomes the analytical signal. Quantitation of the analyte is based on the magnitude of the change in absorbance which is proportional to concentration of analyte in the film which, in turn, is proportional to its concentration in the sample as defined by the partition coefficient.

The significant aspect of this sensor is the three modes of selectivity that are simultaneously achieved. For an analyte to be sensed, it must partition into the selective coating, be electrolyzed at the potential applied to the electrode, and either the analyte or its electrolysis product must absorb light at the wavelength chosen. Selectivity for the analyte relative to other solution components is obtained by choice of coating material, electrolysis potential, and wavelength for optical monitoring. Of the three levels of selectivity, two are tunable, i.e., able to be tuned externally by selecting the potential and wavelength with which to operate the sensor. This concept, therefore, adds an additional component of selectivity compared with many of the existing chemical sensors and is very important for its practical application.

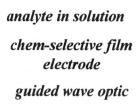 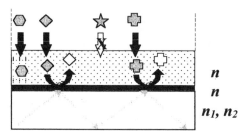

Figure 2: A cross-sectional view of a prototypical spectroelectrochemical sensor is shown consisting of four distinct phases each of which has a refractive index value (n_1, ..., n_4). The vertical arrows denote transport (solid) or prevented transport (unfilled) of analytes into the chemically-selective film. The half-circle arrows denote electrode processes (oxidation, reduction) of analytes that have diffused to the electrode surface. The guided wave optic represents the optical medium in which light is transmitted through the sensor that is diminished by evanescent field interaction with analytes that absorb at the sensor wavelength.

We have developed such a device, the "spectroelectrochemical" sensor, that embodies this general concept. One form of the device consists (*13-18*) of an indium tin oxide coated glass guided wave optical flat that has been over-coated with a sol-gel derived charge-selective thin film or a similar polymer-based coating. Prototype analytes have been used to demonstrate that the change in the absorbance of the guided wave optic resulting from electrochemical oxidation/reduction can be used to quantify an analyte.

Coincident with our work on sensor design, we have developed (*19-24*) two new series of chemically-selective optical materials. These new materials are essential to the functioning of our new spectroelectrochemical sensors. One new series of materials was based on polymer blending in a host of glutaraldehyde cross-linked poly(vinyl alcohol). Chemically-selective dopants in this host demonstrate property-selective separations of chemicals from mixtures. We have optimized the composition, optical properties and the coating procedures for several specific blends for optical sensing. These blends have clear UV and visible spectral regions for direct spectroscopic sensing and they are excellent absorbers of many inorganic and organic charged species from aqueous environments. A second new series consists of polyelectrolyte-containing silica composites prepared by sol-gel processing. The thickness of spin-coated films of these materials on glass can be varied from 0.1 μm to 4 μm. These materials are ion exchangeable and less brittle than the parent silica substrate due to the incorporation of the organic polyelectrolyte. These new composites retain the nano-scale porosity and optical transparency into the ultraviolet of the parent silica sol-gel processed glasses making them attractive host matrices for the immobilization of a variety of chemical reagents.

The details of one form of the spectroelectrochemical sensor are also illustrated in **Figure 2**. This sensor consists of four distinct phases. The first phase consists of an underlying optical element (refractive indices n_1, n_2) that serves to define the channel or layer in which light propagates through the sensor by total internal reflection. This element can be as simple as a microscope slide or as sophisticated as a micro-machined planar waveguide. In general, phase one consists of a supporting optical substrate (n_1) and phase two (n_2) consists of the guided wave layer in which total internal reflection of light ($n_1 < n_2$) is achieved. For the purposes of this paper, we have used commercially available ITO-coated tin float glass where the first two phases are poorly defined due to the graded index of the glass beneath the ITO coating. In the case of tin oxide float glass, it is known that the graded layer of SnO_2 established by the float process can support one or more optical waveguide modes. We have intentionally arranged the optical (prism) coupling to the ITO coated tin float glass to favor the base glass multiple internal reflection transmission of light rather than excite the waveguide modes preferentially. Therefore, in our sensor configuration, one has an optical guiding medium, essentially the tin float glass substrate (base glass 1 mm thick, $n_1 = 1.53$), as phases one and two. (For the sake of simplicity we have ignored the complications that this transparent lossy layer causes in the full optical analysis of the sensor.)

Phases one and two are over-coated (by spin-coating) with phase three (n_3), the chemically-selective film. This third phase is either a sol-gel derived charge-selective film (typically 0.75 μm thick, $n_3 = 1.47$) or, equivalently, a charge-selective polymer blend spin-coated on the ITO layer. Phase 4 (n_4) is the analyte-containing medium which in the case of an aqueous solution has an index $n_4 = 1.33$, i.e., that of water.

Experimental Methods

The following chemicals were used: tetraethoxysilane, TEOS (Aldrich), potassium nitrate (Fisher), potassium nitrite (Fisher), potassium ferrocyanide (Matheson). PDMDAAC (poly(dimethyldiallylammonium chloride)), MW = 200,000 – 350,000, was purchased from Aldrich as a 20% wt/wt solution in water. All reagents were used without further purification. Reagent solutions were all made by dissolving the appropriate amounts of chemicals into 0.10 M potassium nitrate solution (prepared with deionized water from a Barnstead water purification system). Indium tin oxide (ITO) tin float glass (11-50 Ω/sq., 150 nm thick indium tin oxide layer over tin float glass) was purchased from Thin Film Devices. This ITO tin float glass was cut into 1 in × 3 in or 10 mm x 45 mm slides, scrubbed with Alconox, and rinsed thoroughly with deionized water and then 1-propanol prior to use.

Stock silica sols were prepared according to our previously reported protocol (18). The volume ratio of the appropriate polyelectrolyte solution to the silica sol was chosen to control the composition of the composite materials as previously described (19).

The laboratory instrumentation and electrochemistry methods used for these studies have been previously described (13-18) and the reader is referred to these references for further details. In these studies we have used two different types of sensors fabricated from ITO glass. The first is identical to that already described (13-17) and employed 1 in x 3 in optical substrates; the second employed smaller substrates, 10 mm x 45 mm, smaller sample volumes (800 μL) and was operated with a Panasonic blue LED (wavelength maximum, 450 nm) light source. The latter sensor had fewer multiple internal reflections compared to the previously described sensor.

Results and Discussion

The Ferri/ferrocyanide System

To demonstrate the basic functioning and selective sensing capability, we chose the ferri-ferrocyanide couple, $Fe(CN)_6^{3-/4-}$, as an example of an important anion at Hanford. This couple has well-defined, reversible electrochemistry that is accompanied by distinct spectral changes in the visible wavelength region.

$$Fe^{III}(CN)_6^{3-} \; + \; e^- \quad \leftrightarrow \quad Fe^{II}(CN)_6^{4-} \qquad (E° = 0.25 \text{ V at bare ITO vs Ag/AgCl})$$

ferricyanide   *ferrocyanide*
(absorbs at 450 nm)   (transparent at 450 nm)

We have used a sol-gel processed PDMDAAC-SiO$_2$ film as an anion exchange selective film for partitioning of the $Fe(CN)_6^{3-/4-}$ couple. Sol-gel processed PDMDAAC-SiO$_2$ composite films have been shown (19) to possess a series of properties suited to construction of both electrochemical and optochemical sensors, such as ion-exchangeability, nano-scale porosity, high optical transparency, variable thickness (0.1 - 3.0 μm), and physicochemical stability. Most importantly, it was found that this silica-based, polyelectrolyte-containing composite material can also form uniform and well-adhered thin films on ITO glass by the spin-coating method, thereby making it more useful in

ITO-based spectroelectrochemical sensing applications compared with, for example, SiO_2 on gold surfaces. The electrochemical and spectroscopic characteristics of ferri/ferrocyanide at ITO coated with PDMDAAC-SiO_2 have already been described (13,14).

Performance of the Sensor for Determination of Ferrocyanide in Pure Samples

An illustration of the performance of the spectroelectrochemical sensor operating for the detection of $Fe(CN)_6^{4-}$ in pure sample solutions is shown in **Figure 3**. The cyclic potential applied caused periodic changes in absorbance that correspond to the periodic depletion of $Fe(CN)_6^{4-}$ as it is electrochemically oxidized to $Fe(CN)_6^{3-}$, ferricyanide, and reformed. The absorbance values are expected to be proportional to the concentration of $Fe(CN)_6^{4-}$ in the solution.

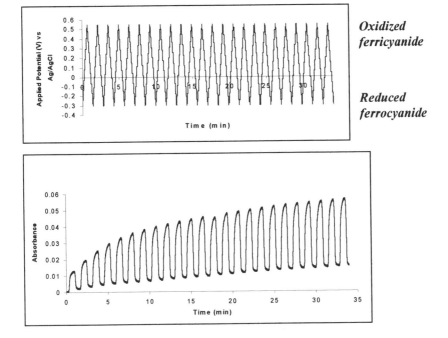

Figure 3. The optical signal modulation at 450 nm (filtered Panasonic LED) for the ferri/ferrocyanide couple incorporated in the SiO₂-PDMDAAC film on ITO-glass with continuous scanning of the potential (top portion of figure). The multiple-internal reflection optic (10 mm x 45 mm) was aligned for 4 reflections.

Demonstration of the Sensor in Hanford Simulants Containing Ferrocyanide

Simulants have been prepared at the Pacific Northwest National Laboratory for the purpose of providing non-radioactive media that can be used to mimic tank waste conditions at Hanford. In the case of simulants for ferrocyanide wastes, the sludge contains insoluble ferrocyanide as $Na_2NiFe(CN)_6$. Since the sensor requires soluble $Fe(CN)_6^{4-}$ for detection, solubilization of the sludge is required for sensor operation. In the case of those wastes containing ferrocyanide, Bryan, et al. (25) have reported a procedure for dissolving simulant ferrocyanide sludges with an elixir composed of aqueous ethylenediamine (EN) and ethylenediaminetetraacetic acid (EDTA) to produce simulant solutions. The sludge and the derived solution that results from treatment of the sludge with elixir (to make the simulant solution) have been provided for these studies. The approximate composition of the simulant solution is given in **Table 1**.

Table 1. Approximate Composition of the Simulant Solution

Species	wt%	Species	wt%
Na^+	0.78	Ni^{2+}	0.04
NO_3^-	1.23	Fe^{2+}	0.04
NO_2^-	0.30	Fe^{3+}	0.12
SO_4^{2-}	0.42	CN^-	0.106
SO_3^{2-}	0.02	EN	4.54
PO_4^{3-}	0.23	EDTA	4.54
NH_4^+	0.01	H_2O	87
Ca^{2+}	0.08		

Note: Cyanide concentration is measured value; Fe^{3+} resulted from Oxidation of added Fe^{2+} and from steel liner corrosion, etc.

The simulant solution, like the actual waste solution resulting from dissolution by elixir, is a complex, high pH (pH = 10) mixture with high concentrations of such oxyions as NO_3^-, NO_2^-, SO_4^{2-}, SO_3^{2-}, and PO_4^{2-}. These ions can be expected to interfere with the detection of ferrocyanide. While cations are excluded by the anion charge-selective PDMDAAC-SiO_2 film, anions such as NO_3^-, NO_2^- are expected to compete with ferrocyanide for the exchange sites in the film. **Figure 4** shows the modulation of the absorbance when the sensor was directly immersed in the simulant solution (Hanford U-Plant-2 Simulant Solution) for approximately 10 min and subjected to cyclic potential modulation. The modulated signal is similar to that in Figure 3, however, absorbance values corresponding to the ferrocyanide present in the simulant solution appear about three times smaller when compared to those obtained for the same ferrocyanide concentration in pure samples with dilute supporting electrolyte in an identical instrument alignment after a similar sensor exposure. Obviously, interferences such as high concentrations of NO_3^- and NO_2^-, which are present in the simulant solution, attenuate the absorbance modulation amplitude. With prolonged exposure of the sensor to the simulant solution, the PDMDAAC-SiO$_2$ film decomposes to such an extent that the modulation amplitude is significantly reduced. For example, after 3 hours of exposure, the amplitude decreased by 60% compared to that at a 10 min exposure. Thus, although the sensor signal corresponding to ferrocyanide in the simulant solution is easily measured with high signal-to-noise, and although it is sufficiently large that the measurements can be performed immediately after sensor exposure to the simulant solution, the magnitude of the signal is reduced by competition of the other oxyanions for exchange sites in the film.

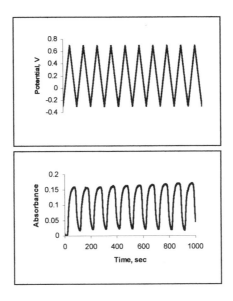

Figure 4. Sensor response signal to simulant solution. Changes in absorbance of the sensor induced by cyclic potential modulation. The applied potential was cycled between – 0.3 and + 0.7 V as noted in the top panel at a scan rate of 20 mV/s. The multiple-internal reflection optic (1 in x 3 in) was aligned for 12 reflections.

Our current procedure is to make sensor measurements in the simulant solution in a very short time period – 1 min exposure. Indeed, the sensor has a

374

sufficient time response and ferrocyanide is present at a relatively high concentration in the simulant solution. **Figure 5** shows sensor calibration curves for high concentrations of $Fe(CN)_6^{4-}$ in 0.7 M KNO_3 and elixir consisting of 5 % EDTA (acid form), and 5 % EN.

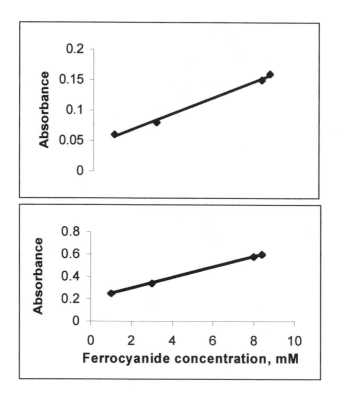

Figure 5. Sensor calibration curves for a thick (top) and a thinner (bottom) PDMDAAC-SiO₂ film for 60 s of sensor exposure. The concentration values of Fe(CN)₆⁴⁻ are 1.0, 3.0, 8.0, and 8.3 mM in 0.7 M KNO₃, 5 % EDTA, and 5 % EN. The multiple-internal reflection optic (1 in x 3 in) was aligned for 12 reflections.

EDTA, and EN are present at the highest concentration and are the major components of the simulant solution. Both calibration curves in either case of film thickness have good linearity ($R^2 = 0.995$ for a thick film and $R^2 = 0.999$ for a thinner film) over the higher concentration range (1.0 mM – 8.3 mM $Fe(CN)_6^{4-}$). These calibration curves can be used to determine the concentration of ferrocyanide in the simulant solution. An absorbance change of 0.16 obtained after 1 min sensor exposure corresponds to 8.3 mM $Fe(CN)_6^{4-}$ for a thick film. For a thinner film, an absorbance change of 0.6 after 1 min also corresponds to 8.3 mM $Fe(CN)_6^{4-}$. This value is in good agreement with the ferrocyanide concentration of 8.27 mM obtained using an FTIR method (25). Note, that the absorbance change at a thinner film is almost four times larger compared to a thick film. Thus, by controlling the film thickness the sensor performance can be optimized.

Design and Fabrication of a Field Portable Spectroelectrochemical Sensor

The instrumentation developed for a portable unit is similar to that described previously (15) and consists of the following components: control electronics for the electrochemistry (potentiostat), electronics for the optical components, and a virtual interface to a computer through which the experiments can be performed and the data acquired and analyzed (**Figure 6**). A "virtual instrument" was assembled with the interface and signal generation being software-based. Instrument front-panels traditionally found have been replaced by software controls that respond to "point and click" commands similar to any Windows interface (**Figure 7**).

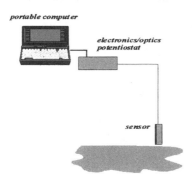

Figure 6. Overview of field-portable spectroelectrochemical sensor.

Remote sensing imposes certain design constraints upon the potentiostat. Long connections between the electrodes and the potentiostat itself can cause serious problems with noise pickup and overall stability of the system. This is usually avoided by locating the potentiostat close to the electrodes, i.e., very near to the sensing region. These considerations also normally apply to the power supply and signal generation source for the potentiostat. In addition, the transducer circuitry becomes relatively elaborate when the primary sensing mode is the electrochemical response. When the electrochemical current response is secondary, as in the case of a spectroelectrochemical sensor, many of these constraints are simplified. The current, primarily double layer charging, is monitored only as an indication that the electrode is still functioning. The analytic signal is the photodiode response. The diode is coupled to the electrode by a length of optic fiber and the variable gain transducer amplifier is located close to the digitization electronics. The potentiostat itself is controlled remotely from the signal generation and digitization electronics by a fiber optic link. Two implementations of this basic concept are under investigation. One uses an analog implementation, which employs a signal generated remotely and then coupled via an analog optical receiver. The other is an all-digital form, which employs small, cheap digital-to-analog converter, analog-to-digital converter and microprocessor on the potentiostat board itself. Both units are battery powered with local regulators and are each under 4 x 2 inches in size. Further size reduction is expected in the surface mounted component forms.

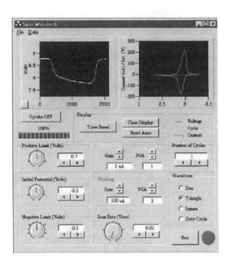

Figure 7. Computer screen display of virtual instrument interface for the portable spectroelectrochemical sensor.

Conclusions

We have demonstrated a new sensor concept which is based on three modes of selectivity, selective partitioning into a thin film, optical selectivity by choice of wavelength, and electrochemical potential selectivity. Advantages of this system include the ability to detect targeted analytes in the presence of direct interferences (*14*). However, there are also some possible disadvantages inherent in the design. These include that the sensor is applicable to only those analytes which are directly electroactive or to analytes which can be coupled to reactions which generate such species. In addition, the electroactive chemical species must have a significant optical extinction coefficient in the region of the spectrum where measurements are made. As a very important application of the sensor, we have demonstrated that it can be directly used to measure ferrocyanide in Hanford waste tank simulants.

Acknowledgement

The authors gratefully acknowledge support from the Environmental Management Science Program of the US Department of Energy, Office of Environmental Management under grant DE-FG07-96ER62311.

References

1. G. Zorpette, G. *Sci. Am.* May 1996, pp. 88-97.
2. D. J. Bradley, D. J.; Frank, C. W.; Mikerin, Y. *Physics Today*, April 1996, pp. 40-45.
3. Campbell, . J. A.; Stromatt, R. W.; Smith, M. R.; Koppenaal, D. W.; Bean, R. M.; Jones, T. E.; Strachan, D. M. ; Babad, H. *Anal. Chem.* **1994**, *66*, 1208A-1215A.
4. Choppin, G. R. *J. Chem. Ed.*, **1994**, *71*, 826-829.
5. U.S. Department of Energy, Office of Environmental Management, *Closing the Circle on the Splitting of the Atom*, second printing, 1996.
6. Gephart, R. E.; Lundgren, R. E. *Report PNL-10773*, Pacific Northwest National Laboratory, Richland, WA, 1995.
7. Burger, L. L.; Reynolds, D. A.; Schulz, W. W.; Strachan, D. M. *Report PNL-7822*, Pacific Northwest Laboratory, Richland, WA, 1991.
8. Meacham, J. R.; Cash, R. J.; Dickinson, D. R.; Reich, F. R.; Grigsby, J. M.; Postma, A. K.; Lilga, M. A. *Report WHC-SD-WM-SARR-038*, Rev. 1, Westinghouse Hanford Company, Richland, WA, 1996.
9. Janata, J.; Josowicz, M.; Vanysek, P.; DeVaney, D. M. *Anal. Chem.* **1998**, *70*, 179R-208R.
10. Vaihinger, S.; Goepel, W. in *Sensors;* Goepel, W.; Hesse, J.; Zemel, J. N., Eds. VCH, Weinheim, Germany, 1991.
11. Kersey, A. D.; Dandridge, A. *IEEE Trans. on Components, Hybrids and Manufacturing Technology* **1990**, *13*, 137-143.
12. Wolfbeis O. *Fiber Optic Chemical Sensors and Biosensors*; CRC Press, Inc., Boca Raton, FL, 1991; Vols. 1 and 2.
13. Shi, Y; Slaterbeck, A. F; Seliskar, C. J; Heineman, W. R. *Anal. Chem.* **1997**, *69*, 3679-3686.

14. Shi, Y; Seliskar, C. J; Heineman, W. R. *Anal. Chem.* **1997**, *69*, 4819-4827.
15. Slaterbeck, A. F; Ridgway, T. H; Seliskar, C. J; Heineman, W. R. *Anal. Chem.* **1999**, *71*, 1196-1203.
16. Gao, L; Seliskar, C. J.; Heineman, W. R. *Anal. Chem.* **1999**, *71*, 4061-4068.
17. Shi, Y; Slaterbeck, A. F; Aryal, S; Seliskar, C. J; Heineman, W. R; Ridgway, T. H; Nevin, J. H. *Proc. SPIE* **1998**, *Series 3258*, pp 56-65.
18. Ross, S. E.; Slaterbeck, A. F.; Shi, Y; Aryal, S; Maizels, M; Seliskar, C. J.; Heineman, W. R.; Ridgway, T. H.; Nevin, J. H. *Proc. SPIE* **1999**, *Series 3537*, 268-279.
19. Shi, Y; Seliskar, C. J. *Chem. Mater.* **1997**, *9*, 821-829.
20. Gao, L; Shi, Y; Slaterbeck, A. F.; Seliskar, C. J.; Heineman, W. R.. *Proc. SPIE* **1998**, *Series 3258*, 66-74.
21. Gao, L; Seliskar, C. J.; Milstein, L. *Appl. Spectrosc.* **1997**, *51*, 1745-1752.
22. Dasenbrock, C. O.; Ridgway, T. H.; Seliskar, C. J., Heineman, W. R.. *Electrochim. Acta* **1998**, *43*, 3497-3502.
23. Gao, L; Seliskar, C. J. *Chem. Mater.* **1998**, *10*, 2481-2489.
24. Hu, Z; Slaterbeck, A. F.; Seliskar, C. J.; Ridgway, T. H.; Heineman, W. R. *Langmuir* **1999**, *15*, 767-773.
25. Bryan, S. A.; Pool, K. H.; Bryan, S. L.; Forbes, S. V.; Hoopes, F. V.; Lerner, B. D.; Mong, G. M.; Nguyen, P. T.; Schiefelbein, G. F.; Sell, R. L.; Thomas, L. M. P. *Report PNL-10696*, Pacific Northwest National Laboratory, Richland, WA, 1995.

Chapter 23

Refinement of Immunochemical Methods for Environmental Analysis of Polycyclic Aromatic Hydrocarbons

Qing X. Li[1], Alexander E. Karu[2], Kai Li[1], and Steven Thomas[1]

[1]Department of Molecular Biosciences and Biosystems Engineering, University of Hawaii at Manoa, Honolulu, HI 96822
[2]Department of Nutritional Sciences, University of California, Berkeley, CA 94720

Polycyclic aromatic hydrocarbons (PAHs) are a large class of combustion byproducts that have similar structures but differ widely in prevalence, human and ecological hazard potential, and the cost and complexity of remediation. Antibody-based methods to detect and recover PAH residues have proven to be reliable and cost-effective for some EMSP needs over the last several years. Wider use of these methods will require improved ways to extract and concentrate parent PAHs, metabolites and conjugates with relatively uniform efficiency, and require little or no adaptation for use with complex matrices. The extracts should be compatible with immunoassay as well as instrumental analysis. This chapter describes the status of our work toward these goals. New PAH haptens and their protein conjugates were synthesized, and ranked with a "recognition index" that identified optimum competitors for immunoassays. Sensitive immunoassays were achieved with broadly cross-reactive monoclonal antibodies and some of the new hapten conjugates as competitors. A Na_4EDTA-assisted CO_2 supercritical fluid extraction (SFE) method was developed, which efficiently recovers apolar parent PAHs and polar PAH metabolites that can be analyzed by immunoassays as well as GC-MS and other instrumental methods. An immunoaffinity column method was developed to separate PAHs from interfering substances in the SFE extracts. Small pilot studies are described, in which these methods were evaluated with surface water, sediment, and marine coral samples from national wildlife refuges in Hawaii.

Introduction

Polycyclic aromatic hydrocarbons (PAHs), their metabolites and protein and DNA adducts are of concern in several EMSP focus areas. PAHs are components of fuels and oils. More than 90% of PAHs in the environment are derived from combustion (1). The U.S. EPA lists 16 PAHs in wastewater, and 24 PAHs in soils, aquatic sediments, hazardous solid wastes and groundwater as priority pollutants (2). Many PAHs are toxic, and several are carcinogens in animals and suspected carcinogens in humans (3). Metabolites of naphthalene, chrysene, pyrene, and benzo[a]pyrene (BaP) form glucuronides and protein and/or DNA adducts that are putative biomarkers of PAH exposure (4-10).

Numerous methods have been developed for the extraction of PAHs in environmental and biological matrices (1, 11-14). Current techniques, that can be automated, used on-site, and minimize organic solvent use, include solid phase extraction (15, 16), accelerated solvent extraction (17, 18), microwave-assisted solvent extraction (19, 20), superheated water extraction (21), and supercritical fluid extraction (SFE). The most common procedures for PAH analyses have been gas chromatography (GC), GC-mass spectrometry (GC-MS), liquid chromatography (LC) and LC-MS (1, 22). Recent advances include capillary electrophoresis (23-25), synchronous fluorescence spectrometry (26), and a combination of hyperthermal surface ionization, supersonic molecular beam GC-MS, and time-of-flight mass analysis (27). Some of these have been miniaturized for on-site use. However, they require skilled personnel and sophisticated data analysis, and the expense and time per sample remain high.

Antibody-based methods offer substantially less time and lower cost per sample, the possibility of on-site use, and potential tools for bioremediation monitoring, and biomarker and worker exposure research. PAHs were among the earlier targets for immunochemical method development. Over the last few years, immunochemical techniques have gained much wider use and acceptance as reliable, cost-effective tools for risk assessment, monitoring, health effect research, and ecotoxicology. Immunoassay kits are commercially available for analyses of PAHs in soil and water. The U.S. EPA adopted official SW-846 method 4035 for immunoassay of PAHs in soils (22). The Department of Energy (DOE) has done much to advance this technology by identifying specific problems that immunochemical methods could at least partly solve, sponsoring validation projects at many sites, and publishing on-line methods for extraction, cleanup, and immunoassay of PAHs in soils (28).

Immunoaffinity chromatography (IAC) offers the advantages of one-step, highly selective cleanup and enrichment of a class or subclass of analytes, metabolites, and adducts for research and regulatory applications. IAC methods have been used for simultaneous recovery of parent PAHs, metabolites, conjugates and adducts from various matrices (29-35). A novel "on-probe" immunoaffinity method for mass spectrometry of PAHs was reported by Liang et al. (36). Figure 1 shows some PAHs with 2-5 fused rings, which are found in a wide variety of environmental and biological matrices, and have vapor pressures and aqueous solubility within the ranges for practical immunoassay and IAC. Their relative abundance provides 'fingerprints' that can be used to identify pollutant sources and monitor the progress of remediation.

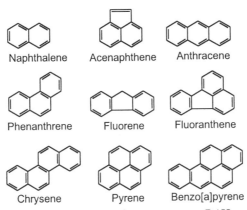

Naphthalene   Acenaphthene   Anthracene

Phenanthrene   Fluorene   Fluoranthene

Chrysene   Pyrene   Benzo[a]pyrene

Figure 1. Structures of some common PAHs.

Our research was motivated by the concept that the next generation of antibody-based methods will use panels of subtly different antibodies to detect or recover several analytes or their analogs (37-41). Two major technical problems must be overcome before this can be realized and used to fill EMSP needs. First, there must be a major advance in the art and science of obtaining antibodies that can distinguish more subtly among PAH subgroups or individual PAHs. In another part of this project, this is addressed with a genetic engineering approach, based upon definition of the PAH-antibody binding mechanisms (Pellequer et al., in this volume). Secondly, sample extraction, cleanup, and recovery methods must be made more efficient, reliable, easier to automate, and compatible with immunoassays. In this chapter, we relate some progress and difficulties encountered in small-scale validation studies toward this goal.

Experimental

Reagents

PAH standards, PAH haptens (the parent compound or a structural analog or mimic that has been derivatized with a spacer arm ending in a functional group that can be linked to protein carriers, enzymes, solid phases, etc.), PAH-protein conjugates, chemicals, solvents and biochemicals were the same as those used in the previous work (42, 43). PAH conjugates used in this study were BSA linked with 6-benzo[a]pyrene isocyanate (BaP-6a-BSA), or with 4-(1-pyrene)butyric acid (PYR-1a-BSA) (42). Hybridoma cell lines producing monoclonal antibodies (mAbs) 4D5 and 10C10, originally derived by Gomes and Santella (12), were generously provided by Dr. Regina Santella (Columbia University, New York, NY).

Field Samples

Fresh water and sediment 'grab' samples were collected from the Pearl Harbor and James Campbell National Wildlife Refuges (NWRs) in Oahu, Hawaii. Corals (*Porites*) were collected from the Kokokahi site in Kaneohe Bay in Oahu, Hawaii, and from Trig Island, French Frigate shoals (44).

Indirect Competitive ELISA (cELISA)

The cELISAs were carried out in 96-well plates (MaxiSorp F96, Nalge Nunc International, Denmark) according to the procedure of Li et al. (42) with the following minor modifications. A microtiter plate was coated with 2.0 ng of PYR-1a-BSA per well in 100 μL of 50-mM carbonate-bicarbonate buffer (pH 9.6) at 37ºC overnight. After the plate was blocked with 0.3% BSA in phosphate buffered saline containing 0.05% of Tween-20 (PBST-BSA), mAb 10C10 culture supernate (1:2000 in PBST-BSA) and biotinylated goat anti-mouse IgG (1:6000 in PBST-BSA) were mixed with an equal volume of analyte (BaP standard in PBST or water sample). The absorbance at 490 nm was read on a Vmax microplate reader, and the data were fitted using Softmax software (Molecular Devices, Sunnyvale, CA).

Immunoaffinity Chromatography (IAC)

4D5 IgG, affinity-purified on Protein A-Sepharose (Hi-Trap, Pharmacia, Piscataway, NJ), was coupled to hydrophilic crosslinked bisacrylamide/azalactone copolymer gel beads according to the manufacturer's procedures (Pierce, Rockford, IL). Purified 4D5 IgG (0.65 mg) in 6 mL of the citrate-carbonate coupling buffer was added to the dry beads (0.25 g) in the vial, and briefly vortexed. Degassed water (6 mL) was added to a polypropylene column (1 x 11 cm), and a porous polyethylene disk was gently pushed through the water to the bottom of the column. The water was then drained, the gel-antibody mixture was added. The column was capped and gently mixed on a hybridization incubator for 1 h at room temperature. The suspension was then washed with PBST (20 mL). To block remaining active groups, six mL of 3-M ethanolamine (pH 9.0) were added to each column and gently mixed for 2.5 h at room temperature. The column was placed in a holder to allow the beads to settle. It was washed with 10 bed volumes (20 mL) of washing solution (1.0 M NaCl), a porous polyethylene disk was placed 1-2 mm above the beads, washed again with three bed volumes of PBST, capped and stored at 4 °C until needed. To determine the coupling efficiency, samples of the buffer from the antibody-gel suspension before and after the reaction, and samples from each subsequent step were analyzed with the BCA protein assay (45).

SFE extracts were subjected to IAC cleanup in a cycle of loading, washing, eluting and regenerating. The columns were pre-washed with 5 mL of PBST at room temperature. A sample extract containing less than 3% methanol (MeOH) in PBST was loaded on the column at a flow rate of 0.5 mL/min. The column was washed with 10 mL of washing solution (acetone: acetonitrile: MeOH: PBST, 1:1:1:7). The bound analytes were eluted with 10 mL of acetone, and collected in a graduated centrifuge tube. The column was regenerated by washing with PBST at a rate of 0.5 mL/min for 45-60 min.

Indirect cELISAs were used to assess the effects of solvents, pH and salts on the binding and release of BaP by 4D5 IgG (42). The binding capacity of 4D5 was determined in wells precoated with BaP-6a-BSA (2.0 ng/100 μL/well) according to the same procedure as the cELISA (42). To assess release of the antibody–PAH complex, 4D5 IgG were incubated in PBST for 1 h at 37 °C with no analytes present in wells precoated with BaP-6a-BSA (2.0 ng/100 μL/well). The plate was washed with PBST, then a test solution (100 μL/well) was added, and the plates were incubated at room temperature for 1 h. The plates were washed, and bound IgG was determined as in the binding reaction. The test solutions were different salts and salt concentrations; buffers at different pH values; and various solvents at different concentrations.

Supercritical Fluid Extraction (SFE) and Solid Phase Extraction (SPE)

An Isco SFX 2-10 extractor (Lincoln, NE) was connected to an Isco 260D syringe pump with a cooling jacket cooled at 10 °C. The extraction parameters were the same as previously reported (43, 46). The samples underwent a 10-min static extraction, followed by a dynamic extraction step to collect 30 mL of SC-CO_2. Sediment and soil samples were adjusted to 15% (w/w) water content, and 5% (w/w) Na_4EDTA (46). Coral samples were air-dried and subjected to SFE with no use of Na$_4$EDTA. The extracts were collected in 10 mL of MeOH. Water samples (0.5-1 L) were extracted with Bakerbond Speedisks™ (C$_{18}$, 50 mm, J. T. Baker, Phillipsburg, NJ) according to the manufacturer's procedure.

GC-MS Analysis

PAHs were analyzed on a HP 6890 series GC equipped with a 5973 MSD, a HP-5 column (0.25 μm x 0.25 mm x 30 m), and a 7683 auto-sampler. Helium carrier gas flow was 36 cm/s. The oven temperature started at 70 °C for 1 min and was increased to 315 °C at 4 °C /min, and held at 315 °C for 5 min. The injector temperature was 300 °C. The injection volume was 1 μL in splitless mode with 1-min delay. The MSD was operated in the selected ion monitoring mode.

Results and Discussion

Optimization of ELISA Procedure

MAbs 4D5 and 10C10 were derived by Gomes and Santella for detection of BaP-protein adducts (12). At the start of this project we synthesized ten new 2-5 fused-ring PAH haptens, and prepared PAH-BSA conjugates for evaluation as coating antigens for indirect ELISAs (42). Using these conjugates, we found that the cross-reactivity of 4D5 and 10C10 with PAHs smaller than BaP was even broader than originally described by Gomes and Santella. In addition, the sensitivity and specificity for different PAHs in cELISAs could be changed with the different conjugates as coating antigens (42). Competitive ELISAs were done with each of the new PAH-BSA conjugates as competitors for 4D5 and 10C10 in wells coated with BaP-6a-BSA which Gomes and Santella used to evoke the antibodies. All of the conjugates competed to different extents that could be ranked as the ratio of I$_{50}$ of the standard conjugate to I$_{50}$ of other conjugates examined (Table I). This ratio, which we defined as a recognition index (RI), represented relative affinities of the mAbs for the conjugates. The ideal RI value would be less than 1 but within a range consistent with optimum sensitivity. Use of PYR-1a-BSA (RI = 0.55) gave the most sensitive cELISA for BaP, with a half-maximal inhibition (I$_{50}$) of about 16 nM (4 ng/mL) BaP (Figure 2) and a working range of about 0.5-500 nM (0.1-125 ng/mL) BaP. This assay proved to be 89-fold more sensitive for BaP with PYR-1a-BSA instead of BaP-6a-BSA as a coating antigen. It was also able to detect several other 2-5 fused-ring PAHs (42). The protocol described is an appropriate method for selecting solid phases and conjugates and optimizing ELISAs.

Table I. Recognition Index of mAb 10C10 to PAH-BSA Conjugates[a]

Conjugate Inhibitor	$I_{50,}$ ng/mL	RI[b]
Pyrene-1a-BSA	52	0.55
Pyrene-1b-BSA	47	0.61
Fluoranthene-BSA	41	0.70
Chrysene-BSA	38	0.75
Anthracene-BSA	30	0.95
BaP-6a-BSA	29	1.0
BaP-6b-BSA	24	1.2
BaP-1a-BSA	15	1.9
Phenanthrene-BSA	13	2.1
Fluorene-BSA	3.1	9.2
Naphthalene-BSA	0.3	95

[a] Data are from ref. 42. The plates were coated with BaP-6a-BSA.
[b] $RI = I_{50}$ of BaP-6a-BSA / I_{50} of a conjugate.

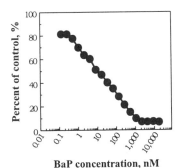

Figure 2. ELISA standard curve of BaP using mAb 10C10.

Surface Water Samples

The improved cELISA was used to screen for PAHs in fresh water. Sample pHs ranged from 6.8 to 8.3 and were not adjusted since they were in the tolerable pH range for ELISA. The assay I_{50} for BaP was approximately 20 ng/mL and the limit of detection (LOD), which was defined as the concentration required for a 20% reduction of signal, was about 0.7 ng/mL. Eleven of 34 samples registered as positive (at least two-fold greater than the LOD, equivalent to $\geq$1.4 ng/mL of BaP). The positive samples were fortified with 3.0 ng/mL of BaP and analyzed by ELISA again. After the 3.0 ng/mL spike was deducted, the estimated original concentration (BaP equivalents) in the samples ranged from 0.04 to 1.01 ng/mL. These results confirmed the results of the first ELISA. The samples tested by ELISA were extracted by SPE, and confirmatory GC-MS analysis was done. However, the ELISA values were higher, up to 10 fold, than those obtained by

386

GC-MS (Figure 3). GC-MS resolves different PAHs, but ELISA does not. The overestimation by ELISA is a common phenomenon, and may be attributed to antibody cross reactivity, nonspecific interference, and calibration PAH standard selected. Cross reactivity and nonspecific interference often biases ELISAs toward higher estimates. In the current ELISA formats using single polyclonal or monoclonal antibodies, results for high concentrations of a weakly cross-reactive PAH or a lower concentration of BaP, may be indistinguishable. Therefore, attention should be paid to selection of appropriate standards, particularly for the ELISAs of multiple analytes such as PAHs and PCBs.

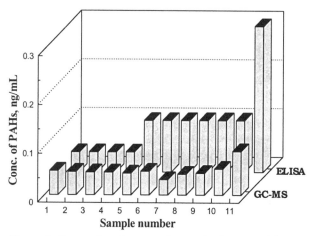

Figure 3. Comparison between the results determined by ELISA and those by GC-MS for PAHs in water from NWFs.

Sediment Samples

Soils and aquatic sediments are among the most heterogeneous matrices for all types of analytical methods, and they are among the largest sinks for PAHs in the environment. Measurements of PAH bioavailability, metabolism, and remediation in soils and sediments are critical for DOE hazard assessment and cleanup needs. There are numerous reliable methods for extraction and cleanup of the apolar parent PAHs (11, 22). However, extraction and cleanup of polar PAH metabolites remain challenging. One of our goals was to devise a way to simultaneously and efficiently extract both the polar and apolar PAH analytes, and to make the procedure as compatible as possible with ELISA and instrumental analyses. One promising technique is the use of supercritical CO_2 with various reagents and/or solvent modifiers to improve recoveries of analytes (22, 47). Published approaches include ion-pair formation (48, 49), in-situ derivatization (50), complexation (51), and formation of an extractable organometallic compound (52) or reverse micelles (53). Recently, we obtained a quantitative recovery of polar aromatic compounds, including xanthenes, phenols, and chlorophenoxyacetic acids, from soil by Na₄EDTA-assisted SFE (43, 46). The sediment extracts obtained by SFE with or without Na₄EDTA were analyzed by GC-MS and ELISA. The total concentrations of PAHs by GC-MS were 166-282 and 209-356 ng/g in the sediments without and with Na₄EDTA, respectively (Figure 4). The PAH concentrations determined by ELISA, expressed as BaP equivalents, were 259-423 and 345-531 ng/g in the sediment samples without and with Na₄EDTA, respectively. The concentrations of total PAHs detected in sediments extracted by Na₄EDTA-assisted SFE were about 20-30% higher than those obtained by SFE without using Na₄EDTA. Confirmatory analyses by HPLC with fluorescence detection revealed that the sediment extracts contained 1-pyrenol at concentrations of 16-30 ng/g. It is known that mAbs 4D5 and 10C10 bind 1-pyrenol and other PAH metabolites (12, 42), so this may account at least partly for the higher estimates in ELISAs.

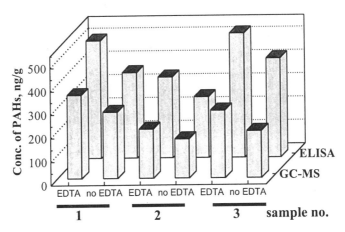

Figure 4. Comparison between results determined by ELISA and those by GC-MS.

To determine the extraction efficiency for 1-pyrenol, sediments were spiked with pyrene and 1-pyrenol (2.75 µg/g), and extracts prepared by SFE with and without Na₄EDTA as a modifier. SFE in the absence of Na₄EDTA quantitatively recovered pyrene, but recoveries of 1-pyrenol averaged 71 ± 6%. SFE with 5% (w/w) Na₄EDTA as a modifier gave quantitative recoveries for pyrene (100%) and 1-pyrenol (101 ± 7%). Use of another modifier, 5% (w/w) Na₂SiO₃, gave 87 ± 8% recovery of 1-pyrenol.

Immunoaffinity Cleanup of Coral Extracts

Coral was chosen as a matrix for the study for several reasons. One was to extend potential uses of SFE and IAC in the analysis of environmental PAHs. Coral reefs and the multitude of life forms they support constitute an ecosystem considered to be as complex and diverse as terrestrial rain forests. In addition to anthropogenic PAH pollution, coral at various locations has been exposed to PAHs from naturally occurring sources such as oil seepage from the ocean floor, and volcanic activity (54). Coral may be a good indicator species for chronic exposure to PAHs from oil spills, coastal runoff and harbor pollution (55-57). Our preliminary GC-MS analysis (total ion monitoring) of PAHs in the crude SFE extracts of corals produced very complex chromatograms with poor separation of the target analytes from interferents. IAC has proven to be an effective cleanup method for PAHs (29-35). Therefore, an IAC method was developed to clean up SFE extracts of coral for PAH analyses.

Indirect ELISAs were used to determine the effects of buffer pH, ionic strength and organic solvents, as a predictor of how PAHs may be bound to and released from the 4D5 IgG-immobilized columns. The 4D5 IgG bound to the BaP-6a-BSA conjugate over the pH range 5.5-8.5, and was released by buffer with pH of 4 or less (Figure 5).

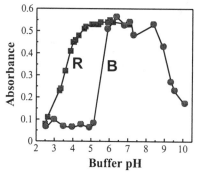

Figure 5. Effects of buffer pH on the binding (B) and releasing (R) reactions between 4D5 IgG and BaP.

Binding of 4D5 to BaP-6a-BSA was progressively reduced by NaCl or $MgCl_2$ at concentrations of 0.4 M or higher in PBST (Figure 6). $MgCl_2$ above 0.8 M strongly disrupted the 4D5–BaP complex. NaCl and KCl had little disruption effects.

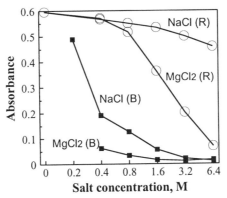

Figure 6. Effects of NaCl and MgCl₂ on the
binding (**B**) and releasing (**R**) reactions.

PAH binding was relatively unaffected by MeOH up to 25% in PBST, reduced by about 25% in PBST containing up to 25% acetone, progressively reduced by acetonitrile, nearly abolished in PBST containing 25% acetonitrile (Figure 7). Acetone at $\geq 60\%$ or acetonitrile $\geq 40\%$ in PBST disrupted the 4D5–BaP complex. MeOH was not an effective disrupter.

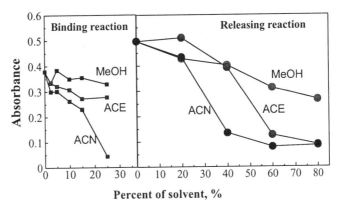

Figure 7. Effects of MeOH, acetone (ACE), and acetonitrile
(ACN) on the binding and releasing reactions.

The results only suggested approximate conditions for IAC because of the differences between microplate and column settings (*e.g.*, non-covalent coating *vs.* covalent attachment).   However, these tests quickly led to workable conditions of IAC loading, washing and elution of SFE extracts of coral for

PAH analysis. These tests have advantages of use of less amount of reagents and quick optimization of various variables for binding and releasing reactions.

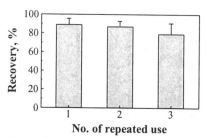

Figure 8. Recovery of IAC for BaP spiked in SFE coral extracts. The column had a theoretical capacity of 8.7 nmol and was loaded with 2 nmol of BaP.

The IAC efficiency was confirmed by analyzing BaP in loading, washing and elution fractions. Good recoveries of BaP (79-89%) were obtained for the three successive cleanup trials when BaP was spiked in SFE extracts of corals (Figure 8). GC-MS chromatograms showed that a simple IAC step reduced the interference and achieved baseline separation of PAHs at ng/g (ppb) levels. PAHs detected in corals from Kaneohe Bay, Oahu Island were anthracene, BaP, benzo[e]pyrene, chrysene, fluoranthene, phenanthrene, pyrene and triphenylene after the SFE extracts of coral samples were cleaned up with the IAC. PAHs detected in corals from Trig Island, North Pacific Ocean, were anthracene, fluoranthene, fluorene, phenanthrene and pyrene. The molar concentration of total PAHs in corals from Kaneohe Bay (1.1 nmoles/g coral or 244 ng/g) was about 5 fold higher than that (0.23 nmoles/g coral or 42 ng/g) from Trig Island. These very limited data were consistent with the coastal uses and pollution histories of both locations.

Summary

The results of these pilot studies point out additional improvements needed to make these immunochemical methods as robust and widely applicable as possible. The degradation products of PAHs prior to mineralization include epoxides, phenols, diols, diol-epoxides, quinones, aromatic acids, and adducts of thiols, glucuronic acid, proteins, and DNA (1, 58). As previously demonstrated (43, 46), SFE with Na_4EDTA as a modifier effectively co-extracted PAHs and their metabolites from soils. In this paper, the same method was shown to work equally well for recovery of PAHs and 1-pyrenol from marine sediments and coral. An ELISA of parent PAHs using the broadly cross-reactive mAbs 4D5 and 10C10 was made almost 90-fold more sensitive using a pyrene-BSA conjugate that the mAbs bound more weakly than BaP-BSA. The cELISA can detect parent PAHs and PAH metabolites while the GC-MS method can only detect parent PAHs. The SFE and ELISA methods can be useful tools in conducting cost-effective site characterization and remediation. The mAb-based IAC columns can be used to recover PAH metabolites formed in remediation processes in addition to initial site assessment for PAH analyses of environmental samples.

Acknowledgments

This work was supported by DOE Grant DE-FG07-96ER-62316. We especially thank R. M. Santella of Columbia University, NY, for providing the 4D5 and 10C10 hybridoma lines.

References

1. *The Handbook of Environmental Chemistry; 3.I. PAHs and Related Compounds: Chemistry.* Neilson, A. H., Ed.; Springer: New York, 1998.
2. Patnaik, P. In *Handbook of Environmental Analysis*; Patnaik, P., Ed.; CRC Press: Boca Raton, Florida, 1997; p165.
3. International Agency for Research on Cancer Monographs on the Evaluation of Carcinogenic Risk of Chemicals to Humans, Vol. 34, *Polynuclear Aromatic Compounds*, Part 3. International Agency for Research on Cancer, Lyon, France, 1984.
4. Autrup, H.; Daneshvar, B.; Dragsted, L. O.; Gamborg, M.; Hansen, A. M.; Loft, S.; Okkels, H.; Nielsen, F.; Nielsen, P. S.; Raffn, E.; Wallin, H.; Knudsen, L. E. Biomarkers for exposure to ambient air pollution – Comparison of carcinogen-DNA adduct levels with other exposure markers and markers for oxidative stress. *Environ. Health Persp.* **1999**, *107*, 233-238.
5. Koganti, A.; Spina, D. A.; Rozett, K.; Ma, B.-L.; Weyand, E. H.; Taylor, B. B.; Mauro, D. M. Studies on the applicability of biomarkers in estimating the systemic bioavailability of polynuclear aromatic hydrocarbons from manufactured gas plant tar-contaminated soils. *Environ. Sci. Technol.* **1998**, *32*, 3104-3112.
6. Lewtas, J.; Dobias, L.; Binkova, B.; Sram, R.; George, M.; Hancock, S.; Williams, R.; Heinrich, U.; Savela, K.; Gallagher, J. Exposure-dosimetry of PAH in humans using DNA adducts as the biomarker of dose and personal exposure monitors for PAH. *Proceedings of the American Association for Cancer Research Annual Meeting* **1994**, *35*, 95.
7. Santella, R. M.; Perera, F. P.; Young, T. L.; Zhang, Y.-J.; Chiamprasert, S.; Tang, D.; Wang, L. W.; Beachman, A.; Lin, J.-H.; Deleo, V. A. Polycyclic

aromatic hydrocarbon-DNA and protein adducts in coal tar treated patients and controls and their relationship to glutathione S-transferase genotype. *Mutation Research* **1995**, *334*, 117-124.

8. Schnell, F. C. Protein adduct-forming chemicals and molecular dosimetry: potential for environmental and occupational biomarkers. In *Reviews in Environmental Toxicology;* Hodgson, E., Eds.; Toxicol. Commun., Inc.: Raleigh, NC, **1993**; *5*, 51-60.

9. Shugart, L.; Theodorakis, C. New trends in biological monitoring: application of biomarkers to genetic ecotoxicology. *Biotherapy* **1998**, *11*, 119-127.

10. World Health Organization. Biomarkers and risk assessment: concepts and principles. World Health Organization: Geneva, Switzerland, 1993.

11. Barcelo, D.; Oubina, A.; Salau, J. S.; Perez, S. Determination of PAHs in river water samples by ELISA. *Anal. Chim. Acta* **1998**, *376*, 49-53.

12. Gomes, M.; Santella, R. M. Immunologic methods for the detection of benzo[a]pyrene metabolites in urine. *Chem. Res. Toxicol.* **1990**, *3*, 307-310.

13. Jones, V. W.; Kenseth, J. R.; Porter, M. D. Microminiaturized immunoassays using atomic force microscopy and compositionally patterned antigen arrays. *Anal. Chem.* **1998**, *70*, 1233-1241.

14. Chen, C. S.; Rao, P. S. C.; Lee, L. S. Evaluation of extraction and detection methods for determining polynuclear aromatic hydrocarbons from coal tar contaminated soils. *Chemosphere* **1996**, *32*, 1123-1132.

15. Michor, G.; Carron, J.; Bruce, S.; Cancilla, D. A. Analysis of 23 polynuclear aromatic hydrocarbons from natural water at the sub-ng/L level using solid-phase disk extraction and mass-selective detection. *J. Chromatogr. A* **1996**, *732*, 85-99.

16. Negrao, M. R.; Alpendurada, M. F. Solvent-free method for the determination of polynuclear aromatic hydrocarbons in waste water by solid-phase microextraction-high-performance liquid chromatography with photodiode-array detection. *J. Chromatogr. A* **1998**, *823*, 211-218.

17. Kenny, D. V.; Olesik, S. V. Extraction of lignite coal fly ash for polynuclear aromatic hydrocarbons: Modified and unmodified supercritical fluid extraction, enhanced-fluidity solvents, and accelerated solvent extraction. *J. Chromatogr. Sci.* **1998**, *36*, 59-65.

18. Kenny, D. V.; Olesik, S. V. Extraction of bituminous coal fly ash for polynuclear aromatic hydrocarbons: Evaluation of modified and unmodified supercritical fluid extraction, enhanced fluidity solvents, and accelerated solvent extraction. *J. Chromatogr. Sci.* **1998**, *36*, 66-72.

19. Tomaniova, M.; Hajslova, J.; Pavelka, J.; Kocourek, V.; Holadova, K.; Klimova, I. Microwave-assisted solvent extraction: A new method for isolation of polynuclear aromatic hydrocarbons from plants. *J. Chromatogr. A* **1998**, *827*, 21-29.

20. Dupeyron, S.; Dudermel, P. M.; Couturier, D. Focused microwave assisted extraction (FMAE) of polynuclear aromatic hydrocarbons from contaminated soil: Role of acetone and water content impact on microwave efficiency. *Analysis* **1997**, *25*, 286.

21. Kipp, S.; Peyrer, H.; Kleiboehmer, W. Coupling superheated water extraction with enzyme immunoassay for an efficient and fast PAH screening in soil. *Talanta* **1998**, *46*, 385-393.

22. *SW-846 Test Methods for Evaluating Solid Waste, Physical/ Chemical methods*, 3rd Ed. US EPA, Office of Solid Waste, Washington, DC, 1996.

23. Brown, R. S.; Szolar, O. H. J.; Luong, J. H. T. Cyclodextrin-aided capillary electrophoretic separation and laser-induced fluorescence detection of polynuclear aromatic hydrocarbons (PAHs). *J. Mol. Recog.* **1996**, *9*, 515-523.

24. Brumley, W. C.; Jones, W. J. Comparison of micellar electrokinetic chromatography (MEKC) with capillary gas chromatography in the separation of phenols, anilines and polynuclear aromatics potential field-screening applications of MEKC. *J. Chromatogr. A* **1994**, *680*, 163-173.

25. Moy, T. W.; Ferguson, P. L.; Grange, A. H.; Matchett, W. H.; Kelliher, V. A.; Brumley, W. C.; Glassman, J.; Farley, J. W. Development of separation systems for polynuclear aromatic hydrocarbon environmental contaminants using micellar electrokinetic chromatography with molecular micelles and free zone electrophoresis. *Electrophoresis* **1998**, *19*, 2090-2094.

26. Zhang, Y.; Zhu, Y.; Xue, X.; Huang, X. Magnetic field effects-polarization-resonant synchronous fluorescence spectrometry for simultaneous analysis of polynuclear aromatic hydrocarbons in mixtures. *Talanta* **1995**, *42*, 1811-1815.

27. Davis, S. C.; Makarov, A. A.; Hughes, J. D. Supersonic molecular beam-hyperthermal surface ionisation coupled with time-of-flight mass spectrometry applied to trace level detection of polynuclear aromatic hydrocarbons in drinking water for reduced sample preparation and analysis time. *Rapid Commun. Mass Spec.* **1999**, *13*, 247-250.

28. U.S. Dept. of Energy; *DOE Methods for Evaluating Environmental and Waste Management Samples (Chapter 8) Method OS060*: Immunoassay for Petroleum Fuel Hydrocarbons in Soil. **1997**; http://www.pnl.gov/methods

29. Bouzige, M.; Pichon, V.; Hennion, M. C. Class-selective immunosorbent for trace-level determination of polycyclic aromatic hydrocarbons in complex sample matrices, used in off-line procedure or on-line coupled with liquid chromatography/fluorescence and diode array detections in series. *Environ. Sci. Technol.* **1999**, *33*, 1916-1925.

30. Cichna, M.; Knopp, D.; Niessner, R. Immunoaffinity chromatography of polycyclic aromatic hydrocarbons in columns prepared by the sol-gel method. . *Anal. Chim. Acta* **1997**, *339*, 241-250.

31. Perez, S.; Ferrer, I.; Hennion, M.-C.; Barcelo, D. Isolation of priority polycyclic aromatic hydrocarbons from natural sediments and sludge reference materials by an anti-fluorene immunosorbent followed by liquid chromatography and diode array detection. *Anal. Chem.* **1998**, *70*, 4996-5001.

32. Bentsen-Farmen, R. K.; Botnen, I. V.; Noto, H.; Jacob, J.; Ovrebo, S. Detection of polycyclic aromatic hydrocarbon metabolites by high-pressure liquid chromatography after purification on immunoaffinity columns in urine from occupationally exposed workers. *Int. Arch. Occup. Environ. Health* **1999**, *72*, 161-168.

33. Cichna, M.; Markl, P.; Knopp, D.; Niessner, R. Optimization of the selectivity of pyrene immunoaffinity columns prepared by the sol-gel method. *Chem. Materials* **1997**, *9*, 2640-2646.

34. King, M. M.; Cuzick, J.; Jenkins, D.; Routledge, M. N.; Garner, R. C. Immunoaffinity concentration of human lung DNA adducts using an anti-benzo[a]pyrene-diol-epoxide-DNA antibody. Analysis by ^{32}P-postlabelling or ELISA. *Mutat. Res.* **1993**, *292*, 113-22.

35. Randerath, K.; Sriram, P.; Moorthy, B.; Aston, J. P.; Baan, R. A.; van den Berg, P. T.; Booth, E. D.; Watson, W. P. Comparison of immunoaffinity chromatography enrichment and nuclease P1 procedures for ^{32}P-postlabelling analysis of PAH-DNA adducts. *Chem. Biol. Interact.* **1998**, *110*, 85-102.

36. Liang, X. L.; Lubman, D. M.; Rossi, D. T.; Nordblom, G. D.; Barksdale, C. M. On probe immunoaffinity extraction by matrix-assisted laser desorption/ionization mass spectrometry. *Anal. Chem.* **1998**, *70*, 498-503.

37. Hock, B. Antibodies for immunosensors. *Anal. Chim. Acta* **1997**, *347*, 177-186.

38. Ekins, R. P.; Chu, F. W. Miniaturized microspot multi-analyte immunoassay systems. In *Immunoanalysis of Agrochemicals: Emerging Technologies;* Nelson, J., Karu, A. E., Wong, R., Eds.; American Chemical Society (Symposium No. 586): Washington D.C., 1995; pp153-174.

39. Rubtsova, M. Y.; Samsonova, J. V.; Egorov, A. M.; Schmid, R. D. Simultaneous determination of several pesticides with chemiluminescent immunoassay on a multi-spot membrane strip. *Food Agric. Immunol.* **1998**, *10*, 223-235.

40. Brecht, A.; Abuknesha, R. Multi-analyte immunoassays: application to environmental analysis. *Trends in Anal. Chem.* **1995**, *14*, 361-371.

41. Dubois, M.; Taillieu, X.; Colemonts, Y.; Lansival, B.; De Graeve, J.; Delahaut, P. GC-MS determination of anabolic steroids after multi-immunoaffinity purification. *Analyst* **1998**, *123*, 2611-6.

42. Li, K.; Chen, R.; Zhao, B.; Liu, M.; Karu A. E.; Roberts, V. A.; Li, Q. X. Monoclonal antibody-based ELISAs for part-per-billion determination of

polycyclic aromatic hydrocarbons: effects of haptens and formats on sensitivity and specificity. *Anal. Chem.* **1999,** *71,* 302-309.

43. Guo, F.; Li, Q.X.; Alcantara-Licudine, J.P. A simple Na$_4$EDTA-assisted sub/supercritical fluid extraction procedure for quantitative recovery of polar analytes in soil. *Anal. Chem.* **1999,** *71*: 1309-1315.

44. Miao, X.-S.; Swenson, C.; Yanagihara, K.; Li, Q.X. Polychlorinated biphenyls and metals in marine species from French Frigate Shoals, North Pacific Ocean. *Arch. Environ. Contam. Toxicol.* **2000,** *38,* 464-471.

45. Smith, P. K.; Krohn, R. I.; Hermanson, G. T. Measurement of protein using bicinchoninic acid. *Anal. Biochem.* **1985,** *150,* 76-85.

46. Alcantara-Licudine, J. P.; Kawate, M. K.; Li, Q. X. Method for the analysis of phloxine B, uranine and related xanthene dyes in soil using supercritical fluid extraction and High-performance liquid chromatography. *J. Agric. Food Chem.* **1997,** *45,* 766-773.

47. Chester, T. L.; Pinkston, J. D.; Raynie, D.E. Supercritical fluid chromatography and extraction. *Anal. Chem.* **1998,** *70,* 301R-319R.

48. Field, J. A.; Miller, D. J.; Field, T. M.; Hawthorne, S. B.; Giger, W. Quantitative determination of sulfonated aliphatic and aromatic surfactants in sewage sludge by ion-pair/supercritical fluid extraction and derivatization gas chromatography/mass spectrometry. *Anal. Chem.* **1992,** *64,* 3161-3167.

49. Jimenez-Carmona, M. M.; Tena, M. T.; Luque de Castro, M. D. Ion-pair-supercritical fluid extraction of clenbuterol from food samples. *J Chromatogr.* **1995,** *711,* 269-276.

50. Hawthorne, S. B.; Miller, D. J.; Nivens, D. E.; White, D. C. Supercritical fluid extraction of polar analytes using in situ chemical derivatization. *Anal. Chem.* **1992,** *64,* 405-412.

51. Liu, Y.; Lopez-Avila, V.; Alcaraz, M.; Beckert, W. F. Off-line complexation/supercritical fluid extraction and gas chromatography with atomic emission detection for the determination and speciation of organotin compounds in soils and sediments. *Anal. Chem.* **1994,** *66,* 3788-3796.

52. Cai, Y.; Alzaga, R.; Bayona, J. M. In-situ derivatization and supercritical fluid extraction for the simultaneous determination of butyltin and phenyltin compounds in sediment. *Anal. Chem.* **1994,** *66,* 1161-1167.

53. Jimenez-Carmona, M. M.; Luque de Castro, M. D. Reverse-micelle formation: a strategy for enhancing CO$_2$-supercritical fluid extraction of polar analytes. *Anal. Chim. Acta* **1998,** *358,* 1-4.

54. Tomascik, T.; Van Woesik, R.; Mah, A. J. Rapid coral colonization of a recent lava flow following a volcanic eruption, Banda Islands, Indonesia. *Coral Reefs* **1996,** *15,* 169-175.

55. Richmond, R. H. Effects of coastal runoff on coral reproduction. In *Proceedings of the Colloquium on Global Aspects of Coral Reefs--Health, Hazards and History*; Ginsburg, R. N. Ed.; University of Miami Rosenstiel

School of Marine and Atmospheric Sciences: Miami, Fla., 1994; pp. 360-364.

56. Hodgson, G. A global assessment of human effects on coral reefs. *Marine Pollut. Bull.* **1999**, *38,* 345-355.

57. Guzman, H. M.; Holst, I. Effects of chronic oil-sediment pollution on the reproduction of the Caribbean reef coral Siderastrea siderea. *Marine Pollut. Bull.* **1993**, *26,* 276-282.

58. Wilson, S. C.; Jones, K. C. Bioremediation of soil contaminated with polynuclear aromatic hydrocarbons (PAHs): a review. *Environ. Pollution* **1993**, *81,* 229-249.

Biochemistry

Chapter 24

Architecture of Antibody Binding Sites for Polynuclear Aromatic Hydrocarbons

Jean-Luc Pellequer[1], Shu-wen W. Chen[1], Ann J. Feeney[2], Bitao Zhao[3],
Hui-I Kao[3], Alexander E. Karu[3], Kai Li[4], Qing X. Li[4],
and Victoria A. Roberts[1]

[1]Departments of Molecular Biology and [2]Immunology, The Scripps Research
Institute, La Jolla, CA 92037
[3]Department of Nutritional Science, University of California, Berkeley, CA 94720
[4]Department of Environmental Biochemistry, University of Hawaii at Manoa,
Honolulu, HI 96822

Polynuclear aromatic hydrocarbons (PAHs), their metabolites, and protein and DNA adducts are of concern in many focus areas of DOE's Environmental Management Science Program. Antibody-based analyses offer many potential advantages, but PAH hydrophobicity, planarity, molecular symmetry, and lack of hydrogen-bonding atoms pose unique problems for developing improved antibodies and assays. This chapter describes molecular models of the binding sites of two similar recombinant Fab antibodies (rFabs), 4D5 and 10C10, and their PAH ligands. Both rFabs bind benzo[a]pyrene in unusually deep antigen-binding pockets, in which a highly conserved framework residue, tryptophan H47, is replaced by valine or leucine. Positively charged side chains flank both sides of the pockets, suggesting that these cationic groups interact with the π electrons of the PAH aromatic rings. This novel mechanism for PAH recognition suggests ways that antibodies could be genetically engineered *in vitro* to produce variants with improved properties for environmental monitoring applications. These findings pose intriguing possibilities for how PAHs may be bound by molecular signaling proteins such as the aryl hydrocarbon receptor and PAH-metabolizing enzymes important for bioremediation.

Introduction

Over the last ten years, antibody-based methods for detection and recovery of small toxic compounds have gained increasing acceptance as reliable, cost-effective, high-throughput analytical tools, suitable for many needs in the Environmental Management Science Program of the Department of Energy (DOE). Immunoassay kits to identify polycyclic aromatic hydrocarbons (PAHs) and total petroleum hydrocarbons (TPH) are commercially available in a variety of formats (coated tubes, cards, magnetic beads, liposomes, etc.) for on-site and point-of-need field use, as well as laboratory analysis (1-3). Several of these products proved successful for compliance monitoring in demonstration projects at DOE and Department of Defense (DOD) sites (4). The U.S. Environmental Protection Agency (EPA) recently promulgated sampling guidelines, data quality objectives, and SW-846 methods 4030 and 4035 for regulatory use of PAH and TPH immunoassay kits (5-9).

Remarkable progress has also been made in developing antibody-based sensors, optical systems, and methods for acquiring and interpreting complex data. It is now feasible to use panels of different antibodies on fiber optic bundles, microarray chips, and other devices for multi-analyte analysis (10-25). Immunoaffinity methods to recover PAHs, their metabolites, and adducts from complex environmental and biological samples have also evolved rapidly (26-30).

The major obstacle to improving and expanding the use of antibody-based analysis for PAH detection is that there are too few antibodies that identify individual analytes, metabolites, and adducts. The two- to five-ring PAHs most amenable to immunoassay are naphthalene, acenaphthene, acenaphthylene, anthracene, phenanthrene, fluorene, fluoranthene, chrysene, pyrene, and benzo[a]pyrene (BaP). All are planar, rigid, and have regions of identical shape and size, making immunological cross-reactivity (binding of different PAH analogs to one antibody) virtually unavoidable. Ideally, it should be possible to create a panel of antibodies from which cross-reactivity data can be interpreted. However, conventional polyclonal and monoclonal antibody methods are not likely to succeed in producing the desired antibodies, because there are very few alternatives for chemical synthesis of PAH haptens different enough to evoke novel antibodies and to serve as competitors in immunoassays.

Applying a new strategy to the problem, we decided that genetic engineering would be the most efficient and productive way to obtain the desired antibodies, and that computational analysis and modeling of PAH-antibody interactions would provide the detailed knowledge needed to guide this effort. This set up a rational design cycle in which recombinant antibody methods were used to clone and express PAH-binding antibodies from hybridoma cells and to select antibodies from large phage display libraries. The binding-site structures of the anti-PAH antibodies have been modeled, and the findings are being used to design antibodies with new selectivities that will be made by *in vitro* mutagenesis. Determining the structural and binding properties of these new antibodies, in turn, will direct further design efforts.

Our first approach is based on hybridoma lines 4D5 and 10C10, which were originally derived by Gomes and Santella from mice immunized with a BaP hapten conjugated to bovine serum albumin (BSA, *31*). The primary purpose of these monoclonal antibodies (MAbs) was to detect BaP metabolites and adducts that may be biomarkers for PAH exposure and effects. We cloned and sequenced the DNA encoding the V_HC_{H1} and V_LC_L domains of 4D5 and 10C10, selected and expressed them as recombinant Fabs (rFabs), and analyzed their affinity, sensitivity, and selectivity (*32*). Like the parent MAbs, the rFabs cross-reacted with eight PAHs and eleven PAH haptens (*32*). Computational models of 4D5 and 10C10 with bound PAHs and PAH haptens revealed novel aspects of the binding site structure, including evidence that a charge interaction may be important for PAH binding (*33*).

In a second approach, we recovered human single-chain Fv antibodies (scFvs) that bound naphthalene and phenanthrene haptens from a large phage display library. Computational models indicated that PAH binding in these scFvs involves a charge interaction similar to that found for mouse rFabs 4D5 and 10C10 (Pellequer, et al., in preparation). Together, these results provide some unexpected fundamental insights into how PAH-binding antibodies with improved properties may be derived and systematically engineered. Our results also impact immunoassay design and data interpretation, as well as sample preparation and cleanup methods. We suggest that these approaches may be generalized to improve immunoassays of other compounds.

Materials and Methods

PAH Haptens and Conjugates

Reagents, synthesis and characterization of the PAH haptens and conjugates, preparation of reference standards, safety precautions, and immunoassay methods have been published (*34*). PAH haptens and conjugates are shown in Table I.

Antibodies

Mouse hybridoma cell lines producing monoclonal antibodies (MAbs) 4D5 and 10C10 were generously provided by Dr. Regina Santella (School of Public Health, Columbia University, New York, NY). Cloning and properties of rFabs 4D5 and 10C10 in the pCOMB3 phage display vector are described elsewhere (*32*). The DNA and deduced amino acid sequences are available from GENBANK. Site-directed mutagenesis was done using a Quick-Change mutagenesis kit (Stratagene, Inc., La Jolla, CA) according to the manufacturer's instructions.

Table I: Synthesized PAHs and PAH haptens

PAH	Hapten		

Naphthalene

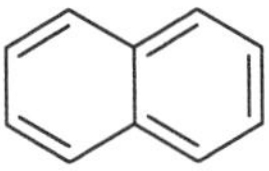

NAP-1a,1b,2a,2b

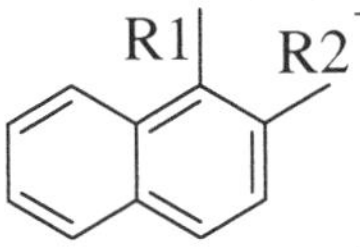

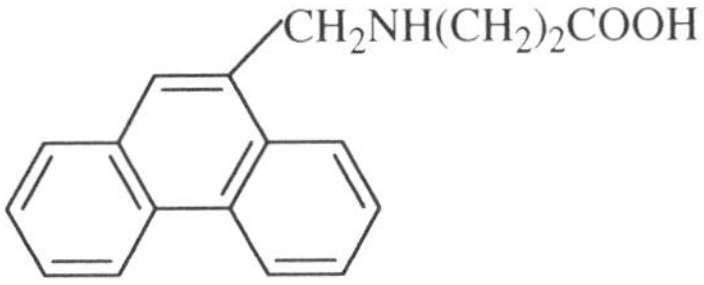

	R1	R2	
	$O(CH_2)_5CO_2H$	H	1a
	$OCHO_2H$	H	1b
	H	$O(CH_2)_5CO_2H$	2a
	H	$OCHO_2H$	2b

Acenaphthene

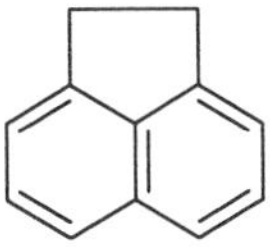

AC-5a

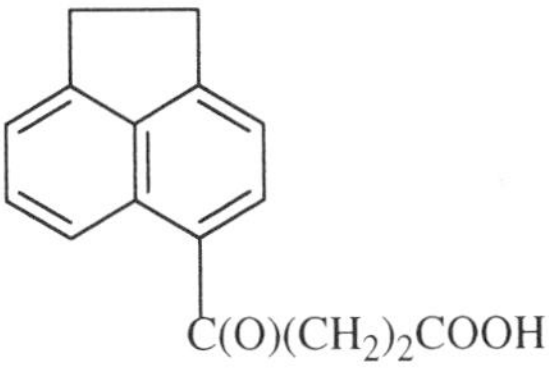

$C(O)(CH_2)_2COOH$

Acenaphthylene

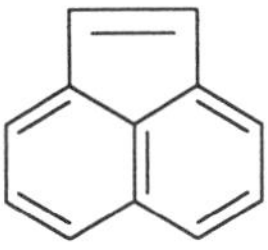

ACN-2a

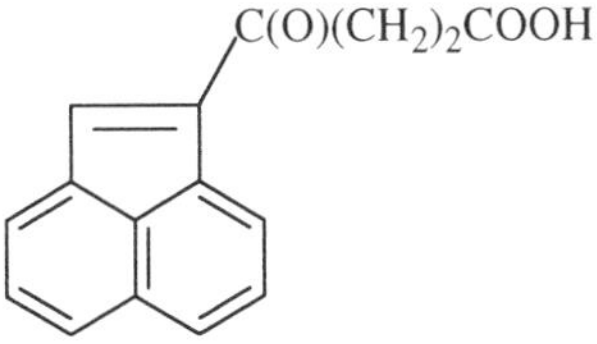

$C(O)(CH_2)_2COOH$

Anthracene

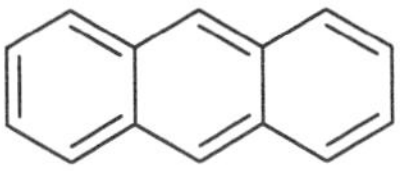

ANT-9a

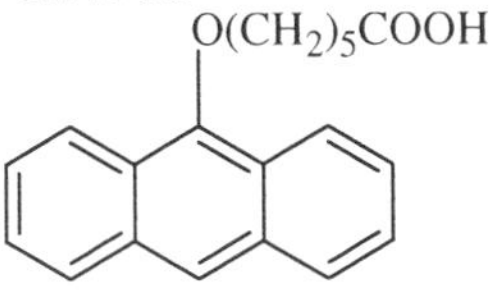

$O(CH_2)_5COOH$

Phenanthrene

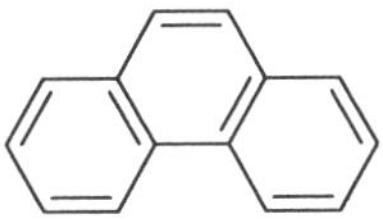

PHE-10a

$CH_2NH(CH_2)_2COOH$

Continued on next page

Table I: Synthesized PAHs and PAH haptens

| PAH | Hapten |

PAH   **Hapten**

Fluorene   FLR-3a

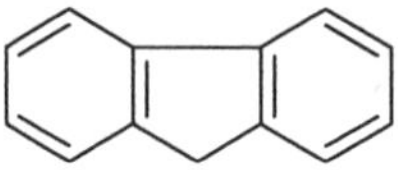

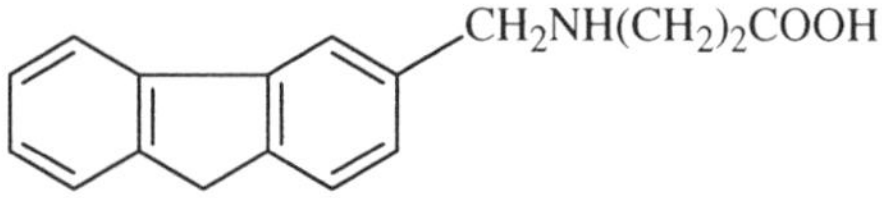

$CH_2NH(CH_2)_2COOH$

Fluoranthene 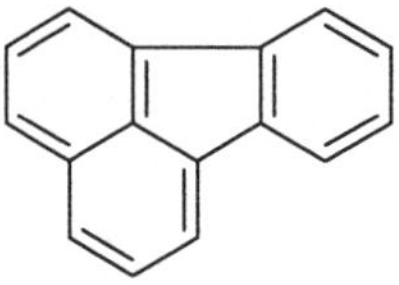   FLA-3a

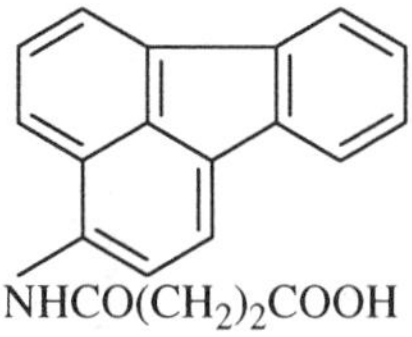

$NHCO(CH_2)_2COOH$

Chrysene 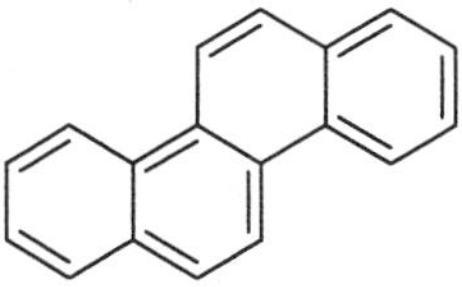   CHR-6a

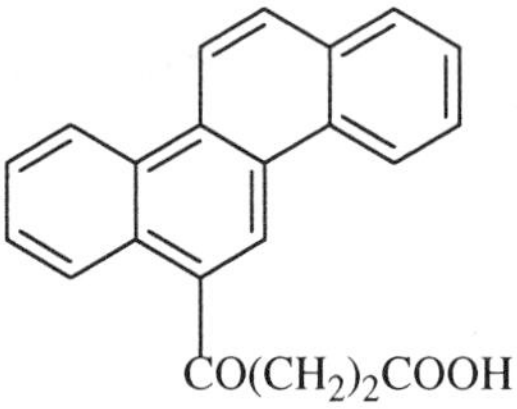

$\overset{|}{C}O(CH_2)_2COOH$

Pyrene 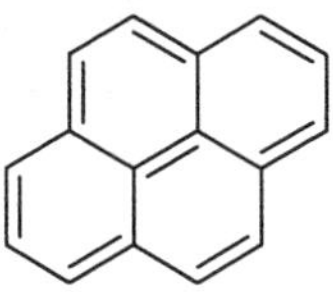   PYR-1a,1b

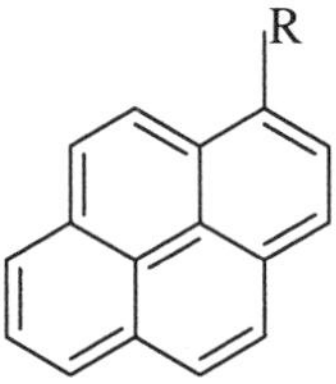

R

$R=(CH_2)_3COOH$ (1a)

CH_2NHCH_2COOH (1b)

Benzo[a]pyrene   BaP-1a,6a,6b

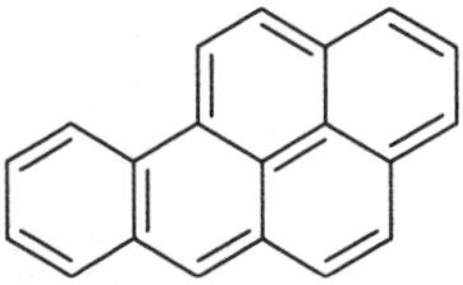

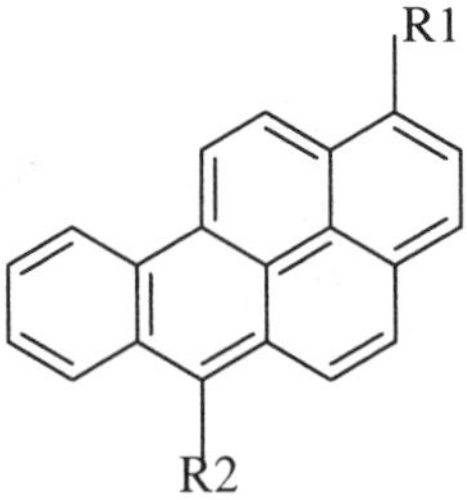

R1

R2

	R1	R2	
	$CO(CH_2)_2CO_2H$	H	1a
	H	N=C=O	6a
	H	$CH_2NH(CH_2)_2$-CO_2H	6b

Combinatorial Antibody Libraries

The Fab (4-22) 2LOX and Nissim human scFv phage display libraries were provided by Dr. Fiona Sait (Dr. Greg Winter's laboratory, Centre for Protein Engineering, Medical Research Council, Cambridge, UK). The libraries were propagated and screened for binding to PAH haptens essentially as described by the developers (*35,36*).

Computational Chemistry and Modeling of PAHs and Haptens

Structural coordinates of PAHs were obtained from the Cambridge Structural Database of small molecules. Geometry optimization and partial atomic charges of PAHs and haptens were determined with the *ab initio* programs Gaussian94 (*37*) and the Amsterdam Density Function (ADF, Vrije Universiteit, Amsterdam, The Netherlands). Molecular orbital calculations were made with Hartree-Fock wave functions at the 6-31G** level in Gaussian94. The total charge of PAHs was set to zero and the multiplicity to one. To calculate the electrostatic potential of each PAH, we used the finite difference Poisson-Boltzmann method in the program Delphi (*38*) and our calculated atomic charges. A 65 x 65 x 65 cubic grid with a spacing of 0.33 Å between each grid point, internal and external dielectric constants of 1.0, and a probe radius of 1.4 Å were used. Atomic radii were 1.7 for C, 1.2 for H, 1.55 for N, and 1.5 for O. The electrostatic potential at the solvent-accessible surface (in kT/e) was mapped onto the molecular surface of PAHs with the program AVS (Plate 1).

Antibody Modeling

Antibody models were built with the aid of the antibody structural database (ASD), which contains superimposed crystallographic structures of antibodies (*39,40*). Structural coordinates of antibodies in the ASD were obtained from the Protein Data Bank (*41*). The modeling was visualized with the program Insight II (Molecular Simulations, Inc., San Diego, CA). The ASD allows inspection of conserved structural features, including backbone conformation and side-chain geometries. Side chains were built using Insight II or Xfit (*42*). Energy minimizations were done with the X-PLOR program (*43*) with the all-atom force field CHARMM22 (*44*). Non-bonded interactions were limited to a radius of 13 Å, with the truncated functions for electrostatic and van der Waals terms set to shifted and switched, respectively, with a cut-on of 9.5 Å and a cut-off of 12 Å.

Conjugate gradient minimizations were considered converged when the gradient reached 0.5 kcal/mol/Å^2 for side-chain optimization with the electrostatic term turned off, or 3.0 kcal/mol/Å^2 for all atoms in a small region with the electrostatic term turned on and a dielectric constant of 1.0. The graphics program Turbo-Frodo (*45*) was used to insert an additional residue in CDRH3. The geometry of the models was evaluated with the PROCHECK program (*46*).

Results and Discussion

Physical Properties of PAHs and Haptens

The electrostatic potential calculated in vacuum for several PAHs (Plate 1) reveals that electron delocalization results in a significant negative electrostatic potential above and below the plane rings for all PAHs. The addition of a methyl-urea linker on BaP, which mimics the conjugated immunogen used for eliciting 4D5 and 10C10, noticeably enhances the negative character of the electrostatic potential. A methyl-succinic acid linker has less effect on the electrostatic potential. Therefore, even though the attached linkers are neutrally charged, they can significantly influence the distribution of electrostatic potential of the PAH ring system. Although the two linkers were attached at different positions, the almost symmetrical shape of BaP allows it to bind in more than one orientation so that both linkers extends out of the antibody binding pocket (*34*).

Modeling 4D5 and 10C10

The murine antibodies 4D5 and 10C10 were raised against BaPisocyan-ate-BSA complex (*31*). Both antibodies have similar binding affinities for BaP and similar cross-reactivity patterns with other PAHs and haptens (*32,34*). The 4D5 variable region (Fv), the domain that binds antigen, has a 90% sequence identity with the 10C10 Fv, differing by 23 residues (Table II), among which are 11 conservative changes. Both antibodies are almost certainly derived from the same germline genes (see below). Although the models of the 4D5 and 10C10 Fv domains were built independently from each other, the same structural templates and model building protocols were used for both. The procedure for building the 4D5 Fv (*33*) is described in detail.

Creating an Fv model requires constructing the variable light chain (V_L) and variable heavy chain (V_H) domains. The antigen-binding site is formed in the V_L/V_H interface by six sequence-variable loops termed the complementarity determining regions (CDRs). Three CDRs are contributed from each chain: CDRs L1, L2, and L3 from V_L and CDRs H1, H2, and H3 from V_H. Templates for V_L and V_H are usually derived from different antibody structures based on sequence identity for each chain. In 4D5, however, both V_L and V_H showed high sequence identity to the corresponding chains of the 17/9 antibody (2 Å resolution, accession code 1HIL, *47*). Therefore, the 17/9 structure was selected as the template for the framework regions of both chains, as well as for CDRs L1, L2, H1, and H2. Table II shows corresponding sequence identities between 17/9 and both target antibodies.

Table II: Sequence identities between template and target antibodies

Domain	Sequence identity (%)		
	4D5 vs. 17/9	10C10 vs. 17/9	4D5 vs. 10C10
V_L FR	88	90	95
CDR L1	59	59	76
L2	86	100	86
L3	37	37	100
V_H FR	83	80	91
CDR H1	80	80	100
H2	65	53	71
H3	25	25	87

Only CDRs L3 and H3 of 4D5 differ significantly from those of 17/9. CDRL3 of 4D5 presents a modeling challenge since it consists of eight residues, one residue shorter than the usual murine CDRL3. Of the four antibody structures in the PDB that have an eight residue CDRL3, and only HyHel-5 (accession code 3HFL, *48*) has Pro at position L95, as is found in 4D5. To incorporate the HyHel-5 CDRL3 into 4D5, the V_L backbone atoms of HyHel-5 were superposed (*40*) onto the 4D5 model and HyHel-5 residues CysL88 to ThrL97 were grafted onto the model. The 4D5 CDRH3 is eight residues long with Arg at position H95, the first residue of CDRH3. The 4D5 CDRH3 is preceded by Arg H94 and also has Asp H101, which is consistent with a β-bulge conformation of CDRH3 (*49*). The antibody TE33 (accession code 1TET, *50*) was chosen as the 4D5 CDRH3 template because it includes Arg H94, Arg H95, and Asp H101, and is just one residue shorter than 4D5. After superposition of V_H backbone atoms (*40*), residues H92 to H105 of TE33 were grafted onto the 4D5 model. An additional residue was built into the 4D5 CDRH3 by computer graphics, resulting in a type I β-hairpin turn, which is a typical turn for the sequence Asp-Tyr-Asp-Ala (*51*). Appropriate side-chain substitutions completed the initial models of the 4D5 V_L and V_H chains. Since 17/9 provided the main template for both chains, the 4D5 V_L and V_H chains were assembled by superposition onto 17/9, providing the V_L/V_H interface. This revealed an antigen-binding site with a deep, well-defined pocket formed by structurally conserved side chains that extend from the β-barrel framework and from regions of the CDRs adjacent to the framework.

Quality Assessment of the 4D5 and 10C10 Models

The predictive value of a molecular model depends on how well its structure and geometry compare with these features in related protein structures. Foremost, it is essential that an antibody model structure maintains the highly conserved antibody β-barrel fold. The root-mean-square deviation (RMSD) between backbone atoms (N, Cα, C) of 4D5 and 17/9 is 0.1 Å for both the V_L and V_H chains (excluding CDRL3 residues L89 to L97 and CDRH3 residues H95-H102). The RMSD between 10C10 and 17/9 is also 0.1 Å. Superposition of the 4D5 model onto the 17/9 template structure reveals the extensive similarity between these two antibodies (Figure 1).

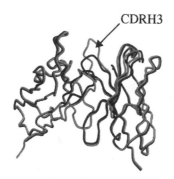

Figure 1: Superposition of the 4D5 (black) model onto the 17/9 template (gray).

Preservation of conserved side-chain geometry is best checked by comparison of the model with related antibodies. This is easily done using the Antibody Structural Database (ASD), which contains superposed V_L and V_H domains of known antibody structures (40). Visual inspection showed that the 4D5 and 10C10 models retain the conserved side-chain geometry found in the framework regions and adjacent CDR residues. Finally, the geometric properties of the model must be reasonable for a protein structure, with main-chain dihedral angles being perhaps the most important indicators (52). In the more than 150 reported crystallographic antibody structures, about 84% of main-chain dihedral angles fall within the "most favored" region as defined by the program PROCHECK (46). In our models of 4D5 and 10C10, 94% and 92% of the main-chain dihedral angles fall in the most favored range (Figure 2).

The outstanding geometric criteria of the 4D5 and 10C10 models are very close to their crystallographic template (17/9) and better than expected for a structure of 2.0 Å resolution, as indicated by the PROCHECK package (Table III).

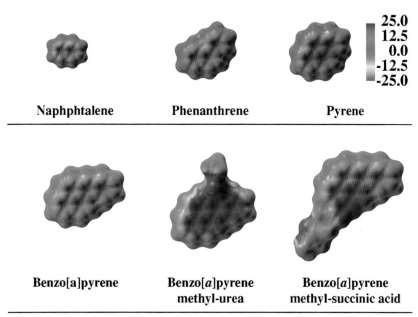

			25.0
			12.5
			0.0
			-12.5
			-25.0

Naphphtalene	Phenanthrene	Pyrene

Benzo[a]pyrene	Benzo[*a*]pyrene methyl-urea	Benzo[*a*]pyrene methyl-succinic acid

Plate 1: Electrostatic potential of PAHs mapped onto their molecular surfaces. Red codes for negatively charged area and blue codes for positively charged area. Units are in kT/e.

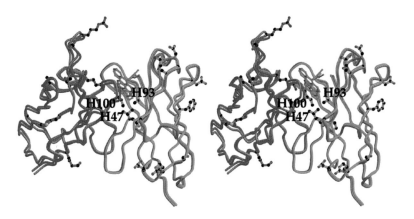

Plate 2: Stereo pair of the variable fragments of 4D5 (magenta for the light chain and cyan for the heavy chain) superposed onto 10C10 (orange) with bound BaP (green for 4D5 and orange for 10C10). Displayed side chains indicate sequence differences between 4D5 and 10C10. Drawing made by Molscript and Raster3D (68,69).

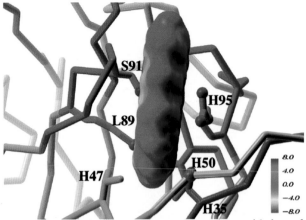

Plate 3: Binding pocket of 4D5, represented as tubes, with bound BaP represented by its molecular surface colored according to electrostatic potential.

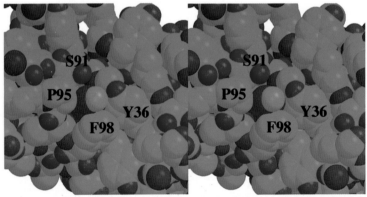

Plate 4: Stereo pair of 4D5 V_L highlighting the buried LysL89 in purple and cyan.

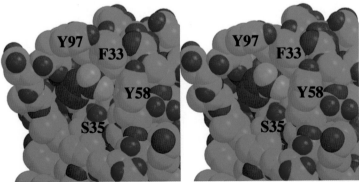

Plate 5: Stereo pair of 4D5 VH highlighting the depression created by small residues (S35, G50) allowing ArgH95 (purple and cyan) to flank the binding pocket.

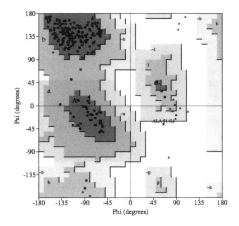

Figure 2: Ramachandran plot of main-chain dihedral angles of 4D5

Table III: Quality assessment of 4D5 and 10C10 models

	Expected	4D5	10C10	17/9
ϕ,ψ in the most favored regions	83.8 +/- 10	94.4	91.3	92.0
ω (peptide bond)	6.0 +/- 3°	6.1°	7.3°	4.8°
ζ (Cα chirality)	3.1 +/- 1.6	1.6	1.7	0.7
$\chi1$ (1st side-chain dihedral angle)	18.2 +/- 4.8°	12.3°	11.1°	11.5°
Bad Contacts	4.2 +/- 10	0	2	0

The most difficult region of an Fv structure to build and evaluate is CDRH3. The crystallographic structure of antibody Mab1-1a (accession code 1A6T, *53*), which became available after completion of the 4D5 model, has a CDRH3 of the same length as 4D5, with 4 of the 8 residues identical to those in the 4D5 CDRH3. The Mab1-1a CDRH3 forms a β-turn in which the AspH96 side chain makes a double hydrogen bond with main-chain nitrogen atoms of residues H98 and H99, a geometry very similar to that built into the 4D5 model (Figure 3). Thus, this recent antibody structure strongly supports our construction of CDRH3, indicating that this region of the model, which is the most variable among Fv structures, has a high likelihood of being accurate.

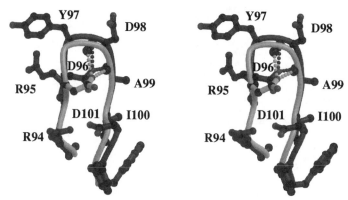

Figure 3: Stereo pair of the superposed CDRH3 loop of 4D5 (dark gray) and Mab1-1a (light gray).

Comparison of 4D5 and 10C10.

As noted above, the Fv domains of 4D5 and 10C10 differ by 23 amino acids. Only three of these differences occur in the binding pocket: residue H47 is Val or Leu, residue H93 is Ala or Gly, and residue H100 is Ile or Val for 4D5 and 10C10, respectively. Position H93 does not contact the bound BaP. The differences at positions H47 and H100 are compensatory, resulting in a conservation of the volume of the binding pocket. These subtle differences result in a slight translational shift of docked BaP in the binding pockets (Plate 2). Among the 20 sequence differences that do not occur in the binding pocket, nine are present in portions of CDRL1 and CDRH2 that do not contact BaP. The others are scattered on the antibody surface and cannot directly affect binding of BaP in 4D5 or 10C10 (Plate 2).

Description of the 4D5 and 10C10 Binding Sites

The shape, charge distribution, and flexibility of an antibody binding site dictates the mode of antigen binding. 4D5 and 10C10 have exceptionally deep pockets (at least 10 Å above the Nζ atom of LysL89, which lies at the bottom of the pocket). The length of the pocket is about 17 Å from the Cβ atom of ValH47 to the Cα atom of TyrH97, and its width is about 7 Å from the Cγ atom of ProL95 to the Cγ atom of ArgH95. Three concurrent criteria are required to build such a deep pocket (*33*, Plate 3): 1) a short framework residue at position H47 (Val in 4D5 and Leu in 10C10), 2) a short CDRL3 (eight residues for both 4D5 and 10C10), and 3) the presence of short side chains around the wall of the pocket (SerL91, SerH35, and GlyH50 for both 4D5 and 10C10, plus IleH100 for 4D5 and ValH100 for 10C10).

Considering the hydrophobic nature of PAHs, it was surprising to find two normally positively charged residues in both the 4D5 and 10C10 binding pockets: LysL89 and ArgH95. On the other hand, the electrostatic potential analyses (Plate 1) indicate substantial negative electrostatic potential above and below the plane of the PAH ring system. This negative charge could be complemented by these two positively charged residues. LysL89 is almost completely buried at the bottom of the binding pocket, and therefore it seems an unlikely candidate (Plate 4). However, ArgH95 flanks the wall of the binding pocket, lying in a depression created by the small amino-acid side chains SerH35 and GlyH50 (Plate 5). Docking BaP-methyl-urea hapten into the binding pocket by computer graphics followed by refinement with rigid minimization using X-PLOR (Plate 2) revealed that the ArgH95 guanidinium group stacks against the planar ring of BaP (33). The manual docking was confirmed (33) by computational docking with the program AUTODOCK (54). In addition, automated docking has shown that BaP itself can dock in multiple orientations in the binding pocket, explaining the similar binding of BaP-1a and BaP-6a to both 4D5 and 10C10 (34). The importance of the positive charge of ArgH95 has been demonstrated by site-directed mutagenesis. Replacement of ArgH95 by Gln totally abolished binding of each and every tested PAH (33). The interaction geometry suggests that the guanidinium group of ArgH95 interacts with the quadrupole moment generated by the π electrons of the BaP ring system (55). In the model, ArgH95 is built in a 'stacked' geometry, as mostly observed in other Arg-aromatic interactions (56-58). This π-cation interaction appears to be essential for stabilizing the bound ligand. Therefore, ArgH95 probably must be retained in our design of new PAH-binding antibodies based on 4D5 and 10C10. The exceptional depth of the binding pocket, however, will allows us to change specificity and affinity using the van der Waals volume exclusion principle. For example, reducing the size of the pocket will reduce binding of larger PAHs.

Genetic origins of PAH binding site structure

We have identified the germline V genes for 4D5 and 10C10 to be the murine kappa light chain gene V_K8-21 and J_K1 for V_L and $V_H7183.10$, DSP2.2, and J_H4 for V_H (Figure 4). Almost all features of the deep binding pocket of 4D5 are encoded by the germline genes except ValH47, which is a somatic mutation of LeuH47 encoded by $V_H7183.10$ (33). 10C10 has LeuH47, reinforcing our confidence that $V_H7183.10$ is the germline gene encoding both 4D5 and 10C10. ArgH95, the first residue of CDRH3, is another common characteristic of 4D5 and 10C10 and is encoded during germline recombination by a two nucleotide addition (GG) at the junction between $V_H7183.10$ and DSP2.2 segments.

Germline genes V_K8-21 and $V_H7183.10$ are seldom observed in B cells. The low usage of V_K8-21 and $V_H7183.10$ may be due to unique features of the

```
1                   5                        10                       15                       20
Asp Ile Val Met Ser Gln Ser Pro Ser Ser Leu Ala Val Ser Ala Gly Glu Lys Val Thr Met Ser Cys
GAC ATT GTG ATG TCA CAG TCT CCA TCC TCC CTG GCT GTG TCA GCA GGA GAG AAG GTC ACT ATG AGC TGC
--G C-C --- --- A-- --- --- --- --- --- --- --- --- --- --- --- --- --- --- --- --- --- ---
--G C-C --- --- A-- --- --- --- --- --- --- --- --- --- --- --- --- --- --- --- --- --- ---

         25              27A 27B 27C 27D 27E 27F      30                  35                  40
Lys Ser Ser Gln Ser Leu Leu Asn Ser Arg Thr Arg Lys Asn Tyr Leu Ala Trp Tyr Gln Gln Lys Pro
AAA TCC AGT CAG AGT CTG CTC AAC AGT AGA ACC CGA AAG AAC TAC TTG GCT TGG TAC CAG CAG AAA CCA
--- --- --- --- --- --- --- --- --- -C- T-G --- --- --- --- --- --- --- --- --- --- --- ---
-TG --- --- --- -C- --- --- --- --- --- --- --- --- --- --- --- --- --- --- --- --- --- --G

         45              50                  55                  60
Gly Gln Ser Pro Lys Leu Leu Ile Tyr Trp Ala Ser Thr Arg Glu Ser Gly Val Pro Asp Arg Phe Thr
GGG CAG TCT CCT AAA CTG CTG ATC TAC TGG GCA TCC ACT AGG GAA TCT GGG GTC CCT GAT CGC TTC ACA
--- --- --- --- --- --A --- --- --- --- --- --- --- --- --- --- --- --- --- --- --- --- ---
--- --- --- --- -C- T-- --- --- --- --- --- --- --- A-- --- --- --- --- -C- --- --- --- ---

     65              70                  75                  80                  85
Gly Ser Gly Ser Gly Thr Asp Phe Thr Leu Thr Ile Ser Ser Val Gln Ala Glu Asp Leu Ala Val Tyr
GGC AGT GGA TCT GGG ACA GAT TTC ACT CTC ACC ATC AGC AGT GTG CAG GCT GAA GAC CTG GCA GTT TAT
--- --- --- --- --- --- --- --- --- --- --- --- --- --- --- --- --- --- --- --- --- --- ---
--- --- --- --- --- --- --- --- --- --- --- --- --- --- G-- --- --- --- --- --- --- --- ---

         90              95
Tyr Cys Lys Gln Ser Tyr Asn Leu Pro Thr (VK8-21)
TAC TGC AAG CAA TCT TAT AAT CTT CCC ACA (VK8-21)
--- --- --- --- --- --- --- T-- --- --G (10C10)
--- --- --- --- --- --- --- T-- --- --G (4D5)


1               5                        10                       15                       20
                                             Val Lys Pro Gly Gly Ser Leu Lys Leu Ser Cys
~~~ ~~~ ~~~ ~~~ ~~~ ~~~ ~~~ ~~~ ~~~ ~~~ ~~~ ~~~ GTG AAG CCT GGA GGG TCC CTG AAA CTC TCC TGT
CAG GTG AAA CTG CTC GAG GAG TCT GGG GGA GGC TTA --- --- --- --- --- --- --- --- --- --- ---
--- --- --- --- --- --- --- --- --- --- --- --- --A --- --- --- --- --- --- --- --- --- ---

         25                  30                  35                  40                  45
Ala Ala Ser Gly Phe Thr Phe Ser Ser Tyr Tyr Met Ser Trp Val Arg Gln Thr Pro Glu Lys Arg Leu
GCA GCC TCT GGA TTC ACT TTC AGT AGC TAT TAC ATG TCT TGG GTT CGC CAG ACT CCA GAG AAG AGG CTG
--- --- --- --T --- --- --- --- --- --- --- -T- --- --- --- --- --- --- --- --- --- --- ---
--- --- --- --A --- --- --- --- --- --- --- -T- --- --- --- --- --- --- --- --- --- --- ---

         50              52A     55                  60                  65
Glu Leu Val Ala Ala Ile Asn Ser Asn Gly Gly Ser Thr Tyr Tyr Pro Asp Thr Val Lys Gly Arg Phe
GAG TTG GTC GCA GCC ATT AAT AGT AAT GGT GGT AGC ACC TAC TAT CCA GAC ACT GTG AAG GGC CGA TTC
--A --- --- --- -GT --- G-- --- -G- --- TA- --A --- --- --- T-- --- --- --- --- --- --- ---
--A --- --- --- -GC --- A-C --- -A- --- GG- --T --- --- --- A-- --- --- --- --- --- --- ---

     70                  75                  80          82A 82B 82C      85
Thr Ile Ser Arg Asp Asn Ala Lys Asn Thr Leu Tyr Leu Gln Met Ser Ser Leu Lys Ser Glu Asp Thr
ACC ATC TCC AGA GAC AAT GCC AAG AAC ACC CTG TAC CTG CAA ATG AGC AGT CTG AAG TCT GAG GAC ACA
--- --- --- --- A-- --- --- --- --- --- --T --T --- --- --- G-- --- G-- --- -GC --- ---
--- --- --- --- G-- --- --- --- --- --- -TC --G --- --- --- A-- --- AG- --- -AG --- ---

     90              95
Ala Leu Tyr Tyr Cys Ala Arg
GCC TTG TAT TAC TGT GCA AGA (VH7183.10)
--- --- --- --- --- -G- --- (10C10)
--- --- --- --- --- -C- --- (4D5)
```

Figure 4: Germline V genes (V_L - top, V_H - bottom) of 4D5 and 10C10

resulting peptide chains (short CDRL3 for V_L; Leu instead of Trp at position H47 in V_H), which in turn may cause poor association with products from other germline genes. The use of these two genes in 4D5 and 10C10 indicates an intense selection by BaP (*33*). In the human germline genes, all V_H genes encode for Trp at H47 and no V_L gene encodes for a nine-residue CDRL3. The absence of these critical features could explain our lack of success in extracting new antibodies against BaP from human phage display libraries. Because CDR maturation techniques applied to these libraries cannot insert or delete residues nor act on framework residues such as H47, it appears that a design strategy based on 4D5 and 10C10 is the best way to engineer antibodies with new specificities and selectivities to various PAHs. Initial mutations will focus on residues present in the germline genes such as introducing the longer MetH100 and altering the H47 side chain to change the depth and size of the binding pocket.

PAH-Binding Antibodies from Combinatorial Libraries

Several research groups worldwide have developed phage display antibody libraries with a diversity orders of magnitude greater than that of the murine and human immune systems. Antibodies with useful affinities and selectivities have been successfully recovered from these libraries. The semi-synthetic human Fab 2LOX library, developed by Winter and co-workers at the Centre for Protein Engineering in Cambridge, England, contains on the order of 10^{12} randomly paired V_H and V_L sequences (*59*). We panned this library on magnetic microbeads or immunoassay microplate wells coated with BSA or cytochrome c conjugates of benzo[*a*]pyrene-6a or 6b, and fluoranthene-3 haptens. After four rounds of phage panning and amplification, phage that bound the haptens were enriched by about 10^6-fold. However, when selected clones were expressed as soluble rFabs, none bound soluble BaP or fluoranthene competitively. Analysis of the library and selected clones by PCR showed that all had V_H sequences, but only about 10% also had a V_L sequence. In earlier experiments, we observed that some phage displaying only a V_H domain nonspecifically bound to BaP and phenanthrene haptens (C.W. Bell and A.E. Karu, unpublished). Consequently, we subtractively panned the library to remove phage that lacked an L chain and phage that bound to plastic or biotin (*60*). The remaining population was then selected on PAH hapten conjugates. Again, no hapten-specific rFabs were recovered, and work with this library was discontinued.

A similar effort was made to derive scFvs that bound naphthalene or phenanthrene haptens from the Nissim library, which consists of approximately 10^8 different human V_H sequences linked to a single Vλ3 light chain from an antibody to BSA (*36*). Two scFvs that bound naphthalene haptens and two that bound phenanthrene haptens were recovered. All of these scFvs had high background binding to the carrier proteins in indirect EIAs, presumably because of the V_L sequence. Only one of the naphthalene-hapten-binding scFvs competitively bound free naphthalene, with I_{50} values of 150-600 ppb in a direct EIA format.

Structural models of the Fv domains of the naphthalene- and phenanthrene-binding scFvs were constructed and compared with the models of the murine 4D5 and 10C10 Fvs. Although the scFvs have much shallower binding pockets, they also have positively charged side chains flanking the binding pockets (Pellequer et al., in preparation).

Significance and Implications

This work was undertaken to alleviate inherent problems of cross-reactivity in anti-PAH antibodies produced by classical immunological methods. PAHs are planar hydrophobic compounds that do not engage in hydrogen bonds, making sub-class distinction a very challenging task. Our strategy was two-fold: a combinatorial approach using phage display libraries to obtain new antibodies, and a rational design approach using computational modeling to determine structural and physical properties of the binding sites of two monoclonal antibodies raised against BaP that in turn could be used for engineering new antibodies. Although, we have not yet made engineered antibodies with improved affinity or selectivity, we have discovered unique binding properties of 4D5 and 10C10 antibodies that indicate how to design antibodies with new specificities. The models of the 4D5 and 10C10 Fvs reveal a novel and unusual π-cation interaction between BaP and a positively charged Arg side chain located on the flank of the binding sites. Similar interactions may occur in the scFv antibodies that bind naphthalene and phenanthrene, suggesting a general protein motif for binding polyaromatic ring systems. These findings pose intriguing possibilities for how PAHs may be bound by molecular signaling proteins such as the aryl hydrocarbon receptor, BaP-binding proteins (*61,62*), and PAH-metabolizing enzymes important for bioremediation.

One conclusion from this study underscores the importance of understanding binding-site architecture for the design and use of combinatorial antibody phage display libraries. It is not sufficient to assume, based on probability alone, that alternative binding motifs for almost any analyte may be found in very large, diverse libraries. Other essential factors include binding-site differences for epitopes on proteins (cup-like), peptides (channel-like), and small haptens (deep cleft) (*63,64*), the possibility of a conformational change upon antigen binding (*47*), genetically conserved preferences among CDR loops (*65*), and differences in structure and expression of V genes from the murine and human repertoires (*66,67*). The latter suggests why the Fab 2LOX and Nissim libraries were not likely to yield desirable antibodies for PAHs. Virtually all known human V_H genes have a Trp, and none have Val or Leu, at position H47. Consequently, the human V gene germline repertoire may not include sequences that can form the unusually deep binding pocket required to bind BaP. By contrast, a library made by mutagenesis of 4D5 or 10C10 should be a good source of useful variants. Construction of this library is now under way in our laboratory.

Acknowledgments

We sincerely thank Christopher W. Bell, Tina Chin, and Yupin Cai for valuable advice and assistance throughout the project. This work was supported by DOE EMSP grant DE-FG07-96ER-62316. AEK was an investigator in the NIEHS Environmental Health Science Center at the University of California, Berkeley (Grant ES01896, B.N. Ames, Director).

Abbreviations.

BaP, benzo[a]pyrene; CDR, complementarity determining region; EIA, enzyme immunoassay; I50, the concentration of analyte that gives half-maximal inhibition in a competition assay; MAb, monoclonal antibody; PAH, polynuclear aromatic hydrocarbon; rFab, recombinant fragment containing disulfide-linked V_H-C_H1 and V_L-C_L polypeptides; V_H, variable region of the immunoglobulin heavy chain; C_H1, first constant domain of the heavy chain; V_L, variable region of the light chain; C_L, constant region of the light chain.

Literature Cited.

1. O'Connell, K. P.; Valdes, J. J.; Azer, N. L.; Schwartz, R. P.; Wright, J.; Eldefrawi, M. E. *J. Immunol. Meth.* **1999**, *225*, 157-169.
2. Abuknesha, R.; Brecht, A. *Biosensors & Bioelectronics* **1997**, *12*, 159-160.
3. Self, C. H.; Cook, D. B. *Curr. Opin. Biotechnol.* **1996**, *7*, 60-65.
4. U.S. Dept. of Energy *DOE Methods for Evaluating Environmental and Waste Management Samples (Chapter 8) Method OS060: Immunoassay for Petroleum Fuel Hydrocarbons in Soil*; DOE Methods online, 1997; Vol. 1999.
5. U.S. EPA Office of Solid Waste *SW-846 On-line Test Methods for Evaluating Solid Waste: Physical/Chemical Methods/4000 Series*; U.S. EPA Office of Solid Waste, 1998; Vol. 1999.
6. U.S. EPA Office of solid waste and emergency response (OSWER) *Soil screening for petroleum hydrocarbons by immunoassay (Method 4030)*; 1996; Vol. 1999.
7. U.S. EPA Office of solid waste and emergency response (OSWER) *Soil screening for petroleum hydrocarbons by immunoassay (Method 4035)*; U.S. EPA Region I Office of Environmental Measurement and Evaluation, 1996; Vol. 1999.

414

8. U.S. EPA Office of solid waste and emergency response (OSWER) *Immunoassay (Method 4000)*; U.S. EPA Region I Office of Environmental Measurement and Evaluation, 1996; Vol. 1999.
9. U.S. EPA Region I - New England Quality Assurance Unit Staff *Immunoassay guidelines for planning environmental projects*; U.S. EPA Region I Office of Environmental Measurement and Evaluation: New England, 1996.
10. Brecht, A.; Abuknesha, R. *Trends Analyt. Chem.* **1995**, *14*, 361-371.
11. Xie, B.; Danielsson, B. *Analyt. Lett.* **1996**, *29*, 1921-1932.
12. Lenigk, R.; Zhu, H.; Lo, T.-C.; Renneberg, R. *Fresenius' J. Analyt. Chem.* **1999**, *364*, 66-71.
13. Cornell, B. A.; Braach-Maksvytis, V. L. B.; King, L. G.; Osman, P. D. J.; Raguse, B.; Wieczorek, L.; Pace, R. J. *Nature* **1997**, *387*, 580-583.
14. Penalva, J.; Gonzalez-Martinez, M. A.; Puchades, R.; Maquieira, A.; Marco, M. P.; Barcelo, D. *Analyt. Chim. Acta* **1999**, *387*, 227-233.
15. Gonzalez-Martinez, M. A.; Puchades, R.; Maquieira, A.; Ferrer, I.; Marco, M. P.; Barcelo, D. *Analyt. Chim. Acta* **1999**, *386*, 201-210.
16. Su, X.; Chew, F. T.; Li, S. F. Y. *Anal. Biochem.* **1999**, *273*, 66-72.
17. Zeravik, J.; Skladal, P. *Electroanalysis* **1999**, *11*, 851-856.
18. Lee, K. S.; Kim, T.-H.; Shin, M.-C.; Lee, W.-Y.; Park, J.-K. *Analyt. Chim. Acta* **1999**, *380*, 17-26.
19. Ekins, R.; Chu, F. *J. Int. Fed. Clinic. Chem.* **1997**, *9*, 100-9.
20. Healey, B. G.; Li, L.; Walt, D. R. *Biosensors and Bioelectronics* **1997**, *12*, 521-529.
21. Fare, T. L.; Cabelli, M. D.; Dallas, S. M.; Herzog, D. P. *Biosensors and Bioelectronics* **1998**, *13*, 459-70.
22. Zhao, C. Q.; Anis, N. A.; Rogers, K. R.; Klines, R. H.; Wright, J.; Eldefrawi, A. T.; Eldefrawi, M. E. *J. Agri. Food Chem.* **1995**, *43*, 2308-2315.
23. Pritchard, D. J.; Morgan, H.; Cooper, J. M. *Analyt. Chim. Acta* **1995**, *310*, 251-256.
24. Narang, U.; Gauger, P. R.; Kusterbeck, A. W.; Ligler, F. S. *Anal. Biochem.* **1998**, *255*, 13-19.
25. Kravec, C. V.; Ghoshal, M.; Rashid, F.; Rashid, S.; Talbot, L. A.; Towt, J.; Tsai, S. C. J.; Wu, R. S.; Salamone, S. J. *Clinic. Chem.* **1997**, *43*, S204-S205.
26. Tierney, B.; Benson, A.; Garner, R. C. *J. Natl. Cancer Inst.* **1986**, *77*, 261-267.
27. Liang, X. L.; Lubman, D. M.; Rossi, D. T.; Nordblom, G. D.; Barksdale, C. M. *Anal. Chem.* **1998**, *70*, 498-503.
28. Cichna, M.; Markl, P.; Knopp, D.; Niessner, R. *Chem. Mat.* **1997**, *9*, 2640-2646.
29. Cichna, M.; Knopp, D.; Niessner, R. *Analyt. Chim. Acta* **1997**, *339*, 241-250.
30. Bentsen-Farmen, R. K.; Botnen, I. V.; Noto, H.; Jacob, J.; Ovrebo, S. *International Arch. Occupat. Environ. Health* **1999**, *72*, 161-168.
31. Gomes, M.; Santella, R. M. *Chem. Res. Toxicol.* **1990**, *3*, 307-310.
32. Bell, C. W.; Li, K.; Zhao, B.; Li, Q. X.; Karu, A. E. *submitted* **2000**.
33. Pellequer, J. L.; Zhao, B.; Kao, H. I.; Bell, C. W.; Li, K.; Li, Q. X.; Karu, A. E.; Roberts, V. A. *Submitted* **2000**.

34. Li, K.; Chen, R.; Zhao, B.; Liu, M.; Karu, A. E.; Roberts, V. A.; Li, Q. X. *Analyt. Chem.* **1999**, *71*, 302-309.

35. Griffiths, A. D.; Williams, S. C.; Hartley, O.; Tomlinson, I. M.; Waterhouse, P. et al. *EMBO J.* **1994**, *13*, 3245-3260.

36. Nissim, A.; Hoogenboom, H. R.; Tomlinson, I. M.; Flynn, G.; Midgley, C.; Lane, D.; Winter, G. *EMBO J.* **1994**, *13*, 692-698.

37. Frisch, M. J.; Trucks, G. W.; Schlegel, H. B.; Gill, P. M. W.; Johnson, B. G.et al. *Gaussian 94*; Gaussian, Inc: Pittsburgh, 1995.

38. Nicholls, A.; Sharp, K. A.; Honig, B. *Proteins* **1991**, *11*, 281-296.

39. Roberts, V. A.; Iverson, B. L.; Iverson, S. A.; Benkovic, S. J.; Lerner, R. A.; Getzoff, E. D.; Tainer, J. A. *Proc. Natl. Acad. Sci. USA* **1990**, *87*, 6654-6658.

40. Roberts, V. A.; Stewart, J.; Benkovic, S. J.; Getzoff, E. D. *J. Mol. Biol.* **1994**, *235*, 1098-1116.

41. Abola, E. E.; Sussman, J. L.; Prilusky, J.; Manning, N. O. *Meth. Enzymol.* **1997**, *277*, 556-571.

42. McRee, D. E. *J. Mol. Graph.* **1992**, *10*, 44-47.

43. Brünger, A. T. *X-PLOR Manual. Version 3.0*; Yale University: New Haven, 1992.

44. Brooks, B.; Bruccoleri, R.; Olafson, B.; States, D.; Swaminathan, S.; Karplus, M. *J. Comp. Chem.* **1983**, *4*, 187-217.

45. Roussel, A.; Cambillau, C. *TURBO-FRODO*; In Silicon Graphics: Mountain View, California, 1989, pp 77-78.

46. Laskowski, R. A.; MacArthur, M. W.; Moss, D. S.; Thornton, J. M. *J. Appl. Cryst.* **1993**, *26*, 283-291.

47. Rini, J. M.; Schulze-Gahmen, U.; Wilson, I. A. *Science* **1992**, *255*, 959-965.

48. Sheriff, S.; Silverton, E. W.; Padlan, E. A.; Cohen, G. H.; Smith-Gill, S.; Finzel, B. C.; Davies, D. R. *Proc. Natl. Acad. Sci.USA* **1987**, *84*, 8075-8079.

49. Morea, V.; Tramontano, A.; Rustici, M.; Chothia, C.; Lesk, A. M. *J. Mol. Biol.* **1998**, *275*, 269-294.

50. Shoham, M. *J. Mol. Biol.* **1993**, *232*, 1169-1175.

51. Hutchinson, E. G.; Thornton, J. M. *Protein Sci.* **1994**, *3*, 2207-2216.

52. EU 3-D Validation Network *J. Mol. Biol.* **1998**, *276*, 417-436.

53. Che, Z.; Olson, N. H.; Leippe, D.; Lee, W.-m.; Mosser, A. G.; Rueckert, R. R.; Baker, T. S.; Smith, T. J. *J. Virol.* **1998**, *72*, 4610-4622.

54. Morris, G. M.; Goodsell, D. S.; Huey, R.; Olson, A. J. *J. Comput.-Aided Mol. Design* **1996**, *10*, 293-304.

55. Luhmer, M.; Bartik, K.; Dejaegere, A.; Bovy, P.; Reisse, J. *Bull. Soc. Chim. Fr.* **1994**, *131*, 603-606.

56. Flocco, M. M.; Mowbray, S. L. *J. Mol. Biol.* **1994**, *235*, 709-717.

57. Mitchell, J. B. O.; Nandi, C. L.; McDonald, I. K.; Thornton, J. M. *J. Mol. Biol.* **1994**, *239*, 315-331.

58. Dougherty, D. A. *Science* **1996**, *271*, 163-168.

59. Medical Research Council (Cambridge England) Centre for Protein Engineering *Instructions for Use of the Human Synthetic Fab (4-22) 2Lox Library*, 1995.

60. Adey, N. B.; Mataragnon, A. H.; Rider, J. E.; Carter, J. M.; Kay, B. H. *Gene* **1995**, *156*, 27-31.
61. Lesca, P.; Pineau, T.; Galtier, P.; Peryt, B.; Derancourt, J. *Biochem. Biophys. Res. Commun.* **1998**, *242*, 26-31.
62. Ogawa, H.; Gomi, T.; Imamura, T.; Kobayashi, M.; Huh, N. *Biochem. Biophys. Res. Commun.* **1997**, *233*, 300-4.
63. Lesk, A. M.; Tramontano, A. *Antibody structure and structural predictions useful in guiding antibody engineering*; Lesk, A. M.; Tramontano, A., Ed.; W.H. Freeman Co.: New York, 1992, pp 1-38.
64. Padlan, E. A. *Molec. Immunol.* **1994**, *31*, 169-217.
65. Vargas-Madrazo, E.; Lara-Ochoa, F.; Almagro, J. C. *J. Mol. Biol.* **1995**, *254*, 497-504.
66. Lara-Ochoa, F.; Almagro, J. C.; Vargas-Madrazo, E.; Conrad, M. *J. Mol. Evol.* **1996**, *43*, 678-84.
67. Almagro, J. C.; Hernandez, I.; del Carmen Ramirez, M.; Vargas-Madrazo, E. *Molec. Immunol.* **1997**, *34*, 1199-214.
68. Kraulis, P. J. *J. Appl. Cryst.* **1991**, *24*, 946-950.
69. Merritt, E. A.; Bacon, D. J. *Meth. Enzymol.* **1997**, *277*, 505-524.

Chapter 25

Genotypic Influence on Metal Ion Mobilization and Sequestration via Metal Ion Ligand Production by Wheat

Teresa W.-M. Fan[1], Fabienne Baraud[1], and Richard M. Higashi[2]

[1]Department of Land, Air and Water Resources, and [2]Crocker Nuclear Laboratory, University of California, One Shields Avenue, Davis, CA 95616–8627

Intracellular production and root exudation of metal ion ligands (MIL) are keys to metal ion mobilization and sequestration by vascular plants. This is well-exemplified by the exudation of phytosiderophores such as mugineic acid and derivatives in the acquisition of Fe (III) and Zn (II) by graminaceous plants. However, the genetic and biochemical mechanism(s) for the mobilization and sequestration of most pollutant metal ions remain unknown, due largely to a lack of knowledge in MIL involved in the process. This information is critically needed for developing and engineering plant-based remediation of metal contamination at DOE, DOD, and other industrial facilities. Here, we investigated the influence of elevated metal (Zn, Cu, Ni, Mn, and Cd) ion treatment on root exudation, tissue MIL profiles, and metal accumulation in Chinese spring (CS) wheat, its 7 genotypes prepared from disomic addition of the wheatgrass chromosomes to CS, and the amphiploid between the two species. Broad-screen and structure elucidating nuclear magnetic resonance spectroscopy and gas chromatography-mass spectrometry were employed to obtain comprehensive profiles of MIL in root exudates and tissues, difficult to achieve with conventional approaches. In addition, thiol-rich peptide profiles were acquired using fluorescent bromobimane tagging and sodium dodecyl sulfate-polyacrylamide gel electrophoresis, while comprehensive metal ion profiles were obtained using X-ray fluorescence. Genotypic covariation was noted between metal ions (Mn, Fe, Cu, and Zn) and such MIL as 2'-deoxymugineic acid and acetate in root exudates and citrate and malate in roots. Profiles of phytochelatin-like peptides also displayed genotypic variations and strong correlation with Cd accumulation. These results suggest possible chromosomal location(s) of gene(s) governing metal ion mobilization and sequestration in wheat.

Introduction

Vascular plants are known to utilize two different strategies to mobilize Fe(III) from soils (*1*). The so-called Strategy II plants (graminaceous monocots) release powerful iron chelators into root exudates to complex Fe(III) for uptake. These chelators, which are termed phytosiderophores (PS) when applied strictly to Fe acquistion, consist principally of mugineic acid and its derivatives, 2'-deoxymugineic acid (2'-DMA) and 3-*epi*-hydroxymugineic acid (3-*epi*-DMA) (*2*). In addition to Fe(III), PS also complex with other micronutrients such as Zn(II) and Cu(II) to facilitate their uptake by plants (*3,4*) However, it is unclear whether PS are involved in the mobilization and uptake of pollutant metal ions (e.g. Cd(II), Pb(II), Sr(II), Cs(II)) commonly found in contaminated soils including those at DOE facilities. Also uncertain is the involvement of exudate components other than PS in metal ion mobilization, for lack of a comprehensive knowldege of root exudate composition. Moreover, once absorbed, it is largely unknown how these pollutant ions (usually toxic to plants) are translocated to shoots or sequestered inside root cells. Insights into these questions are crucial to the design and implementation of plant-based remediation of metal contamination from soils and sediments.

Without prior knowledge of chemical composition, analysis of crude mixtures such as plant root exudates requires a profiling approach, for which we developed a combined nuclear magnetic resonance (NMR) spectroscopy and gas chromatography-mass spectrometry (GC-MS) method (*5*). We have used this approach for simultaneous determination of known, unexpected, and even unknown metal ion ligands (MIL), in addition to PS, directly from crude exudates (*5*). With this tool in hand, it is now possible to examine environmental and genetic factors influencing intracellular MIL and root exudation profiles, and their role in metal ion acquisition. Among known PS-releasing plants, wheat ranks as one of the highest in terms of PS production (*6*). The principal PS released by common wheat is 2'-DMA which has a very high stability constant ($K_d \approx 10^{33}$) towards Fe(III) (*6,7*). It seems reasonable to expect that some of the wheat relatives, in particular those inhabiting Fe-deficient environments, may have evolved even higher PS production. In the present study, we exploited the wealth of genetic stocks that have been developed for *Lophopyrum elongatum* (Host) Love, a diploid wheatgrass closely related to wheat, for study of the production of 2'-DMA and possibly other MIL, and in turn the consequences of such production on metal ion uptake. With these genetic stocks, we examined the influence of each chromosome of this species on root exudation, metabolite profiles, and metal accumulation under normal and elevated metal treatments. Covariation of 2'-DMA, thiol-rich peptides, and other MIL with acccumulation of Mn, Fe, Cu, Zn, and Cd by these genetic stocks made it possible to acquire insights into relationships between these compounds and the uptake and sequestration of specific heavy metals and potential chromosomal location of genes regulating these processes.

Materials and Methods

Plant growth. An octaploid amphiploid with three genome pairs of the hexaploid *Triticum aestivum* L. cv. Chinese Spring (2n = 42, genomes AABBDD) and one pair of genomes of dipoid wheatgrass, *Lophopyrum elongatum* (Host) Love (2n = 14, genomes EE) was produced by Rommel and Jenkins (1959) (*8*) and supplied by J. Dvorak, Department of Agronomy, University of California, Davis. In addition to the amphiploid and Chinese Spring, a complete set of disomic addition lines of the *L. elongatum* chromosomes in Chinese Spring were investigated. The lines were produced by backcrossing the amphiploid to Chinese Spring and ultimately selecting plants with each of the seven chromosome pairs of *L. elongatum* added to the chromosome complement of Chinese Spring (*9-11*). These genetic stocks were designated here as follows: CS, the parental bread wheat; AgCS, the amphiploid; DA1E, DA2E, DA3E, DA4E, DA5E, DA6E, and DA7E, disomic addition lines with added chromosomes 1E, 2E, 3E, 4E, 5E, 6E, and 7E, respectively.

Seeds of the 8 genotypes plus CS were sterilized in 1:10 diluted commercial bleach, germinated in 0.5 mM $CaSO_4$ on sterile germination paper at room temperature for 3-4 days, transplanted to six 10-L black polyethylene containers with 3 genotypes per container, and grown in one-half strength Hoagland solution as modified from Epstein (1972) (*12*). The seedlings were aerated and maintained in a growth chamber (Sanyo) with a cycle of 16/8h light/dark periods, day/night temperature of 25/19°C, and humidity regulated at 70%. The pH of the nutrient solution was adjusted to 6 daily.

On day 10, metal treatment was initiated on one set of 9 genotypes (in 3 containers) by changing the nutriet solution to one-half strength Hoagland solution plus elevated levels of (in mg/L) of Mn (0.275), Ni (0.15), Cu (0.08), Zn (0.33) plus Cd (at 1 mg/L as $CdSO_4$) while the Fe concentration remained unchanged. The other set was unchanged and used as control. On day 20, the plants were harvested, separated into roots and shoots, lyophilized, and dry weight determined. The lyophilized tissues were pulverized into < 3 μm particles and stored at −70°C before analysis.

Root exudate collection. Three days before harvest, root exudates were collected 3 times each starting 2 h after the onset of the photoperiod, as previously described (*5*). The three collections were combined, lyophilized, and dry weight determined.

Tissue metal contents. Total shoot and root metal contents were determined by energy dispersive X-ray fluorescence spectroscopy (ED-XRF, JVAR Inc. EX3600 spectrometer). Tomato leaf tissues, including #1573a from the National Institute of Standards and Technology, spiked with appropriate concentrations of relevant elements were used as calibration standards. The element concentrations of selected wheat tissues were also analyzed by ICP-AES after $HClO_4/HNO_3$ (1:5) digestion to confirm the XRF analyses.

SH-rich proteins and peptides. Sulfhydryl (SH)-rich proteins and peptides from wheat tissues were extracted and derivatized simultaneously using a method modified from Fahey and Newton, 1987 (*13*), and subjected to SDS-PAGE (polyacrylamide gel electrophoresis) (modified from ref *14*). After PAGE, the gel was immediately fixed in 40% methanol plus 10% acetic acid and fluorescence-imaged using 365nm-excitation. Loss of peptides of ≤1 kDa would occur with longer soaking in the fixing solution. After fluorescence imaging, the gel was stained in Coomassie blue G-250 (BioRad protocol) for comparison with the fluorescence pattern.

^{1}H NMR and GC-MS of root metabolites and exudates. Root metabolites were extracted from pulverized root tissues and analyzed by GC-MS as described previously (*5,15,16*). The lyophilized exudates were redissolved in 1 mL of DDI water and subjected to both GC-MS and ^{1}H NMR analysis as previously described (*5*), except that the ^{1}H NMR was performed at 11.75 Tesla on a Varian Unity Plus spectrometer housed in the DOE Environmental Molecular Science Laboratory (EMSL), Richland, WA. Peak assignments were based on previously assigned standards (*5,17*). Acetate, 2'-DMA, glycinebetaine (GB), and an unknown compound in root exudates were quantified from the 1-D ^{1}H NMR spectra (cf. Figure 2) based on the intensity of peaks at 1.91, 4.09, 3.26, and 3.35 ppm, respectively, against that of the CH_3 resonance at 1.47 ppm of an external Ala standard (*5,18*). The quantification for the unknown was only tentative and the number of protons represented by the peak at 3.35 ppm was assumed to be 2 based on the methylene assignment for this peak.

Results and Discussion

Biomass production.

Under control conditions, the parent CS had the highest root and shoot biomass while DA2E, DA3E, and AgCS were among the lowest (data not shown). Under elevated metal treatment, the dry weight of both shoots and roots decreased for all wheat genotypes relative to the respective control, with the reduction ranging from 23 to 67% (data not shown). The biomass of DA3E, DA4E, DA6E, and DA7E was less impacted than that of AgCS and DA2E.

Biomass reduction of a similar extent was observed with Cd treatment alone (Fan and Higashi, unpublished), so the reduction observed here was presumably due to Cd toxicity while other metal ions (i.e. Mn, Cu, Ni, and Zn) may have had only minor effects.

Compared to the CS parent, DA4E, DA6E, and DA7E also exhibited less metal-induced reduction in both root and shoot biomass, suggesting that addition of these *L. elongatum* chromosomes to CS confers some degree of tolerance to the metal treatment. The % change in the shoot/root dry weight ratio between treated and control plants differed among genotypes. DA1E to DA6E showed a negative change while DA7E, AgCS, and CS exhibited a positive change (data not shown), suggesting that genetic influence on shoot/root ratio may be altered by chromosomal additions of 1E to 6E.

Tissue metal profiles.

A notable genotypic variation in the Mn, Fe, and Cu content in both control and metal-treated roots was observed, with variation for Zn and Ni less pronounced (Figure 1A). In addition, metal treatment elicited a large increase in the concentrations of Zn, Fe, Cu, and Ni but a slight reduction for Mn content in roots of all lines, with Zn showing the highest increase (up to 10 fold) (Figure 1A). This level of Zn accumulation was disproportionately greater than the Zn elevation in the treatment, which could be related to the co-treatment with Cd. Cd treatment alone caused a > 2-fold increase in Zn uptake into wheat roots under control conditions (Fan and Higashi, unpublished data). Since Zn and Cd are both Group IIB elements, it is possible that they share a similar uptake mechanism.

At 1 mg/L, Cd treatment led to 80-200 fold accumulation of Cd in the roots of the 9 lines, with AgCS and DA3E showing the highest extent of accumulation (Figure 1A). This level of accumulation argues for a selective uptake mechanism for Cd, while the lack of enhanced Mn accumulation suggests feedback control for Mn uptake. Although DA2E, DA3E, and the amphiploid had regularly higher root concentrations of Mn, Fe, and Zn, these genotypes also had the lowest biomass production, which may be a consequence of accumulating "toxic" levels of these metal ions. More importantly, the result suggests that the gene(s) governing the uptake of these metal ions may reside on L. elongatum chromosomes 2E and 3E.

In contrast to roots, Mn, instead of Fe, was the most concentrated metal ion in shoots of all genotypes under control growth (Figure 1B). This suggests a higher translocability of Mn in wheat. Shoots of all genotypes also exhibited variations in their metal content, but to a lesser extent than roots, except for Ni and Cu under elevated metal conditions. In particular, DA3E had an unusually high Ni content in the treated shoot relative to the control shoot while DA5E showed a similar behavior for Cu. These results suggest that the mechanism(s) for Ni and Cu translocation to shoots under elevated metal conditions may be altered by the addition of chromosomes 3E and 5E, respectively.

It is also interesting to note that relative to control conditions, little accumulation or even suppression of shoot content in Mn, Fe, and Cu was

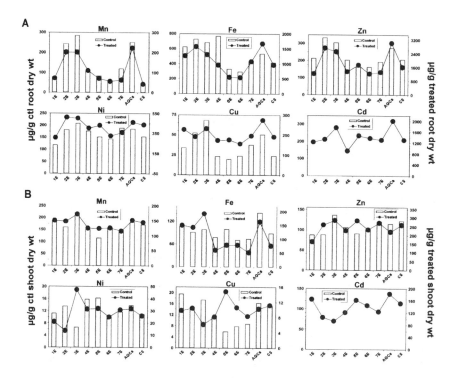

Figure 1. Genotypic variations in wheat tissue metal content under control and elevated metal treatments. Metal contents of roots (A) and shoots (B) were acquired using ED-XRF as described in Materials and Methods. Each graph represents results for an element. The gray bar (left ordinate) and black circle (right ordinate) respectively denote control and metal-treated tissues, and the abscissa for each graph are the nine genotypes 1E, 2E, 3E, 4E, 5E, 6E, 7E, AgCS, and CS.

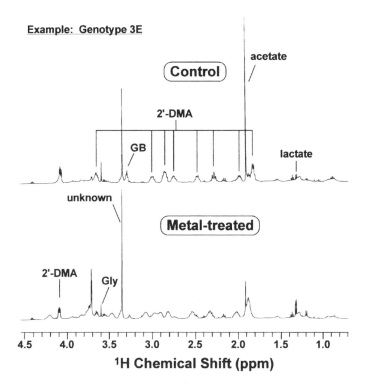

Figure 2. ^{1}H NMR profiles of example wheat root exudates from control and metal treatments. The ^{1}H NMR spectra were acquired from control and metal-treated DA3E as described in Materials and Methods. Besides the major components that are labeled, a number of less abundant components were confirmed from GC-MS analysis (data not shown) including alanine, valine, leucine, isoleucine, succinate, proline, fumarate, serine, threonine, phenylalanine, glutamate, and citrate.

424

observed for most genotypes under elevated metal conditions (Figure 1B). The lack of excess Mn accumulation in shoots presumably reflected the lack of excess uptake into roots (Figure 1A). However, this was not the case for Fe and Cu since a substantial accumulation of the two metal ions did occur in roots. It is possible that the Fe and Cu translocation to shoot was either disturbed by the treatment or highly regulated in wheat. Although a significant accumulation of Zn, Ni, and Cd was observed in shoots of all genotypes, the extent of the accumulation was much less than that in roots. This may be again related to the perturbation in translocation capacity and/or feedback control.

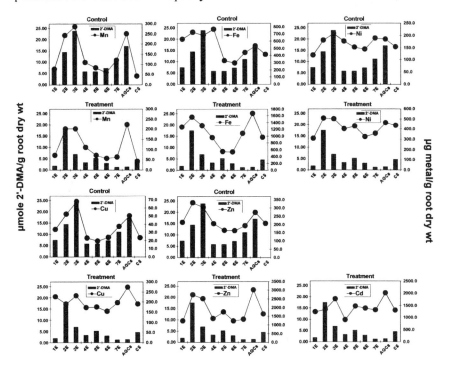

Figure 3A. Genotypic variations in 2'-DMA exudate (gray bars, left ordinate) and Mn, Fe, Ni, Cu, Zn, and Cd content (black circles, right ordinate) of control and metal-treated wheat roots. Exudate components and metals were quantified ass described in Materials and Methods. The abscissa for each graph are the nine genotypes 1E, 2E, 3E, 4E, 5E, 6E, 7E, AgCS, and CS.

Exudates production and their profiles.

Under control conditions, the parent CS produced the highest amount of root exudates per g root mass while DA2E, DA3E, and AgCS were among the lowest in exudation. This suggests that root exudation was perturbed by the addition of *L. elongatum* chromsomes 2E and 3E. In contrast, under metal treatment conditions, DA2E, DA3E, and AgCS produced higher amounts (50-70%) of exudates per g root mass than the parent CS. DA2E and AgCS also had the highest biomass reduction due to the metal treatment, which could be related to a resource diversion to root exudation, in addition to the toxic effect of metal accumulation (Figure 1A).

Analysis of the crude exudates by ^{1}H NMR revealed the abundance of components including the phytosiderophore 2'-deoxymugineic acid (2'-DMA), lactate, acetate, glycinebetaine (GB), glycine (Gly), and an unknown compound (Figure 2). Identification of these components by 2-D ^{1}H total correlation spectroscopy (TOCSY), ^{1}H-^{13}C heteronuclear single quantum coherence spectroscopy (HSQC), and heteronuclear multiple bond correlation spectroscopy (HMBC) (5) will be presented elsewhere.

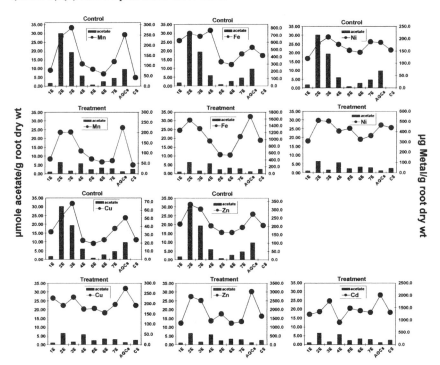

Figure 3B. Genotypic variations in acetate exudate (gray bars, left ordinate). All other aspects of this figure is identical to Figure 3A.

The concentrations of exudate components, 2'-DMA (Figure 3A), acetate (Figure 3B), and the unknown (Figure 3C), followed the same pattern across all lines as the root concentration of Mn, Fe, Cu, and Zn under control conditions.

DA2E, 3E and AgCS ranked among the highest for both exudate concentrations and metal content. This is in contrast to their low total exudation by dry weight (see above), which illustrates the necessity for determining specific exudate profile, not just total exudation, for a functional understanding of metal ion mobilization and sequestration. Since 2'-DMA is known to be involved in Fe and Zn uptake into wheat roots (*18*), such genotypic correlation suggests that the gene(s) involved in 2'-DMA production and thus Fe and Zn uptake may be located on *L. elongatum* chromosomes 2E and 3E. This is consistent with the earlier conclusion based on root metal content (see above). Whether the same case holds for acetate and the unknown awaits further investigation.

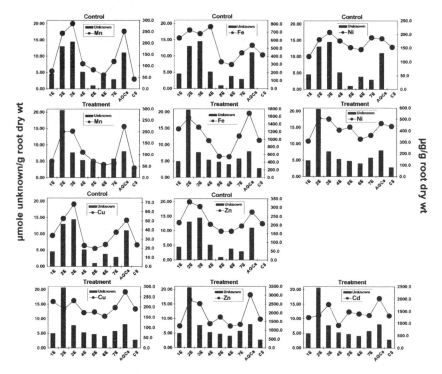

Figure 3C. Genotypic variations in the unidentified exudate (gray bars, left ordinate). All other aspects of this figure is identical to Figure 3A.

In addition, PS has been shown to mobilize Mn and Cu from soil and the MnPS and CuPS complexes are taken up by graminaceous roots albeit less preferentially than the uptake of FePS (*20*). Thus, the covariation of Mn and Cu with 2'-DMA concentrations was to be expected. However, the relationship for Ni and the three exudate components was ambiguous (e.g. Figure 3), which suggests that Ni may be absorbed by a mechanism different from that via these components. It should also be noted that the amplitude of genotypic responses in root metal content (Figures 1A and 3) may be attenuated to some extent, due to the accessibility of exudate components (therefore their metal ion complexes) by each set of three genotypes (DA1E-3E, DA4E-6E, and DA7E/AgCs/CS) grouped in the same container.

Compared to control conditions, the genotypic covariation for 2'-DMA, Mn, Fe, Zn, and Cu became much less discernable under metal treatment (Figure 3). This was also the case for Cd. It is possible that the excess free metal ions under the elevated metal treatment is readily available to plant roots, which in turn requires less assistance from exudation of MIL such as 2'-DMA. Consequently, metal uptake into roots becomes decorrelated with the level of exudation for these components. This account is also consistent with an overall reduction in the exudation of 2'-DMA, acetate, and the unknown induced by the metal treatment.

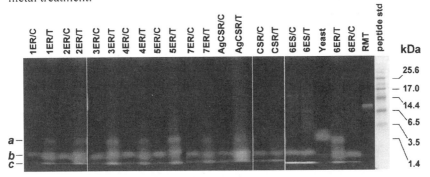

Figure 4. SDS-PAGE pattern of thiol-rich peptide extracts of wheat genotypes. The thiol-rich peptides from 9 wheat genotypes were extracted, derivatized, separated by electrophoresis, and imaged as described in Materials and Methods. R/C, R/T, S/C, and S/T denote root/control, root/metal treated, shoot/control, and shoot/metal treated, respectively. Rabbit metallothionein (RMT) and Cd-treated yeast (Saccharomyces pombe, obtained from Dr. D. Ow) were also derivatized and electrophoresed along with the wheat extracts and peptide standards (BioRad). The peptide standards were stained with Coomassie blue G-250. The dominant yeast band of > 1.4 kDa was presumably PC₄ (phytochelatin with chain length = 4) or PC with higher chain lengths (23). The bands from wheat are labeled as a, b, and c.

Intracellular MIL: SH-rich peptides and proteins.

It is well-recognized that thiol-rich peptides such as metallothioneins (MT) and phytochelatins (PC, also known as class III metallothioneins) are induced for intracellular metal sequestration, particularly for Cd (*21-23*). Through complexation with metal ions, these peptides help improve biological tolerance to metal accumulation. However, the analysis of these peptides has been relatively labor-intensive and limited in scope. Here, we present a preliminary method that combined bromobimane (BrB) fluorescent tagging with polyacrylamide gel electrophoresis to greatly simplify the analysis of all thiol-containing proteins and peptides. Figure 4 illustrates the fluorescence gel patterns of BrB-derivatized root and shoot extracts of different wheat genotypes, along with those of the rabbit MT and Cd-exposed yeast extract. At least three bands in the < 3.5 kDa region were observed in the yeast extract which may correspond to PC_2, PC_3, and PC_4, known to be produced under Cd treatment (22).

The gel patterns of wheat roots were distinctly different between control and metal treatments, with an additional band of approximately 1.4 kDa (a) appeared in all metal treated but not in control roots. This band had an apparent molecular weight similar to the calculated weight of BrB-derivatized PC_3. The two bands below the 1.4 kD band (b, c) also shows different intensity between control and treated roots. Moreover, there were genotypic differences in the band pattern, and in particular band b of metal-treated AgCS root (AgCSR/T) had an unusually high intensity. It appeared that these peptides were induced by the metal treatment and that they are likely to be phytochelatins. We are currently conducting further analysis of these peptides to confirm their identity. It is also interesting to note that wheat shoots (e.g. 6ES/C and 6ES/T) was very low in band a and that no significant difference in bands b and c was observed between control and metal treatments. This is correlated with the much lower level of metal (Fe, Ni, Cu, Zn, and Cd) accumulation in metal-treated shoots than roots. These results suggest that band a may be important in sequestering excess metal ions to reduce their toxicity in wheat roots. Moreover, the thiol-rich peptide pattern of wheat differed somewhat from that of yeast, where the yeast peptide pattern was more enriched in a higher molecular weight component than that of wheat. This could reflect a difference in the biosynthesis of these peptides.

Other MIL profiles in wheat roots.

In addition to the thiol-rich peptides, the root tissues were analyzed broadly for MIL by GC-MS and ¹H NMR. Figure 5 shows the genotypic profile of some of the MIL obtained from the GC-MS analysis along with the root Fe profile. The identity of these compounds was also confirmed by the ¹H NMR analysis (data not shown).

Under control conditions, no discernable relationship was observed between these MIL and Fe content across all lines. However, under elevated metal conditions, a genotypic covariation between Fe, citrate, and malate was evident while an overall negative correlation was noted between Fe and lysine for DA1E through DA6E (Figure 5). A similar trend also applied to Zn and Mn (cf. Figures

1A and 3). Moreover, the concentrations of these MIL increased many fold in response to the metal buildup in roots, except for that of asparagine in AgCS (the data for CS was missing due to lack of tissue for analysis). Among all lines, DA2E and DA3E demonstrated the highest accumulation of malate and citrate. Therefore, these MIL and the SH-rich peptides above may collectively participate in the binding of excess metal ions, presumably to reduce their toxicity to roots. The genetic component(s) involved in this response may be located on chromosomes 2E and 3E.

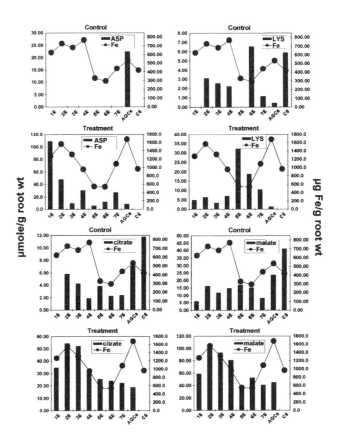

Figure 5. Genotypic variations in other metal ion ligands and Fe content of control and metal-treated wheat roots. Metal ion ligands (MIL) and other root metabolites were analyzed as described in the text. Gray bar and filled circle denote MIL and Fe contents, respectively. The abscissa for each graph are the nine genotypes 1E, 2E, 3E, 4E, 5E, 6E, 7E, AgCS, and CS.

In conclusion, a comprehensive screening for element and metabolites by XRF, ^{1}H NMR, GC-MS, and SDS-PAGE revealed significant differences in metal ion accumulation, root exudation of MIL, and buildup of endogenous MIL in roots among different genetic stocks of wheat. This approach provided clues to the biochemistry and genetic components involved in metal mobilization and sequestration as well as metabolic adaptation to metal toxicity in plants. Such fundamental knowledge should facilitate the design and efficacy of plant-based remediation of metal contamination.

Acknowledgement

This work was supported by the DOE/EMSP grant #DE-FG07-96ER20255) and EPA grant #R825960010. We are grateful to Dr. J. Dvorak for his comments on the manuscript, Dr. D. Ow for supplying the yeast culture, R. Kaufman for assistance in thiol-rich peptide analysis, and T.A. James for assistance on ED-XRF method development. We also wish to acknowledge Dr. Nancy Isern for her assistance with the NMR measurements and Environmental Molecular Science Laboratory (Richland, WA) for the NMR access.

References

1. Marschner, H. in Plant roots : the hidden half; Waisel, Y.; Eshel, A.; Kafkafi, U., Eds; M. Dekker: New York, **1991**, pp 503-526.
2. Kawai, S.; Takagi, S.; Sato, Y. J. Plant Nutr. **1988**, 11, 633-642.
3. Romheld, V. Plant and Soil **1991**, 130, 127-134.
4. Hopkins, B. G.; Whitney, D. A.; Lamond, R. E.; Jolley, V. D. J. Plant Nutr. **1998**, 21, 2623-2637.
5. Fan, T. W. M.; Lane, A. N.; Pedler, J.; Crowley, D.; Higashi, R. M. Analyt. Biochem. **1997**, 251, 57-68.
6. Takagi, W.-I. In Iron chelation in plants and soil microorganisms; Barton, L.; Hemming, B. C., Eds; Academic Press: San Diego, 1993, pp 111-131.
7. Nomoto, K.; Yoshioka, H.; Arima, M.; Fushiya, S.; Takagi, S.; Takemoto, T. Chimica **1981**, 35, 249-250.
8. Rommel, R.; Jenkins, B. C. Wheat Inf. Serv., **1959**, 9-10, 23.
9. Dvorak, J. Canad. J. Genet. Cytol. **1980**, 22, 237-259.
10. Dvorak, J.; Chen, K. C. Canad. J. Genet. Cytol. **1984**, 26, 128-132.
11. Tuleen, N. A.; Hart, G. E. Genome **1988**, 30, 519-524.
12. Epstein, E., Mineral Nutrition of Plants: Principles and Perspectives; John Wiley and Sons: New York, **1972**.
13. Fahey, R.C.; Newton, G.L. Methods in Enzymology 1987, 143, 85-96.
14. Schagger, H.; von Jagow, G. Analyt. Biochem. **1987**, 166, 368-379.

15. Fan, T. W. M.; Higashi, R. M.; Frenkiel, T. A.; Lane, A. N. J. Exp. Bot. **1997**, 48, 1655-1666.
16. Fan, T. W.-M.; Lane, A. N.; Martens, D.; Higashi, R. M. Analyst **1998**, 123, 875-884.
17. Fan, T. W.-M. Progr. Nucl. Magn. Reson. Spectrosc. **1996**, 28, 161-219.
18. Fan, T. W. M.; Colmer, T. D.; Lane, A. N.; Higashi, R. M. Analyt. Biochem. **1993**, 214, 260-271.
19. Zhang, F. S.; Romheld, V.; Marschner, H. Soil Sci. Plant Nutr. **1991**, 37, 671-678.
20. Römheld, R. Plant and Soil **1991**, 130, 127-134.
21. Kaigi, J., Schaffer, A Biochemistry **1988**, 27, 8509-8515.
22. Grill, E.; Winnacker, E.-L.; Zenk, M. H. Science **1985**, 230, 674-676.
23. Rauser, W.E. Annu. Rev. Biochem. **1990**, 59, 61-86.

Author Index

Subject Index

More Best Sellers from ACS Books

Microwave-Enhanced Chemistry: Fundamentals, Sample Preparation, and Applications
Edited by H. M. (Skip) Kingston and Stephen J. Haswell
800 pp; clothbound ISBN 0–8412–3375–6

Designing Bioactive Molecules: Three-Dimensional Techniques and Applications
Edited by Yvonne Connolly Martin and Peter Willett
352 pp; clothbound ISBN 0–8412–3490–6

Principles of Environmental Toxicology, Second Edition
By Sigmund F. Zakrzewski
352 pp; clothbound ISBN 0–8412–3380–2

Controlled Radical Polymerization
Edited by Krzysztof Matyjaszewski
484 pp; clothbound ISBN 0–8412–3545–7

The Chemistry of Mind-Altering Drugs: History, Pharmacology, and Cultural Context
By Daniel M. Perrine
500 pp; casebound ISBN 0–8412–3253–9

Computational Thermochemistry: Prediction and Estimation of Molecular Thermodynamics
Edited by Karl K. Irikura and David J. Frurip
480 pp; clothbound ISBN 0–8412–3533–3

Organic Coatings for Corrosion Control
Edited by Gordon P. Bierwagen
468 pp; clothbound ISBN 0–8412–3549–X

Polymers in Sensors: Theory and Practice
Edited by Naim Akmal and Arthur M. Usmani
320 pp; clothbound ISBN 0–8412–3550–3

Phytomedicines of Europe: Chemistry and Biological Activity
Edited by Larry D. Lawson and Rudolph Bauer
336 pp; clothbound ISBN 0–8412–3559–7

For further information contact:
Order Department
Oxford University Press
2001 Evans Road
Cary, NC 27513
Phone: 1-800-445-9714 or 919-677-0977

Highlights from ACS Books

Desk Reference of Functional Polymers: Syntheses and Applications
Reza Arshady, Editor
832 pages, clothbound, ISBN 0–8412–3469–8

Chemical Engineering for Chemists
Richard G. Griskey
352 pages, clothbound, ISBN 0–8412–2215–0

Controlled Drug Delivery: Challenges and Strategies
Kinam Park, Editor
720 pages, clothbound, ISBN 0–8412–3470–1

Chemistry Today and Tomorrow: The Central, Useful, and Creative Science
Ronald Breslow
144 pages, paperbound, ISBN 0–8412–3460–4

A Practical Guide to Combinatorial Chemistry
Anthony W. Czarnik and Sheila H. DeWitt
462 pages, clothbound, ISBN 0–8412–3485–X

Chiral Separations: Applications and Technology
Satinder Ahuja, Editor
368 pages, clothbound, ISBN 0–8412–3407–8

Molecular Diversity and Combinatorial Chemistry: Libraries and Drug Discovery
Irwin M. Chaiken and Kim D. Janda, Editors
336 pages, clothbound, ISBN 0–8412–3450–7

A Lifetime of Synergy with Theory and Experiment
Andrew Streitwieser, Jr.
320 pages, clothbound, ISBN 0–8412–1836–6

For further information contact:
Order Department
Oxford University Press
2001 Evans Road
Cary, NC 27513
Phone: 1-800-445-9714 or 919-677-0977
Fax: 919-677-1303